Climate Change Management

Series Editor

Walter Leal Filho, International Climate Change Information and Research Programme, Hamburg University of Applied Sciences, Hamburg, Germany

The aim of this book series is to provide an authoritative source of information on climate change management, with an emphasis on projects, case studies and practical initiatives – all of which may help to address a problem with a global scope, but the impacts of which are mostly local. As the world actively seeks ways to cope with the effects of climate change and global warming, such as floods, droughts, rising sea levels and landscape changes, there is a vital need for reliable information and data to support the efforts pursued by local governments, NGOs and other organizations to address the problems associated with climate change. This series welcomes monographs and contributed volumes written for an academic and professional audience, as well as peer-reviewed conference proceedings. Relevant topics include but are not limited to water conservation, disaster prevention and management, and agriculture, as well as regional studies and documentation of trends. Thanks to its interdisciplinary focus, the series aims to concretely contribute to a better understanding of the state-of-the-art of climate change adaptation, and of the tools with which it can be implemented on the ground.

Notes on the quality assurance and peer review of this publication

Prior to publication, the quality of the works published in this series is double blind reviewed by external referees appointed by the editor. The referees are not aware of the author's name when performing the review; the referees' names are not disclosed.

Walter Leal Filho · Marina Kovaleva ·
Maria Alzira Pimenta Dinis · Johannes M. Luetz ·
Fátima Alves · Gustavo J. Nagy · Sidat Yaffa ·
Desalegn Yayeh Ayal · Jokastah Kalungu
Editors

Climate Change Mitigation and Adaptation in Practice

Springer

Editors
See next page

ISSN 1610-2002 ISSN 1610-2010 (electronic)
Climate Change Management
ISBN 978-3-031-85219-0 ISBN 978-3-031-85217-6 (eBook)
https://doi.org/10.1007/978-3-031-85217-6

This Springer imprint is published by the registered company Springer Nature Switzerland AG
The registered company address is: Gewerbestrasse 11, 6330 Cham, Switzerland

Editors
Walter Leal Filho
Faculty of Life Sciences
Hamburg University of Applied Sciences
Hamburg, Germany

Maria Alzira Pimenta Dinis
Fernando Pessoa Research, Innovation and Development Institute (FP-I3ID)
University Fernando Pessoa (UFP)
Porto, Portugal

Marine and Environmental Sciences Centre (MARE)
University of Coimbra
Coimbra, Portugal

Fátima Alves
Universidade Aberta
Lisbon, Portugal

Centre for Functional Ecology
University of Coimbra
Porto, Portugal

Sidat Yaffa
School of Agriculture and Environmental Sciences
University of the Gambia
Kanifing, Gambia

Jokastah Kalungu
KITUI
South Eastern Kenya University
Kitui, Kenya

Marina Kovaleva
Faculty of Life Sciences
Hamburg University of Applied Sciences
Hamburg, Germany

Johannes M. Luetz
Graduate Research School
Alphacrucis University College
Brisbane, QLD, Australia

Gustavo J. Nagy
Instituto de Ecología y Ciencias Ambientales (IECA)
Universidad de la Republica
Montevideo, Uruguay

Desalegn Yayeh Ayal
Addis Ababa University
Addis Ababa, Ethiopia

Preface

Climate change is one of the most pressing challenges of our time, with profound implications for ecosystems, economies, and societies worldwide. The scientific consensus is clear: human activities, primarily through the burning of fossil fuels and deforestation, are driving unprecedented changes in the Earth's climate. As global temperatures rise, we are witnessing more frequent and severe weather events, rising sea levels, and disrupted natural systems. These changes not only threaten biodiversity but also pose significant risks to food security, water resources, public health, and economic stability.

While the magnitude of the problem can seem overwhelming, there is hope. Climate change is a complex issue that requires a two-pronged approach: mitigation and adaptation. Mitigation efforts aim to reduce or prevent the emission of greenhouse gases, thereby addressing the root cause of the problem. Adaptation, on the other hand, involves adjusting our practices, processes, and infrastructures to better cope with the changing climate. Both strategies are essential for building resilience and ensuring a sustainable future.

The book *Climate Change Mitigation and Adaptation in Practice* brings together an interdisciplinary collection of research, case studies, and practical solutions that address these dual approaches. This book is designed to provide a comprehensive understanding of the scientific, policy, and social dimensions of climate change. It delves into strategies for reducing emissions, enhancing carbon sequestration, and implementing adaptive measures across different sectors such as agriculture, energy, urban planning, and water management.

The book emphasises practical solutions and showcases successful examples of climate action from around the world. By highlighting real-world applications, it aims to inspire readers to implement changes in their own communities, businesses, and governments. The diverse perspectives presented underscore the importance of collaborative and innovative approaches to climate action, acknowledging that the path to a sustainable future will require the involvement of all sectors of society.

The contributors to this book come from various fields, including environmental science, engineering, economics, social sciences, and public policy. Their combined expertise provides a holistic view of the challenges and opportunities in climate change mitigation and adaptation. By addressing both technical solutions and policy frameworks, the book seeks to equip readers with the knowledge and tools necessary to contribute effectively to the global effort.

As we confront the realities of a warming world, this book aims to serve as a valuable resource for students, researchers, practitioners, and policymakers who are dedicated to tackling climate change. It is our hope that the insights and lessons shared within these pages will not only inform but also inspire transformative action.

We thank all authors and reviewers for their valuable contributions. We hope that this book will inspire further works in this field.

Ultimately, the fight against climate change is not merely about preserving the environment; it is about safeguarding our future, ensuring social justice, and fostering a world where both people and nature can thrive. The time to act is now, and this book seeks to be a catalyst for the change we urgently need.

Hamburg, Germany — Walter Leal Filho
Hamburg, Germany — Marina Kovaleva
Porto/Coimbra, Portugal — Maria Alzira Pimenta Dinis
Brisbane, Australia — Johannes M. Luetz
Porto/Lisbon, Portugal — Fátima Alves
Montevideo, Uruguay — Gustavo J. Nagy
Kanifing, Gambia — Sidat Yaffa
Addis Ababa, Ethiopia — Desalegn Yayeh Ayal
Kitui, Kenya — Jokastah Kalungu

Contents

Chapter 1
Climate Change: Assessment of Adaptation Intervention Projects, Challenges and the Need for Climate Change Education in Nigeria

Kemi Funlayo Akeju

Abstract Nigeria is currently recognised as a part of the principal susceptible nations to climate change in Africa. The most obvious climatic changes in Nigeria are reflected in change in rainfall, rising temperatures and precipitation that have adverse impact on agricultural outputs, water pollution, environmental landscapes and livelihoods. Climate change's effects on human lives and livelihoods demonstrate the necessity of establishing adaptation and resilience to these changes. This study considers progress made via financing of notable intervention schemes launched by international organization such as African Union, AFDB, World Bank etc. on climate adaptation projects in recent years towards the adoption of adaptation strategies in Nigeria. It presents the challenges to these projects implementation which are associated with the failure to align climate discourse with national framework. It further showcases the necessity of incorporating climate change education into both secondary and tertiary educational curricula to enhance adequate knowledge of the ecosystem and training by larger societies. Suggestions on ways of aligning climate change programmes, strategies and education with local community members and stakeholders at local and national level are presented.

Keywords Climate change · Adaptation intervention programmes · Challenges · Climate change education · Community inclusion

Introduction: Climate Change Issues in Nigeria

Changes in the climate is anticipated to push many people in Sub-Saharan Africa (SSA) into further poverty if no adequate and concrete measures for climate and development action take place by 2050 (Jafino et al., 2020; Sono et al., 2021).

K. F. Akeju (✉)
Department of Economics, Faculty of the Social Sciences, Ekiti State University, Ado-Ekiti, Ekiti, Nigeria
e-mail: kemi.akeju@eksu.edu.ng

W. Leal Filho et al. (eds.), *Climate Change Mitigation and Adaptation in Practice*, Climate Change Management, https://doi.org/10.1007/978-3-031-85217-6_1

Nigeria continues to be one of the major African nations where individuals residing in unstable and and ravaged environments may be most severely affected by the massive climate change's impacts (Osayemwenre et al., 2023). Nigeria, with an estimated 211 million inhabitants, it is African's continent most inhabited nation according to World Bank estimates from 2021 (Abdulkarim, 2023; Omolola et al., 2023). The country is widely regarded as possessing an elevated level of exposure to climate risks and facing a variety of difficulties, including risks to its physical well-being and means of subsistence due to its large population's heavy reliance on agriculture, a sector that is highly dependent on the climate (Chen et al., 2022; Echendu, 2020). Lower industrial output and an impact on agricultural output are caused by rising temperatures (Chen et al., 2019). Meteorologists and scientists note that along with rising temperatures, extreme precipitation is becoming more intense in much of the northern portion of Nigeria (Dike et al., 2020; Shiru et al., 2020), whereas precipitation patterns are becoming less extreme in other areas (Ogolo & Matthew, 2022).

The widely held perception is that Africa is excessively affected by climate change (Nyiwul, 2021), as the continent is experiencing a rate of increasing temperatures and the resulting impacts of coastal loss, degradation, biodiversity waste and seepage of saltwater compared to the rest of the world (Ndabula et al., 2021; World Bank & WHO, 2020). Nigeria's adaptation capacities and susceptibility to climate catastrophes were assessed by the Notre Dame Global Adaptation Initiative (ND-GAIN) in 2021 (Andrijevic et al., 2020) and the country was placed 161 out of 182 nations (Prince et al., 2023). Nigeria is particularly susceptible to the consequences of climate change because of its extensive coastline, and high population (Shiru et al., 2020), little funding for climate change from governmental and private sector organisations (Raimi et al., 2021), and a lack of information about adaptation (Adeagbo et al., 2021). The nation is now dealing with complicated issues that have both indirect as well as direct effects, such as hunger, relocation, war, and poor health (George & Adelaja, 2022; Mfon, 2023). In particular, local populations that practise agriculture with irrigation have been severely impacted by droughts and degradation in the arid and semi-arid regions of the northern part of Nigeria (Awolala et al., 2022; Ifeanacho & Okudu, 2020; Lenshie et al., 2022).

Rising temperatures in Africa have been linked to erratic precipitation (more intense and unpredictable rainfall) (Fu, 2022), severe occurrences (extreme weather, fires, flooding, heat waves, tropical cyclones, and excessive rainfall) (Clarke et al., 2022; Di Capua & Rahmstorf, 2023), and increasing sea levels increase the risk of coastal erosion and flooding, resulting in physical harm and destruction, endangering public health with illnesses related to water, and contaminated water supply and farmland with salt (Majedul Islam, 2022; Singer et al., 2022). These elements affect the lives of people and overall well-being as reported by Murray and Ebi (2012) in the Intergovernmental Panel on Climate Change and literature (Atwoli et al., 2022; Semenza et al., 2022). Nigeria is among the major food producers in Africa, but study suggests that the danger presented by rising temperatures might worsen the food shortages brought on by the COVID-19 lockdown's detrimental effects on agricultural output (Ayo-Lawal et al., 2022; Balana et al., 2023; Omotayo et al., 2022).

Local communities in northern Nigeria's semi-arid and dry regions have been disproportionately affected by desertification and droughts (Adewuyi & Ezeamaka, 2023). Due to droughts and desertification, indigenous crop-farming groups and nomadic animal-rearing populations are at odds over dwindling resources, causing the former to migrate from the north to the south (Bala & Tar, 2021). Also as the herders migrate, illnesses that may be made worse by climate change are brought by these itinerant livestock herders into host communities (Elelu et al., 2019).

According to recent research, rising global temperatures and more erratic rainfall patterns are already endangering population health and well-being (Challinor et al., 2014; Mora et al., 2017; Raimi et al., 2021), whose effect jeopardises the nutritional well-being and food availability for children (Anderko et al., 2020; Brenton et al., 2022). The need for proper nutrition for under-five kids is acknowledged on a global, national, and individual level as essential to human growth (McCoy et al., 2022), it has been demonstrated that prolonged starvation in early life damages a child's physical and mental growth irreversibly, stalling their growth and developmental changes (Pereira et al., 2023; Shabangu, 2023). Concerns are becoming apparent as the environment continues to produce increasingly frequent and violent extreme weather events, thus making it critical to comprehend the intricate relationships between nutrition, climate, and societal vulnerability in areas where there are nutritional shortages (Carr, 2023).

A plethora of research over the last ten years has demonstrated the significant and quantifiable effects that global climate change has on health (Agache et al., 2022; Atwoli et al., 2022). Evidence suggests that the effects of climate change will manifest more rapidly and virulently in the nations closest to the equator (Kruidbos, 2022). Research has shown that environmental stressors, such as heat stress, failing agriculture, disease transmission, and unstable household finances, can have a detrimental impact on nutritional health in early childhood (Challinor et al. 2014). Adopting robust adaptation measures has become more urgent as the effects of climate change on Nigeria become more severe and the sector's threats increase (Akpabio et al., 2023; Kala et al., 2023).

There is mounting evidence that education has played a significant role in disseminating information about solutions to deal with the risks and issues raised by climate change in industrialized nations around the world (Ajwang & Nambiro, 2022; Alam & Kumar, 2023; Ogadimma et al., 2023; Swain, 2022). To better prepare youth to respond to the complex interplay of science, society, politics, and ethics, there has been a need for participative, multidisciplinary, creative, and affect-driven methods for climate change education. In Nigeria, however, the application of climate change education and knowledge sharing across schools, communities, and universities is not well defined and taken into consideration, despite the continuous evidence on how climate change is affecting society as a whole and what trade–offs it may have on society's welfare, including poverty, conflict and income inequality (Taconet et al., 2020).

Nigeria's revised Nationally Determined Contributions 2021 (NDCs), the 2021 Climate Change Act, the National Adaptation Plan (NAP), and other Federal Government frameworks and initiatives for adaption have been set in place (Orie, 2021;

Ozor et al., 2020). These frameworks also offer guidelines for developing adaptation practices that are consistent with the context and economic objectives of Nigeria. A gender-responsive NAP process, community and ecosystem-based adaptation, awareness of climate change as a cross-cutting issue with potential trade-offs, and many more subsequent co-benefits are among the important considerations that come with the Framework (Malley et al., 2021). Although there is a NAP framework that the Nigerian government has created, there is no documentation regarding its application or implementation in terms of lowering vulnerability at the local level (Leal Filho et al., 2019).

In evaluating the adaptation intervention programmes in Nigeria, this study takes into account the advancements achieved through the funding of significant intervention initiatives introduced by international organisations such as the World Bank, African Union, and African Development Bank on climate adaption projects in the past few years. It presents the challenges to adaptation projects implementation which are associated with the failure to align climate discourse with national framework. It further highlights the necessity of including climate change education in secondary and postsecondary educational curriculum.

The remaining section of this paper is structured in this sequence, Section 'Methodology' present the method of assessment of adaptation intervention, Sect. 'Climate Change Intervention: Adaptation Strategies' describes the climate change adaptation interventions with consideration on both the locally led and internationally financed adaptation projects in Nigeria, Section 'Barriers to Climate Change Adaptation Strategies Implementation' presents the barriers and challenges to adaptation interventions, Sect. 'Conclusion' present the study conclusion, Sect. 'Recommendation' provides the study recommendations while the limitation of the study is presented in Sect. 'Study Limitation'.

Methodology

A qualitative research technique involving the use of documentary analysis was adopted for this paper. Documents were gathered from public and educational sources to provide support claims and evidence of established climate adaptation interventions in Nigeria. A literature review exploring previous studies on locally adaptation strategies and climate adaptation intervention projects put in place by international organizations in Nigeria was carried out. The steps for eligibility criteria are; having document with discussion on climate adaptation strategies in Nigeria, studies with identified climate intervention programmes in Nigeria and studies with on-going funded climate change adaptation projects in Nigeria. These studies were collated using search strategy and their outcome and challenges drawn out.

Climate Change Intervention: Adaptation Strategies

Adapting to climate change has the ability to enhance livelihoods, health, and community resilience in Nigeria, and it is just now becoming recognized as a significant area of activity (Onyimadu, 2023). Because Nigeria has not yet communicated the impact of climate change to the greater public through schools, colleges, and institutions, interventions on climate change concerns have not been adequately provided in Nigeria. It hasn't been acknowledged in the public dialogue as a serious issue needing immediate attention to stop the threat it poses to health, food security, animals, and the spread of disease. The necessary frameworks that are essential for offering direction and methodical assistance for organising, carrying out, and managing adaptation practises are still in the early phase of development, and individuals at all levels lack access to sufficient information (Akpan et al., 2023). When there isn't a widely recognised adaptation process—if not a universal one—adaptation practise is defined by lone efforts by government agencies, international organisations, and local groups.

Considering that climate change is inevitable, adaption measures are required to make sure that societies are able to withstand negative effects and seize new possibilities (Malhi et al., 2020). Although the term "adaptation" is often used, no single definition is accepted as universal. Crucially, adaptation is defined as "modifications in systems of nature or humanity in reactions to current or projected climate stimuli or their consequences, which regulates damages or leverages advantageous potential" by the Intergovernmental Panel on Climate Change (IPCC, 2001). Adaptation refers to variant adjustments in human activities or practices which are structured in response to anticipated or changes in climate with the main aim of addressing the challenges brought about by the influence of the climate change or maintaining the capacity to deal with current and future changes towards taking advantage of new opportunities that may be presented (Owen, 2020). Such action includes activities that are taken before the occurrences of an event or before the impacts of change becomes observatory and after the impacts have emerged or taken place. Adaptation also refers to the actions that individuals, groups, companies, organisations, and governments take to get ready for and adapt to a changing climate (Dapilah et al., 2020).

Understanding local hazards is essential, particularly the risks that rural farmers would face in the event that a similar disaster recurred and how to lessen its impact (Dadzie, 2023). In Nigeria, rural farmers plays a major role as they are involved in the provision of food and the creation of job opportunities, therefore understanding the local ways at which climate change affect their crop plants, farmland, rainfall supply, poultry birds and fish ponds are essential. In order to lessen the speed and severity of climate change, measures are required. Adaptation to the predicted climate change is crucial, in addition to the strategies being developed to counteract it. Adaptation is crucial to lessen the harms from climate change that cannot be prevented.

Locally Led Adaptation Strategies in Nigeria

In Nigeria, a number of entities have created adaptation plans as well as implementation techniques to lessen the impact of climate change in various industries. At the individual and communal levels, adaptation techniques have been implemented to decrease the effect of catastrophes in riverine areas and places vulnerable to floods, therefore lowering loss of life and property. The majority of initiatives focus on preventing disasters that might harm people and have financial repercussions later on by taking measures prior to the next rainy or dry season. Nigerians are suffering from extensive human, economical, and societal repercussions as a result of the recent flood calamity (Echendu, 2020). Floods ruined property worth billions of naira, carried away numerous farmlands and agricultural goods, inundated several settlements, and left thousands of people homeless.

Some local adaptation measures implemented by both people and communities to cushion the effect of the changing climate include:

(i) Farmer's adjustment to changing weather using replanting of fast maturing varieties mechanism following the harvest season to adapt to changing climatic conditions.
(ii) The use of irrigating techniques to supply fields with water if rainfall is scarce.
(iii) Wide drainage channels are being built to prevent floods and erosion, and all drainage paths are being cleared to facilitate the easy flow of water across waterways in farmlands and villages.
(iv) providing farmers with grassroots education on climate change in their native tongues through social gatherings and meetings.
(v) Planting of trees trees arranged in rows to create wind blockers further to monitor erosion in villages and the dissemination of information and supply of seedlings to rural farmers.
(vi) To prevent land degradation, advocate against overstocking animals and overgrazing a portion of land. Farmers may also help restore the soil by using crop rotation practises.
(vii) Forecasting seasonal rainfall and using local organisations and farmers associations to spread knowledge about climate change across the nation. As climate change involves changing weather, early dissemination of change in planting season help inform the rural farmers about the necessary method of adaptation.
(viii) Launching a forceful education campaign in advance of and after natural catastrophes by informing the public about imminent dangers using community radios and other indigenous communication methods.
(ix) The most common farming practises are the planting of cover crops and the use of improved crop varieties (hybrid and pollinated-openly), which have been selected for traits like resistance to pests and diseases, tolerance to drought, with early maturation and large yields.

(x) Houses relocation/movement from water channels to new locations. Internal migration of farmers and herders take place in a bid to address climate change with relocation to new environment.

Assessment of Adaptation Intervention Projects in Nigeria

Climate change adaptation is becoming more prevalent in Nigeria across a range of industries, having the capacity to enhance livelihoods, promote health, and create healthy societies. Nigeria updated its NDC in July 2021, with many progressive modifications and a stronger aim with the update added HFCs to the range of greenhouse gases it addressed, as well as, and it now takes strategies in the waste industry into account (Federal Government of Nigeria, 2021). The Nigeria's revised Nationally Determined Contributions 2021 (NDCs), which are included in the 2021 Climate Change Act, include a number of federal programmes and agendas that serve as guidelines for developing adaption practises that are consistent with the country's economic objectives (Ajayi, 2022). A gender-responsive National Adaptation Plan (NAP) approach, community- and ecosystem-based adaptation, the acknowledgment of climate change as a multifaceted problem with possible trade-offs, and many more subsequent co-benefits are included in this framework. NAPS seeks to: i. lessen susceptibility to the to the consequences of global warming by enhancing resilience and adaptive capability. ii. assisting in the cogent integration of adaptation to climate change into significant new and ongoing policies, programmes, and initiatives—more especially, strategies and processes for development planning—across all relevant sectors and at all levels as needed.

Even with the adoption of the NAPS agenda and regulations, funding remains a challenge to their execution. Globally, financing adaptation is a crucial problem, and Nigeria is no exception. International institutions have provided the majority of funding for climate adaptation programmes in Nigeria (Williams et al., 2021). Nigeria, like most of the nations in Sub-Saharan Africa, depends more than 50% of its climate adaptation efforts on grants, loans, and other forms of foreign assistance. The World Bank, GIZ, FCDO, and USAID are the primary multilateral financial institutions (MFIs) that bear a substantial portion of the project expenditures.

Climate Change Adaptation projects in Nigeria include:

1. The five-year Building Nigeria's Response to Climate Change (BNRCC) initiative, which started in 2007 with pilot projects on vulnerability and its awareness, was supported by the Canadian International Development Agency (CIDA). The pilot programmes' objectives were to provide stoves made of wood that use less fuel, suggest aquaculture as a substitute source, and use improved varieties to increase food security. In cooperation with the Federal Ministry of Environment, it led to the creation of the National Adaptation Strategy and Plan of Action on Climate Change for Nigeria (NASPA-CCN) (Woodley, 2011).

2. In order to address the threat of gully erosion in southeast and considering further land degradation practices in northern Nigeria,, the Federal Ministry of Environment and the World Bank collaborated on the design and approval of the Nigeria Erosion and Watershed Management Project (NEWMAP) in 2012 (Fadeyi & Maresova, 2020). The Gully Rapid Action and Slope Stabilisation (GRASS), Integrated Watershed Management, and Adaptive Livelihoods programmes that are being implemented throughout Nigeria's nineteen (19) states are the focal points of the project.
3. With funding from the Deutsche Gesellschaft für Internationale Zusammenarbeit (GIZ), the Lake Chad Basin is adapting to climate change (Shaibu et al., 2020). The project's main goal is to give 1100 communities in the surrounding nations and the area best practises and solutions for adaption. A specific goal of the five-year initiative, which ran from 2013 to 2018, was to increase food security.
4. The United States Agency for International Development (USAID) and the UK Foreign, Commonwealth and Development Office (FCDO) have also provided funding for the creation of country reports, knowledge products, resources on climate risk, and research on climate change throughout Nigeria. The World Bank introduced the Agro-Climatic Resilience in Semi-Arid Landscapes (ACReSAL) programme in northern Nigeria in 2021 as a replacement for NEWMAP (Butu et al., 2022). In addition to strengthening institutional capability as well as enabling the institutional and policy foundation for multi-sectoral climatic resilience along with collaborative ecosystem management, this project also took into account the difficulties posed by large-scale watershed degradation in northern Nigeria. Finally, it installed a financing mechanism that is accessible to borrowers in investment project financing operations, allowing for the prompt use of uncommitted money toto address qualifying catastrophes or tragedies.
5. The Integrated Climate Change Action Plan (ECCAP) of the Yobe state government campaign to restore degraded lands to be able to address the climate change's effects and desertification on the people of Yobe State and restore the forest landscape in various locations within the state are among the other noteworthy projects proposed by the state government. A reforestation project involving up to 20,000,000 nursed seedlings of various fruit trees, including Doum palm (*Hyphaene thebaica*), Kenaf (*Hibiscus cannabinus*), Date palm (*Phoenix dactylifera*), Mango (*Mangifera indica*), and Tamarind (*Tamarindus indica*), will be implemented in approximately seventeen local government areas that have been chosen.

Barriers to Climate Change Adaptation Strategies Implementation

It is critical that climate adaptation be appropriately understood and managed due to Nigeria's high level of vulnerability and the acknowledged restrictions in resource allocation. Nigeria's economic development would stall in the absence of a strong reaction to climate change, as the nation will have to deal with a plethora of externalities that will probably lead to complicated social problems, including those at the local and regional levels. Although a number of players have created adaptation policies and implementation plans, it is still unclear if these plans and policies are adequate or effective enough to support further climate action.

Barriers and limitations to the implementation of climate change adaptation policies in Nigeria are:

(i) Economic and Financial Restraint: Despite the implementation of the National Adaptation Plan (NAP), the National Climate Change Policy, and the 2021 Climate Change Act, these measures have not received sufficient funding (Brigid et al., 2022; Chete & Chete, 2021; Raimi et al., 2021). Since the majority of these policies are handled centrally, their implementation has not expanded to other states or local government units. Nigeria has a sizable population, thus it is crucial that climate change policies and programmes are distributed at the state and municipal levels in order to guarantee a comprehensive impact at the lower level. When it comes to sustainable development, money is necessary for getting things done and implementing changes including climate change adaptation in Nigeria's development master plans and economic blueprints as a critical strategic move.

(ii) Social and Cultural Practises: Social barriers consist of a range of processes associated with normative and cognitive constraints that hinder individuals or groups from identifying the best means of adapting (Biesbroek et al., 2013; Moser & Ekstrom, 2010; Wilson et al., 2020) It has to do with the cultural perspective and idea that the effects of climate change may be unpredictable and that adaptation measures need not be taken right away, as well as the inclination to put off taking action until after the effects are felt. There are many different types of social institutions, such as collective ownership rights over forest resources, local farmer collectives, and indigenous knowledge institutions. Barriers are also created by constrictive cultural conventions, such as the limited role that women play in the home and community and the reliance on conventional methods of coping with climatic hazards (Grashuis & Dary, 2021; Makate, 2020). Example is how land tenure and fragmentation systems are used throughout Africa, which may impede farmers' ability to adapt to climate change, particularly for women farmers. Farmland is not owned by the majority of African peoples; rather, it is held in trust for their future generations by the current generation. This restricts the amount of money that individual farmers can invest in the development of their property, and the fragmented character of farmland may make it more difficult for farmers to implement the novel agricultural techniques that they may need in order to adjust to climate change (Jayne et al., 2019).

(iii) Information and Human Knowledge: policymakers have inadequate knowledge of the effects of climate change, and there is a dearth of funding and support to enable adaptation measures (Owen, 2020; Sharifi, 2021). In order to inform adaptation policies, many Sub-Saharan African nations lack access to a multitude of observational data and climate modelled forecasts. Since education provides individuals with information, and awareness,the knowledge of the impact of climate change in addition to solutions for it adaptability is essential in tackling these concerns (Cordero et al., 2020).

Conclusion

The most recent predictions state that Africa will continue to see rising inequality as well as threats to food security, water quality, and human health due to the effects of climate change (Hadley et al., 2023; Naheed, 2023; Nyiwul, 2021). Nonetheless, African nations are making a greater effort to address climate change, as seen by their ratification of the Paris Agreement. These efforts are to be complemented by factors that can help address various challenges to the implementation and adjustment to climate adaptation and strategies. Efforts on promoting adaptation strategies are important in providing support for gender mainstreaming into the climate change policy and agenda. It is no news that African nations are also having difficulty adapting to gender-based power disparities that exacerbate pre-existing vulnerabilities, especially those faced by women (Kovaleva et al., 2022).

This study considers the progress made with the adoption of various developmental programmes and policies to address climate change impact in Nigeria. It presents the challenges to these projects implementation which are associated with funding, social barriers, inadequate information and the failure to align climate discourse with national framework. It reveal that the primary focus of government should be on assisting farmers and households in better ways of adapting to climatic variability, as there are significant negative effects of climate variability on economic livelihoods and food security in Nigeria. Programmes implemented are to be strengthened with education and training of the people on climate change and various adaptation processes from the early years through classroom, media Nigeria towards enhancing people's adaptive ability in order to deal with the harms brought on by climate change. When a harvest fails, a lot of Nigerian farmers frequently end up absolutely powerless, establishing of educational programmes has the potential to inform them on way out and also strengthen their knowledge for future occurrences.

In view of the aforementioned, this research highlighted a few adaptation techniques that can help Nigerians deal with the effects of climate change on food security. This chapter affirms the existence of climate change and the extreme vulnerability of Nigeria to it. Nigeria's low agricultural output is evidence that climate change is having a detrimental influence on the country's food security. A significant portion of Nigeria's population continues to suffer from different health issues, hunger, starvation, and malnourishment as a result of food insecurity brought on by climate change. Nigeria's food security should significantly improve and the nation's susceptibility to climate change should decrease if the recommendations stated in this article are successfully implemented.

Recommendation

To maintain food security in the nation, Nigeria must implement certain adaptation measures that will help her address the consequences of climate change. Climate change educational regulations are desperately needed in Nigeria at the federal, state, and local levels of government in order to accomplish this. To create transdisciplinary climate change knowledge in scientific research, everyday life, the workplace, and study, climate change education must be needs to be incorporated into primary, secondary and higher education programmes as well as into daily communication. Policymakers benefit from climate change education by better understanding the necessity and significance of putting improved national climate change measures into action. Barriers and restrictions are mostly the result of cultural norms and institutions, especially at the local level. However, it's crucial to remember that these restrictions can only be acknowledged, impacted, and removed through educational interventions. Communities learn about the effects of climate change on themselves, how to address those effects, and how to reduce their own carbon footprint. Particularly, the dissemination of climate change education information contributes to the development of resistance building of already vulnerable societies as well as those most likely to suffer negative consequences from climate change.

Study Limitation

The assessment of adaptation interventions and the challenges discussed in this chapter is limited to the use of documentary analysis. The evaluation is based on a sample of the available, evaluated documents rather than a comprehensive qualitative research methodology. To fully understand the impact of other recent climate adaptation intervention projects in Nigeria, more research is required.

References

Abdulai, I. A., Enu-Kwesi, F., & Agyenim Boateng, J. (2022). Landowners' willingness to supply agricultural land for conversion into urban uses in peri-urban Ghana. *Local Environment, 27*(2), 145–159.

Abdulkarim, Y. (2023). A systematic review of investment indicators and economic growth in Nigeria. *Humanities and Social Sciences Communications, 10*(1), 1–13.

Adeagbo, O. A., Ojo, T. O., & Adetoro, A. A. (2021). Understanding the determinants of climate change adaptation strategies among smallholder maize farmers in South-west, Nigeria. *Heliyon, 7*(2).

Adewuyi, T. O., & Ezeamaka, C. K. (2023). Ecological disasters, climate change, and rural violence in Nigeria. In *Rural Violence in Contemporary Nigeria* (pp. 87–108).

Agache, I., Sampath, V., Aguilera, J., Akdis, C. A., Akdis, M., Barry, M., Bouagnon, A., Chinthrajah, S., Collins, W., Dulitzki, C., Erny, B., Gomez, J., Goshua, A., Jutel, M., Kizer, K. W., Kline,

O., LaBeaud, A. D., Pali-Schöll, I., Perrett, K. P., … Nadeau, K. C. (2022). Climate change and global health: A call to more research and more action. *Allergy, 77*(5), 1389–1407.
Ajayi, O. O. (2022). Examining Nigeria's institutional response to climate change. *Journal of Law, Policy & Globalization, 126*, 20.
Ajwang, S. O., & Nambiro, A. W. (2022). Climate change adaptation and mitigation using information and communication technology. *International Journal of Computer Science and Research, 6*, 1046–1063.
Akpabio, E. S., Akeju, K. F., Omotoso, K. O., Ohunakin, F., & Ogundokun, R. O. (2023, April). Climate smart agriculture and Nigeria's SDG 2 prospects: A review. In *2023 International conference on science, engineering and business for sustainable development goals (SEB-SDG)* (Vol. 1, pp. 1–6). IEEE.
Akpan, G. E., Ndukwu, M. C., Etim, P. J., Ekop, I. E., & Udoh, I. E. (2023). Food safety and agrochemicals: Risk assessment and food security implications. In *One health implications of agrochemicals and their sustainable alternatives* (pp. 301–333). Springer Nature Singapore.
Alam, M. A., & Kumar, V. (2023). Assessment of community-based risk (CBR) and indigenous knowledge on climate change adaptation: An overview. In *Indigenous knowledge and disaster risk reduction: Insight towards perception, response, adaptation and sustainability* (pp. 539–566).
Anderko, L., Chalupka, S., Du, M., & Hauptman, M. (2020). Climate changes reproductive and children's health: A review of risks, exposures, and impacts. *Pediatric Research, 87*(2), 414–419.
Andrijevic, M., Crespo Cuaresma, J., Muttarak, R., & Schleussner, C. F. (2020). Governance in socioeconomic pathways and its role for future adaptive capacity. *Nature Sustainability, 3*(1), 35–41.
Atwoli, L., Muhia, J., & Merali, Z. (2022). Mental health and climate change in Africa. *Bjpsych International, 19*(4), 86–89.
Awolala, D. O., Mutemi, J., Adefisan, E., Antwi-Agyei, P., & Taylor, A. (2022). Profiling user needs for weather and climate information in fostering drought risk Preparedness in Central-Southern Nigeria. *Frontiers in Climate, 4*, 787605.
Ayo-Lawal, R., Ilevbare, O. E., Omotoso, K. O., Omimakinde, E. A., & Ukwuoma, O. (2022). Social assistance and food security during covid-19 pandemic lock-down: insights from Nigeria. *Future of Food: Journal on Food, Agriculture and Society, 10*(1).
Bala, B., & Tar, U. A. (2021). Lake Chad Basin: Transnational insurgency, counter- insurgency and the proliferation of small arms. In *The Palgrave handbook of small arms and conflicts in Africa* (pp. 927–950).
Balana, B. B., Ogunniyi, A., Oyeyemi, M., Fasoranti, A., Edeh, H., & Andam, K. (2023). COVID-19, food insecurity and dietary diversity of households: Survey evidence from Nigeria. *Food Security, 15*(1), 219–241.
Biesbroek, G. R., Klostermann, J. E., Termeer, C. J., & Kabat, P. (2013). On the nature of barriers to climate change adaptation. *Regional Environmental Change, 13*, 1119–1129.
Brenton, P., Chemutai, V., & Pangestu, M. (2022). Trade and food security in a climate change-impacted world. *Agricultural Economics, 53*(4), 580–591.
Brigid, O. I., Okwuchukwu, E. E., & Marcel, J. A. (2022). The political economy of climate change in Nigeria. *Scholars Journal of Arts Humanities and Social Sciences, 7*, 324–338.
Butu, H. M., Okeke, C. U., & Okereke, C., (2022). Climate change adaptation in Nigeria: Strategies, initiatives, and practices. Working Paper No. 3. APRI.
Carr, C. (2023). Repair and care: Locating the work of climate crisis. *Dialogues in Human Geography, 13*(2), 221–239.
Challinor, A. J., Watson, J., Lobell, D. B., Howden, S. M., Smith, D. R., & Chhetri, N. (2014). A meta-analysis of crop yield under climate change and adaptation. *Nature Climate Change, 4*(4), 287–291.
Chen, X., & Yang, L. (2019). Temperature and industrial output: Firm-level evidence from China. *Journal of Environmental Economics and Management, 95*, 257–274.

Chen, Y., Zhang, D., Wu, F., & Ji, Q. (2022). Climate risks and foreign direct investment in developing countries: The role of national governance. *Sustainability Science, 17*(5), 1723–1740.

Chete, O. B., & Chete, L. N. (2021). Climate change and food security in smallholder farming households of Kogi State, Nigeria. *Handbook of climate change management: Research, leadership, transformation* (pp. 285–304). Springer International Publishing.

Clarke, B., Otto, F., Stuart-Smith, R., & Harrington, L. (2022). Extreme weather impacts of climate change: An attribution perspective. *Environmental Research: Climate, 1*(1), 012001.

Cordero, E. C., Centeno, D., & Todd, A. M. (2020). The role of climate change education on individual lifetime carbon emissions. *PLoS ONE, 15*(2), e0206266.

Dadzie, S. K. N. (2023). Maize farmers' adaptation to drought: Do risk attitudes and perceived risk probability matter? *Cogent Food & Agriculture, 9*(2), 2266197.

Dapilah, F., Nielsen, J. Ø., & Friis, C. (2020). The role of social networks in building adaptive capacity and resilience to climate change: A case study from northern Ghana. *Climate and Development, 12*(1), 42–56.

Di Capua, G., & Rahmstorf, S. (2023). Extreme weather in a changing climate. *Environmental Research Letters, 18*(10), 102001.

Dike, V. N., Lin, Z. H., & Ibe, C. C. (2020). Intensification of summer rainfall extremes over Nigeria during recent decades. *Atmosphere, 11*(10), 1084.

Echendu, A. J. (2020). The impact of flooding on Nigeria's sustainable development goals (SDGs). *Ecosystem Health and Sustainability, 6*(1), 1791735.

Elelu, N., Aiyedun, J. O., Mohammed, I. G., Oludairo, O. O., Odetokun, I. A., Mohammed, K. M., Bale, J. O., & Nuru, S. (2019). Neglected zoonotic diseases in Nigeria: Role of the public health veterinarian. *Pan African Medical Journal, 32*(1).

Fadeyi, O., & Maresova, P. (2020). Stakeholders' perception of climate actions in some developing economies. *Climate, 8*(6), 66.

Federal Republic of Nigeria. (2021). *Petroleum Industry Act, 2021*. https://placng.org/i/wpcontent/uploads/2021/08/Petroleum-Industry-Act2021.pdf

Fu, R. H. Y. (2022). Impacts of climate change and vulnerability of African pastoralists—Trend and pattern of local climate indices. *Journal of Arid Land Studies, 32*(S), 211–215.

George, J., & Adelaja, A. (2022). Armed conflicts, forced displacement and food security in host communities. *World Development, 158*, 105991.

Grashuis, J., & Dary, S. K. (2021). Design principles of common property institutions: The case of farmer cooperatives in the upper west region of Ghana. *International Journal of the Commons, 15*(1).

Hadley, K., Talbott, J., Reddy, S., & Wheat, S. (2023, October). Impacts of climate change on food security and resulting perinatal health impacts. In *Seminars in perinatology* (p. 151842). WB Saunders.

Ifeanacho, M. O., & Okudu, H. O. (2020). Climate change and nutrition security in Nigeria. *Journal of Applied Sciences and Environmental Management, 24*(11), 1853–1860.

IPCC. (2001). *Climate change 2001: Impacts, adaptation, and vulnerability, summary for policymakers and technical summary of the working group II report*. Intergovernmental Panel on Climate Change.

Jafino, B., Walsh, B., Rozenberg, J., & Hallegatte, S. (2020). Revised estimates of the impact of climate change on extreme poverty by 2030. Policy Research Working Paper No. 9417. World Bank, Washington, D.C.

Jayne, T. S., Muyanga, M., Wineman, A., Ghebru, H., Stevens, C., Stickler, M., Chapoto, A., Anseeuw, W., van der Westhuizen, D., & Nyange, D. (2019). Are medium-scale farms driving agricultural transformation in sub- Saharan Africa? *Agricultural Economics, 50*, 75–95.

Kala, N., Balboni, C., & Bhogale, S. (2023). Climate adaptation. *Voxdevlit, 7*, 3.

Kovaleva, M., Leal Filho, W., Borgemeister, C., & Kalungu, J. W. (2022). Understanding needs and potentials for gender-balanced empowerment and leadership in climate change adaptation and mitigation in Africa. *Sustainability, 14*(15), 9410.

Kruidbos, F. (2022). Ecology of zoonotic pathways indicating conflict and mass migration. *The climate-conflict-displacement nexus from a human security perspective* (pp. 251–291). Springer International Publishing.

Leal Filho, W., Balogun, A. L., Olayide, O. E., Azeiteiro, U. M., Ayal, D. Y., Muñoz, P. D. C., Nagy, G. J., Bynoe, P., Oguge, O., Toamukum, N. Y., Saroar, M., & Li, C. (2019). Assessing the impacts of climate change in cities and their adaptive capacity: Towards transformative approaches to climate change adaptation and poverty reduction in urban areas in a set of developing countries. *Science of theTotal Environment, 692*, 1175–1190.

Lenshie, N. E., Ojeh, V. N., Oruonye, E. D., Ezeibe, C., Ajaero, C., Nzeadibe, T. C., Celestine, U. C., & Osadebe, N. (2022). Geopolitics of climate change-induced conflict and population displacement in West Africa. *Local Environment, 27*(3), 287–308.

Majedul Islam, M. M. (2022). Threats to humanity from climate change. *Climate change: The social and scientific construct* (pp. 21–36). Springer International Publishing.

Makate, C. (2020). Local institutions and indigenous knowledge in adoption and scaling of climate-smart agricultural innovations among sub-Saharan smallholder farmers. *International Journal of Climate Change Strategies and Management, 12*(2), 270–287.

Malhi, Y., Franklin, J., Seddon, N., Solan, M., Turner, M. G., Field, C. B., & Knowlton, N. (2020). Climate change and ecosystems: Threats, opportunities and solutions. *Philosophical Transactions of the Royal Society B, 375*(1794), 20190104.

Malley, C. S., Omotosho, D., Bappa, B., Jibril, A., Tarfa, P., Roman, M., Hicks, W. K., Kuylenstierna, J. C. I., de la Sota Sandez, C., & Lefèvre, E. N. (2021). Integration of climate change mitigation and sustainable development planning: Lessons from a national planning process in Nigeria. *Environmental Science & Policy, 125*, 66–75.

McCoy, D. C., Seiden, J., Cuartas, J., Pisani, L., & Waldman, M. (2022). Estimates of a multi-dimensional index of nurturing care in the next 1000 days of life for children in low-income and middle-income countries: A modelling study. *The Lancet Child & Adolescent Health, 6*(5), 324–334.

Mfon, U. Y. (2023). Climate change and farmers-pastoralists conflict in Nigeria: A development-centered analysis. *Global climate change and environmental refugees: Nature, framework and legality* (pp. 121–136). Springer International Publishing.

Mora, C., Dousset, B., Caldwell, I. R., Powell, F. E., Geronimo, R. C., Bielecki, C. R., Counsell, C. W. W., Dietrich, B. S., Johnston, E. T., Louis, L. V., Lucas, M. P., McKenzie, M. M., Shea, A. G., Tseng, H., Giambelluca, T. W., Leon, L. R., Hawkins, E., & Trauernicht, C. (2017). Global risk of deadly heat. *Nature Climate Change, 7*(7), 501–506.

Moser, S. C., & Ekstrom, J. A. (2010). A framework to diagnose barriers to climate change adaptation. *Proceedings of the National Academy of Sciences, 107*(51), 22026–22031.

Murray, V., & Ebi, K. L. (2012). IPCC special report on managing the risks of extreme events and disasters to advance climate change adaptation (SREX). *Journal of Epidemiology and Community Health, 66*(9), 759–760.

Naheed, S. (2023). An overview of the influence of climate change on food security and human health. *Life, 3*, 15.

Ndabula, C., Terdoo, F., & Jidauna, G. G. (2021). Evidence of climate and environmental change in Nigeria: Synthesis from the driving force, pressure, state, impact and response (DPSIR) framework.

Nyiwul, L. (2021). Climate change adaptation and inequality in Africa: Case of water, energy and food insecurity. *Journal of Cleaner Production, 278*, 123393.

Ogadimma, E. C., Raza, S. H., Khan, S. W., & Mohamad, B. (2023). Dearth of Media reportage of environmental issues in Nigeria: Need for innovation and adoption of green culture to curb environmental decay. *Information Development*, 02666669231187364.

Ogolo, E. O., & Matthew, O. J. (2022). Spatial and temporal analysis of observed trends in extreme precipitation events in different climatic zones of Nigeria. *Theoretical and Applied Climatology, 148*(3–4), 1335–1351.

Omolola, O. E., Elijah, O. A., & Ede, O. J. (2023). Effects of population dynamics on economic growth among the world most populous countries. *African Journal of Economic Review, 11*(3), 69–90.

Omotayo, A. O., Omotoso, A. B., Daud, S. A., Omotayo, O. P., & Adeniyi, B. A. (2022). Rising food prices and farming households food insecurity during the COVID-19 Pandemic: Policy implications from SouthWest Nigeria. *Agriculture, 12*(3), 363.

Onyimadu, C. O. (2023). Climate change adaptation and wellbeing among smallholder women farmers in Gwagwalada and Kokona, Nigeria. *Futures, 153*, 103238.

Orie, E. G. (2021). Climate change adaptation mechanism for sustainable development goal 1 in Nigeria: Legal imperative. *African handbook of climate change adaptation* (pp. 1–21). Springer International Publishing.

Osayemwenre, O. B., Olayemiomolola, C. O., & Faith, M. (2023). Nigeria pattern of migration: Issues on human sustainable development. *Journal of Techno-Social, 15*(1), 1–15.

Owen, G. (2020). What makes climate change adaptation effective? A systematic review of the literature. *Global Environmental Change, 62*, 102071.

Ozor, N., Nyambane, A., Onuoha, C. M., Makokha, M. O., & M'mboyi, F. (2020). Nationally determined contributions (NDCs) implementation index, monitoring and tracking tools for selected countries in Africa.

Pereira, S. P., Grilo, L. F., Tavares, R. S., Gomes, R. M., Ramalho-Santos, J., Ozanne, S. E., & Matafome, P. (2023). Programming of early aging. In *Aging* (pp. 407–431). Academic Press.

Prince, A. I., Badewole, A. O., & Omotoso, J. A. (2023). Investigating the adaptive practices implemented by rural communities in Nigeria to alleviate the impacts of climate change on agricultural sectors. *International Journal of Agriculture and Earth Science (IJAES), 9*(4). E-ISSN 2489-0081, P-ISSN 2695-1894.

Raimi, M. O., Vivien, O. T., & Oluwatoyin, O. A. (2021). Creating the healthiest nation: Climate change and environmental health impacts in Nigeria: A narrative review. *Scholink Sustainability in Environment*.

Semenza, J. C., Rocklöv, J., & Ebi, K. L. (2022). Climate change and cascading risks from infectious disease. *Infectious Diseases and Therapy, 11*(4), 1371–1390.

Shabangu, G. B. (2023). *Food safety practices, nutrition knowledge and menu adequacy at non-registered Early Childhood Development (ECD) centres in the Ntuzuma Area, Durban, South Africa* [Doctoral dissertation].

Shaibu, M. T., Omoyele, B. H., & Raphael, O. O. (2020). Climate change and trans-border migration from lake Chad to Nigeria: Are there policy responses towards a sustainable lake? *International Journal of Engineering Applied Sciences and Technology, 4*, 37–44.

Sharifi, A. (2021). Co-benefits and synergies between urban climate change mitigation and adaptation measures: A literature review. *Science of the Total Environment, 750*, 141642.

Shiru, M. S., Shahid, S., Dewan, A., Chung, E. S., Alias, N., Ahmed, K., & Hassan, Q. K. (2020). Projection of meteorological droughts in Nigeria during growing seasons under climate change scenarios. *Scientific Reports, 10*(1), 10107.

Singer, M., Shoreman-ouimet, E., & Graham, A. L. (2022). Climate change and health: Anthropology and beyond. In *A companion to medical anthropology* (pp. 429–441).

Sono, D., Wei, Y., & Jin, Y. (2021). Assessing the climate resilience of Sub-Saharan Africa (SSA): A metric-based approach. *Land, 10*(11), 1205.

Swain, K. A. (2022). Media framing of climate change mitigation and adaptation. *Handbook of climate change mitigation and adaptation* (pp. 3295–3363). Springer International Publishing.

Taconet, N., Méjean, A., & Guivarch, C. (2020). Influence of climate change impacts and mitigation costs on inequality between countries. *Climatic Change, 160*, 15–34.

Williams, P. A., Simpson, N. P., Totin, E., North, M. A., & Trisos, C. H. (2021). Feasibility assessment of climate change adaptation options across Africa: An evidence-based review. *Environmental Research Letters, 16*(7), 073004.

Wilson, R. S., Herziger, A., Hamilton, M., & Brooks, J. S. (2020). From incremental to transformative adaptation in individual responses to climate-exacerbated hazards. *Nature Climate Change, 10*(3), 200–208.

Woodley, E. (2011). Building Nigeria's response to climate change: Pilot projects for community-based adaptation in Nigeria. In *Experiences of climate change adaptation in Africa* (pp. 297–315).

World Bank, & WHO. (2020). Access to clean fuels and technologies for cooking (% of population)—Nigeria | Data. https://data.worldbank.org/indicator/EG.CFT.ACCS.ZS?locations=NG

Chapter 2
Higher Education Alignment with the Climate Action in Mozambique: A Review of Research and Outreach University Functions

Malaquias Zildo António Tsambe, Herieta Massango, Patrício Vitorino Langa, and Luís Jorge Ferrão

Abstract This study analyzes the alignment of higher education institutions' research and outreach activities with Mozambique's Climate Actions. More specifically, the article examines Goal 13 pertaining to Climate Action. A content analysis, using NVivo 12 software, of scientific reports and interviews with main key actors engaged in Climate Action projects was conducted. The assessment tool for the analysis was based on UN goal 13: Climate Action SDG and on Mozambique's indicators to the goal. Six universities from a total of twenty-one were selected, two in each Mozambique's three (southern, central and northern) regions. The study reveals that universities are carrying out different research and outreach projects on Climate Action. However, more than 50% of these projects do not align with Mozambique's Climate Action indicators and Climate Action targets not, even with local indicators. There is misalignment between UN SDG goal 13 indicators and local indicators reported in the activities carried out by Mozambican Higher Education Institutions.

Keywords Climate change (CC) · Climate change action (CCA) · Higher education (HE)

M. Z. A. Tsambe (✉) · H. Massango · P. V. Langa
Institute for Post School Studies, University of Western Cape, Cape Town, South Africa
e-mail: mtsambe@up.ac.mz

M. Z. A. Tsambe · H. Massango · L. J. Ferrão
Faculty of Natural Science and Mathematics, Universidade Pedagógica de Maputo, Maputo, Mozambique

P. V. Langa
Faculty of Education, Universidade Eduardo Mondlane, Maputo, Mozambique

W. Leal Filho et al. (eds.), *Climate Change Mitigation and Adaptation in Practice*, Climate Change Management, https://doi.org/10.1007/978-3-031-85217-6_2

Introduction

Mozambique is one of the countries that has a very insignificant contribution to climate change duo to the country's level of industrialization. Mozambique's contribution to global warming in 2019 was about 0.21% of global emissions (The World Bank Group, 2023). Nevertheless, as a consequence of the country geographical location it's one of the most vulnerable to climate change impacts. This makes the country prone to extreme climate change hazards such as cyclones, droughts and floods, which contribute to the increase of tropical diseases and humanitarian crisis (MFAN, 2018). The vulnerability is heightened not only by the country's coastline, which is about 2470 km, but also by its socioeconomic fragility and weak institutional development. More than 60% of the Mozambican population lives in low-lying coastal areas. In these areas, intense storms and rising sea levels are frequent and this puts infrastructure, coastal agriculture, key ecosystems, and fisheries at risk (Uamusse et al., 2020). The climate change effects lead to the actual level of poverty and many other sustainability challenges. The Notre Dame Adaptation Initiative ranked Mozambique vulnerability to the climate change 154 position out of 181 countries (ND-GAIN, 2021). The UNHCR (2022) referee that Mozambique is the 35th most vulnerable and 24th least ready country, meaning that it is vulnerable to, yet unready to address climate change effects. The country has 45% of population living below the poverty line and in addition to that, 70% depends on agriculture production for their livelihood (McCowan, 2020; Uamusse et al., 2020).

The current challenges of climate variability and weather risk are compounded by climate change. This situation suggests a set of local actions that leads the country to develop polices on mitigation, adaptation, and resilience to guarantee community capabilities at the same time, the achievement of the precepts defined by SDG 13. Beyond policies, to address climate change, there is a need to have strong popular commitment to decisive actions, as well as the knowledge and skills to navigate the complex interlocking systems of human societies and natural environment (McCowan, 2021). This requires climate literacy, where various sectors of society come together and build an environmental consciousness on climate change. Across all the sectors, HEIs play a key role in developing and disseminating knowledge.

This research has its focus on evaluating the Mozambican Higher Education Sector level of contribution on Climate Action through rise public perception, values, attitudes and behaviors on climate change mitigation, adaptation and resilience. It analyzes the alignment of higher education institutions' research and outreach activities with the precepts of the United Nations Sustainable Development Goals (SDG). More specifically, the article examines Goal 13: Climate Action.

Higher Education's Role for Climate Change

The implementation and materialization of Education for Sustainable Development (ESD), in different level of education, particularly in tertiary education, where contents related to climate change are addressed, is a consequence of previous movements to use education as an instrument to achieve a sustainable world (Cheeseman et al., 2019; UNESCO, 2014). The assumption that education is important to a better present and future, is found in SDG 4, "ensure that all learners acquire the knowledge and skills needed to promote sustainable development" (UN, 2015).

Education is seen as a pillar to emancipate local communities to face climate change. This offers practical skills, knowledge, adherence to values of interaction between human and nature, change personal lifestyle and collective modes of organization that can bring changes (Dyer & Andrews, 2011; McCowan, 2021).

Historically, there has been strong acknowledgement that Higher Education sector can and does play an important part in the process of nation building and rebuilding, as custodians and generators of awareness and cultures (Beall, 2021). There is a clear ongoing critique on management of higher education on how it promotes Climate Action awareness and responsibility (Ghoshal, 2005; Snelson-Powell et al., 2016).

The call for a greater attention to climate change on HEIs has been welcomed and it has historically been included transversally or integrated into specific activities, either formally, through the teaching–learning and research process, or informally through the engagement activities in the communities (Leal Filho et al., 2018; McCowan, 2021).

Around the world, HHEIs have taken a leadership role in Climate Action, showing a clear trend on sustainability moving from isolated programs to a core mandate of operations, ensuring training, research and community engagement (Dyer & Andrews, 2011; Prudente et. al., 2015). HEIs also have a duty to shape new ways of educating citizens and delivering knowledge regarding mitigation, adaptation and resilience-building, taking into account the local realities of the country (Hovmöller et. al., 2019).

In Mozambique, different stakeholders are taking several actions to the climate change. Mozambican HEIs are called to take part in the materialization, according to the local context, the 13 SDG targets and indicators. Actually, a number of HEIs in Mozambique, from the three pillars (teaching, research and outreach), develop their activities focused mainly on two: teaching and research. Nevertheless, the results of this knowledge and research do not have a significant impact on the community due to the absence of the third pillar, outreach.

Theoretical Perspective

Higher Education is recognized as formalized organization with a formal structure arising from rationalized institutional rules, in a neo-institutional perspective,

(DiMaggio & Powell, 1983; Hasse & Krücken, 2013; Isomura, 2020; Koenig & Dierkes, 2011; Li & Du, 2016; Meyer & Rowan, 1977). These organizations operate in an environment dominated by generalized rules, beliefs, conventions and agenda, which defines what, constitutes acceptable form of being integrated (DiMaggio & Powell, 1983; Li & Du, 2016). In this sense, HEIs are situated in and influenced by other organizations and wider social forces. HEIs as organizations are driven to incorporate global practices and institutionized procedures prevailing in the society, guided by an universal comparative instrument of global integration (Hasse & Krücken, 2013; Koenig & Dierkes, 2011, Massango et al., 2022; Meyer & Rowan, 1977).

According to Akrivou and Bradbury-Huang (2015), organizations need to reflect on their sustainability values that include behaviour changes to the Climate Action. Therefore, it suggests an urgent need by the HEIs to became isomorphic as a way of legitimacy in a globalized word (Hasse & Krücken, 2013; Koenig & Dierkes, 2011; Li & Du, 2016). Legitimacy can be stands for generalized perceptions that the actions of an entity or organization are desirable and appropriate within a system estabilished by socially constructed norms, values, beliefs and definitions (Ndibuuza & Langa, 2020). This movement is associated with practices, political influence, regulatory agencies, and the resources providers such as government and other research financial agencies (DiMaggio & Powell, 1983; Hasse & Krücken, 2013; Ndibuuza & Langa, 2020). The HEI challenges suggest a relational interconnectedness of involved processes through which development outcomes in one place are sharped through linkage in other places, which characterize the north–south global interconnection (Horner, 2020). The perspective of global development raises questions for approaches, which see development solely, shaped as a challenge just for, the global south (Hasse & Krücken, 2013).

Questions arise from the north–south global interconnection, need to look the local challenges on climate changes. UN established the PRME (Principles of Responsible Management Education) initiative to help promote and deliver sustainable thinking. However, the integration of sustainability, and climate change particularly, into education management remains limited as a consequence of ongoing pressure on organizations to align their policies, actions and practices in a contemporary concept of decoupling (Bromley & Powell, 2012; Snelson-Powell et al.,; 2016; Wall et al., 2022; Westphal & Zajac, 2001). This perspective suggests a common understanding of decoupling or rupture as a gap between policy, action and practice in an organization (Westphal & Zajac, 2001).

Methodology

The study is based on mapping research and outreach university functions that aim to materialize the scope of SDG 13—Climate Action. A review on HE and Climate Action was conducted to identify and evaluate the evidences on how HEIs contribute with research and outreach activities on climate change. The research focuses on a

Table 2.1 Selected HEI for study (*Source* Authors)

Country regions	Name of the HEI	Registered year	Number of students (2022)
South	Eduardo Mondlane University (UEM)	1962	49,678
	Pedagogical University of Maputo (UP-Maputo)	1985	14,047
Center	Catholic University of Mozambique (UCM)	1995	33,527
	Zambeze University (UniZambeze)	2007	8029
North	Lúrio University (UniLúrio)	2006	4888
	Rovuma University (Uni-Rovuma)	2019	14,069

comparative implementation of Goal 13 targets by HEIs as a way to achieve the goal, taking into account the local context.

Six (five public and one private) from the total number of 21 universities in Mozambique were selected. Three criteria were conjugated to select the HE, namely: (1) regionalization (two in each southern, central, and northern region); (2) number of students (two more populated HEI in each region), and (3) year of implantation (two oldest institutions by region). Exceptionally, for Rovuma University only the two first criteria were taken into account. These six universities covered about 58.1% (124,238 students) of the total number of HE students (Table 2.1).

Data Source and Analyses Procedure

The data collection was based on field survey, where institutional reports from 2019 to 2022 were analyzed, face-to-face interviews with the main actors of Climate Action projects were conduct and observation of research and outreach activities were made. The set of information regarding the research and extension activities developed within the scope of the actions that materialize the objective 13 was collected.

In order to ensure clarity related to these dimensions of existence, the data were also analyzed via thematic content method proposed by Bardin (2011), using NVivo software 12. This method use a set of analitical technics to transform singular contents into quantities through the presence or absence of a set of characteristics in message fragments (Bardin, 2011). NVivo was used to support the content analisys, coding keywords that caracterize each target indicator for the Goal or by synonymous keywords.

The data collection tool was developed based on the five targets from UN Goals 13—Climate Action: (i) strengthen resilience and adaptive, (ii) integrate climate change measures in policies, strategies and planning, (iii) improve education, awareness-raising and human and institutional capacity, (iv) implement the commitment of mobilizing jointly all sources to address the needs of meaningful mitigation actions, (v) promote mechanisms for raising capacity for effective climate change-related planning and management (UN, 2015), conjugated with the Mozambique Commitment Action for this Goal taking considering the local indicators defined by the Mapping of Monitoring Instruments and Systems on SGDs and the National (2013–2025) Strategy for Adaptation and Mitigation of Climate Change (MEF, 2017; MICOA, 2012; UN, 2015; VNR, 2020) (Table 2.2).

Table 2.2 Tools components (targets for the climate action and respective local indicator) (*Source* Authors)

Target for the climate action (UN, 2015)	Local indicator (MICOA, 2012; MEF, 2017; VNR, 2020)
13.1 Strengthen resilience and adaptive capacity to climate-related hazards and natural disasters in all countries	13.1.1. Minimize the number of dead, missing persons, affected by disasters per 100,000 people
	13.1.2. Ensure community resilience on Climate Change impacts
13.2 Integrate climate change measures into national policies, strategies and planning	13.2.1. Response to HEI policy, strategy and planning regarding to Climate Change
	13.2.2. Enables the improvement of institutional policies, strategies and planning to achieve the goal
13.3 Improve education, awareness-raising and human and institutional capacity on climate change mitigation, adaptation, impact reduction and early warning	13.3.1. Ensure students and community awareness by integrate mitigation, adaptation and resilience on curriculum
	13.3.2. Develop institutional capacity on research issues related to mitigation, adaptation and resilience
13.a Implement the commitment undertaken by developed-country parties to the United Nations Framework Convention on Climate Change to a goal of mobilizing jointly $100 billion annually by 2020 from all sources to address the needs of developing countries in the context of meaningful mitigation actions and transparency on implementation and fully operationalize the Green Climate Fund through its capitalization as soon as possible	13.a.1. Ensure continued mobilization of resources for research and outreach on Climate Change
	13.a.2. Promote meaning mitigation, adaptation and resilience actions on Climate Change in communities;
13.b Promote mechanisms for raising capacity for effective climate change-related planning and management in least developed countries and small island developing States, including focusing on women, youth and local and marginalized communities	13.b.1. Ensure acquisition of technologies to face Climate Change impact,
	13.b.2. Promote the presence of youth and woman on Climate Change activities

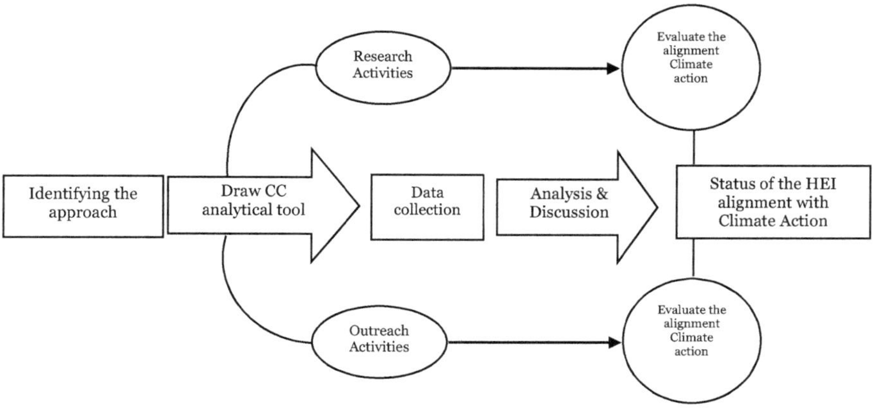

Fig. 2.1 Targets for the climate action (*Source* Authors, adapted from UN, 2015; VNR, 2020)

This tool was used to identify and analyze in which dimensions each HEI activity responds to the challenges imposed by the 2030 Agenda, with regard to Climate Action (Fig. 2.1). It was designed with a view to guarantee a qualitative assessment characterized by credibility, replicability, and reliability factors (Massango et al., 2022; Velasco, 2009; Wright et al., 2021).

The overview of the methodology approach used in the present research is represented by Fig. 2.2.

Results and Discussion

The UN Climate Action (UN Goals 13) defines a set of targets which aims to face climate change and its impact at global and local level. The goal aims to ensure the achievement, at different level of a sustainable world, the development of capacities to face the impacts of climate change. There are considerable uncertainties over future climate change. In this sense, HE sector is called to develop local knowledge and capabilities to face the changes (Arndt et al., 2011). The goal emphasizes the need for global cooperation to reduce greenhouse gas emissions, increase resilience and adaptability to climate related disasters, and promote sustainable practices in industries, communities, and governments. From this, we can assume that education is in the center of all the stakeholders, since besides being a goal, it is a means to reach all other goals specially creating a conscious invidious who is capable to face clime change adversities. In this section, the major findings on how universities are aligned with Goal 13 in terms of research and outreach activities are presented. It also presented a comparative perspective on how universities respond, at the national level, to climate adversities regardless of their geographical location.

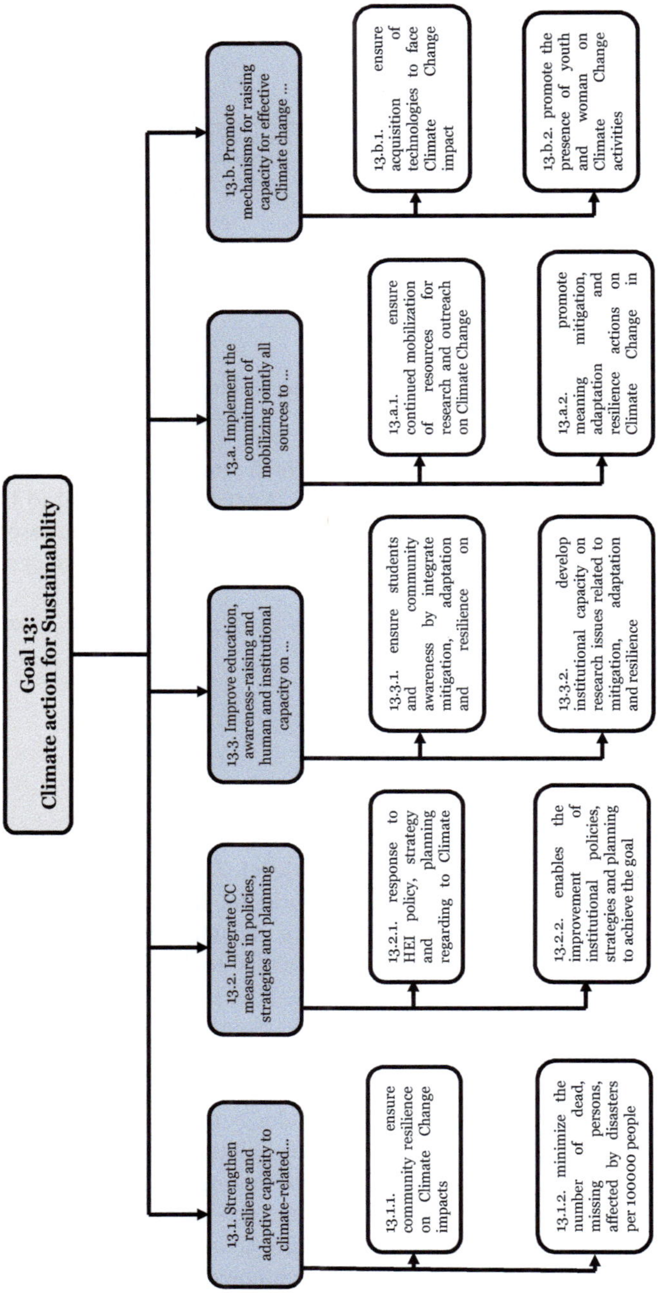

Fig. 2.2 Framework of analysis (*Source* Authors)

Mozambican Higher Education and Climate Action

The study results indicate that universities are engaged with research and outreach activities to face Climate Adversities. Between 2019 and 2022, universities conduct research and outreach activities to find the best ways of approaches to respond to Climate issues according to the different Mozambican realities. Table 2.3 presents, in number for each university the activities related to Climate Action by UN Goal 13 targets conjugated with the Mozambique Commitment action for this Goal.

The six universities from this study, cover around 58.1% (124,238 students) of the all student population from Mozambican HE sector. This represents a potential impact on the communities in which the different actors from these universities are inserted. As mentioned by Aarts et al. (2020) Kioupi and Voulvoulis (2020) and Massango et al. (2022) there is a challenge of integrate sustainable goals in HE sectors, particularly in African countries. Including research and outreach activities in the daily dynamics of university activities can contribute to minimizing this gap.

Findings associated with HE intervention and alignment with Goal 13 in terms of research and outreach activities are displayed in Fig. 2.3. The results indicate that HEIs are engaged with research and outreach activities to face Climate Adversities. The research and outreach activities are translated in 63 projects implemented and being implemented by HEIs. Of these, 43 are research projects which represents 68.3% and 20 are outreach projects that represents 31.7%. However, the presence of more research than outreach projects, in some extent, is not a surprise since HEI in Mozambique have been so far engaged more in teaching and research activities. The outreach activities are the application of the results from the acquired knowledge and research results. One of the universities roles is to influence the social lifestyle of a community by its engagement as consequence of research activities and outreach (Chankseliani & McCowan, 2020; Massango et al., 2022; Wals, 2015). However, we continue to witness the almost absence of universities in the face of population crises caused by climate effects in Mozambique. At some point the question arises "what is the contribution of the academic community to this crisis?". A lot can be done from technology transfer to training provision. HEIs play a crucial role in transferring technology and knowledge to local communities. HEIs provide training and capacity-building programs to local communities. These programs aim to enhance the skills and knowledge of individuals and organizations to better address development challenges including climate change.

Climate Action by Higher Education Institutions

From the regional distribution on climate action by HEIs, South region has more research projects 47% compared to the central and north regions 28% and 26% while for outreach projects, the three regions seem to be balanced, South and North regions having 35% projects each and Central region 30%, Fig. 2.3. The HEIs have

Table 2.3 Distribution of climate action projects in each university, location and climate action perspective (*Source* Authors)

Name of higher education institution	Type of projects		Project location						Project dimension in climate action perspective*				Number of projects
	Research	Outreach	South		Central		North						
			Research	Outreach	Research	Outreach	Research	Outreach	Vulnerability	Mitigation	Adaptation	Resilience	
UEM	19	4	15	2	2	1	2	1	8	3	6	6	23
UP-Maputo	9	5	5	5	2	0	2	0	5	2	3	4	14
UniZambeze	5	4	0	0	5	4	0	0	2	3	2	2	9
UCM	5	2	0	0	3	1	2	1	2	2	2	1	7
UniLurio	3	2	0	0	0	0	3	2	0	1	2	2	5
UniRovuma	2	3	0	0	0	0	2	3	1	1	1	2	5
Partial total	43	20	20	7	12	6	11	7	18	12	16	17	63

*The Climate action dimension was identified by the aims and the activities of each project. The Climate action dimension in this study define: vulnerability dimension—as the degree to which a system is susceptible to, and unable to cope with, adverse effects of climate change, including climate variability and extremes. Mitigation dimension—as efforts to reduce or prevent emission of greenhouse gases. Adaptation dimension—as the process taken to adjust to the actual or expected climate and its effects. Resilience dimension—as the capacity of social, economic and ecosystems to cope with hazardous event or trend or disturbance, responding or reorganizing in ways that maintain their essential function, identity and structure (IPCC, 2007)

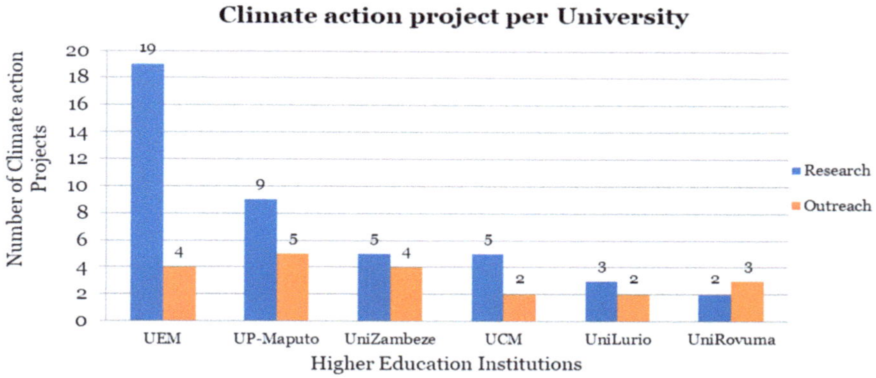

Fig. 2.3 Climate action projects per university (*Source* Authors)

in total more projects (Research and Outreach) on Climate Action in south region of Mozambique. The fact that HEIs located in south have more projects can be justified by the fact that these universities, in addition to being public, were one of the first established in the country. This gives them a certain degree of recognition. These universities have been so far recognized in the contribution for the country development (Fig. 2.4).

The distribution of Climate Action projects by Higher Education Institutions in each region is shown in Fig. 2.5. The two Universities UEM and UP-Maputo located in south, have projects in all three regions (south 74%, central 13% and north 13%) of the country. UCM is located in the central region, has projects in central and north regions (57% in central and 43% in north regions). However, UniLurio and UniRovuma located in the north region have projects only in the north.

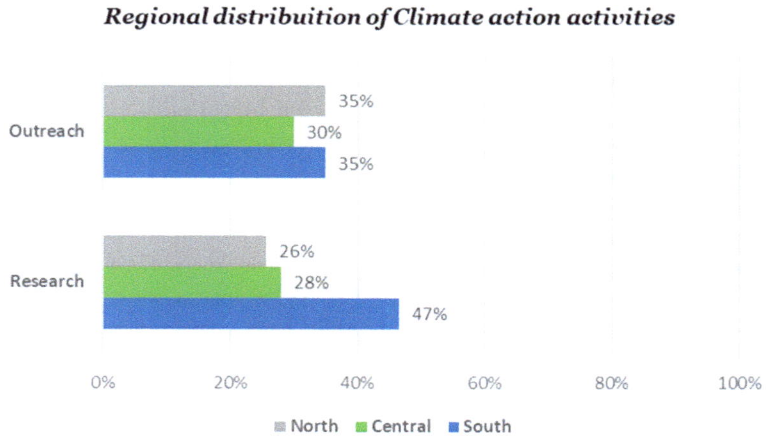

Fig. 2.4 Climate action activities in each Mozambique region (*Source* Authors)

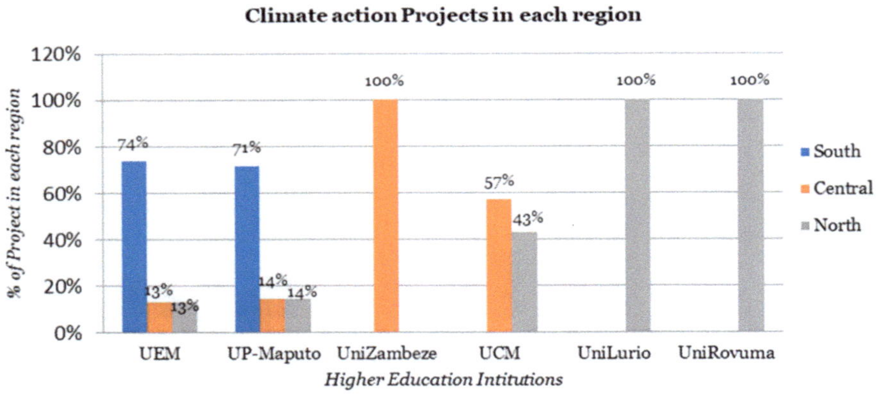

Fig. 2.5 Climate action activities by higher education institutions in each Mozambique region (*Source* Authors)

In general, the projects carried out by HEI contribute to improving citizen knowledge about climate change among youth people in Mozambique. As mentioned by Arnld et al. (2011), there is a need to enhance education investment to accelerate human capital accumulation on Climate Actions. The country registered an annual growth rate in tertiary education of about 3%, which indicates an educated skilled Citizen who will be able to face the world's adversity including climate change and its impact (Arnld et al., 2011). Youths are referred by these authors as future decision makers and they shall become responsible for implementing climate action, presently and in the future. Hence, their involvement in HE Climate Action projects enables them the international acquisition of necessary knowledge, capability and skills that can help the next generation face climate change (Tunji-Olayeni et al., 2021).

The fact that UEM and UP-Maputo have projects in all three regions of Mozambique, is due to the factors mentioned before adding the fact that these Universities have well stablished partnerships and more experience in obtaining competitive funds for research and outreach activities compared to the newest universities, such as UniRovuma and Unilúrio. The actual education scenario, in HE sectors, also demonstrate how it impacts the development of human capability to intervene to the regeneration of damaged areas in order to reclaim the loses caused by the impacts of climate change. Many projects are concentrated in central region of Mozambique. This region showed itself to be more susceptible to cyclones and extreme weather events. In recent years, powerful cyclones like Cyclone Idai and Cyclone Kenneth have caused widespread devastation, including flooding, destruction of homes, infrastructure, and agricultural lands, leading to loss of lives and displacement of populations. However, recently these climate events tend to be felled in all the country. Massango et al. (2022) draw attention to the need to ensure alignment between the Mozambican HE policies and their actions in term of research and outreach activities, taking into account that this sector has an important role to play on the contribution for achievement of sustainability.

Conclusion

There are several Climate Action activities that are being carried out by higher education institutions in Mozambique both by extension and research projects in other to development climate change capabilities in mitigation, adaptation, and resilience. In terms of regionalization, it is worth mentioning that the distribution presented in the study indicates an absence of institutional guidance between higher education activities and policies, from an up to bottom perspective. The oldest HEI in Mozambique is characterized by aggregate a greater number of research and outreach activities on climate action. The six Mozambique HEIs, that represent 58.1% (124,238 students) of the total number of higher education students, showed a significant presence of Climate Action in their research and outreach activities, even without a legal mandate for that by the universities police, which does not aspects reelected to climate action and its effect in local communities in north, center and south regions. Generally, Mozambique HEIs showed many challenges on communication about their research and outreach activities on climate change between scientists, communities, and other non-expert audience.

Recommendation

There is a need to develop specific research on how Mozambique HEIs plan and align the research and outreach activities with formal curricula. Many aspects related to climate change such as implementation of sustainability initiatives, for example on campus research to reduce carbon footprint and promote environmental stewardship.

Another aspect as to do with and how the Mozambique HEIs delivery knowledge to communities, taking into account the complex scientific language on climate change. One can be done by organizing outreach events, public lectures, and awareness campaigns in schools, villages, and urban areas to educate people about climate change and promote sustainable practices.

In addition, there is a need to develop research on how HEIs act to develop adaptability capability in communities that suffer with different climate change events (cyclone—Idai, Kennedy etc. and dry season) and what can be done to develop mitigation capability in vulnerable communities.

References

Aarts, H., Greijn, H., Mohamedbhai, G., & Jowi, J. O. (2020). The SDGs and African higher education. In M. Ramutsindela & D. Mickler (Eds.), *Africa and the sustainable development goals* (pp. 231–242). Springer International Publishing. https://doi.org/10.1007/978-3-030-14857-7

Akrivou, K., & Bradbury-Huang, H. (2015). Educating integrated catalysts: Transforming business schools toward ethics and sustainability. *Academy of Management Learning & Education, 14*(2), 1–19. https://doi.org/10.5465/amle.2012.0343

Arndt, C., Strzepeck, K., Tarp, F., Thurlow, J., Fant IV, C., Wright, L. (2011). Adapting to climate change: An integrated biophysical and economic assessment for Mozambique. African regional perspectives. *Sustainability Science, 6*, 7–20. https://doi.org/10.1007/s11625-010-0118-9. Springer International Publishing.

Bardin, L. (2011). *Content analysis* (p. 70). Ediçõess.

Beall, J. (2021). Levelling up international higher education: Universities, nations and global goals. In H. Land, A. Corcoran, & D. Lancu (Eds.), *The promise of higher education* (pp. 107–112). Springer. https://doi.org/10.1007/978-3-030-67245-4

Bromley, P., & Powell, W. W. (2012). From smoke and mirrors to walking the talk: Decoupling in the contemporary world. *Academy of Management Annals, 6*(1), 1–48.

Chankseliani, M., & McCowan, T. (2020) Higher education and sustainable development goals. *Higher Education*, 1–8. https://doi.org/10.1007/s10734-020-00652w

Cheeseman, A., Wright, T. S., Murray, J., & McKenzie, M. (2019). Taking stock of sustainability in higher education: a review of the policy literature. *Environmental Education Research*, 1–16. https://doi.org/10.1080/13504622.2019.1616164

DiMaggio, P. J., & Powell, W. W. (1983). The iron cage revisited: Institutional isomorphism and collective rationality in organizational fields. *American Sociological Review, 48*(2), 147–160. https://doi.org/10.2307/2095101

Dyer, G., & Anderws, J. (2011) Higher education's role in adapting to charging climate. Education for sustainability. American College & University. www.secondnature.org

Ghoshal, S. (2005). Bad management theories are destroying good management practices. *Academy of Management Learning and Education, 4*(1), 75–91.

Hasse, R., & Krücken, G. (2013). Competition and Actorhood: A further expansion of the neo-institutional agenda. *Sociologia Internationalis, 51*(2), 1–18.

Horner, R. (2020). Towards a new paradigm of global development? Beyond the limits of international development. *Progress in Human Geography, 44*(3), 415–436. https://doi.org/10.1177/0309132519836158

Hovmöller, E. E., Lagerlöf, D., & Aldén, S. (2019) *Rethinking higher education: Inspired by the sustainable development goals.* Karolinska Institutet Campus Solna.

IPCC. (2007). *Climate change 2007: The physical science basis.* Contribution of Working Group I to the IPCC's Ar4.

Isomura, K. (2020) The theory of formal organization. In *Organization theory by Chester Bernard. Springer Briefs in Business.* Springer. https://doi.org/10.1007/978-981-15-92065_3

Kioupi, V., & Voulvoulis, N. (2020). Sustainable development goals (SDGs): Assessing the contribution of higher education programmes. *Sustainability, 12*, 6701. https://doi.org/10.3390/su12176701, www.mdpi.com/journal/sustainability.

Koenig, M., & Dierkes, J. (2011). Conflict in the world polity—Neo-institutional perspectives. *Acta Sociologica, 54*(1), 5–25. https://doi.org/10.1177/0001699310392600

Leal Filho, W., Brandli, L., Becker, D., Skanavis, C., Kounani, A., Sardi, C., & Marans, R. (2018). Sustainable development policies as indicators and pre-conditions for sustainability efforts at universities. *International Journal of Sustainability in Higher Education, 19*(1), 85–113. https://doi.org/10.1108/IJSHE-01-2017-0002

Li, J., & Du, J. (2016). Globalization and decentralization forces in China's higher education administration and management reform (1953–2015): A neo-institutional analysis. *US-China Education Review, 6*(1), 1–19. https://doi.org/10.17265/2161-6248/2016.01.001

Massango, H., Tsambe, M. Z. A., Geschwind, L., & Langa, P. V. (2022). An assessment of higher education system and institutional policies alignment with sustainable development goal in Mozambique. *International Journal of Higher Education and Sustainability, 4*(2), 115–133.

McCowan, T. (2020). *The impact of universities on climate change: a theoretical framework.* Centre for Global Higher Education. ISSN 2398-56X.

McCowan, T. (2021). Climate change in higher education: a curriculum topography approach. Transforming universities for a changing climate, Working Paper Series Nr. 6. ISSN: 2754-0308.
MEF—Ministério da Economia e Finanças. (2017). Mapeamento dos Instrumentos e Sistemas de Monitoria sobre os Objectivos de Desenvolvimento Sustentável [Fote, Isidro; Salência, Celestino; Afonso, Felisberto (eds.)], Maputo, Moçambique.
Meyer, J. W., & Rowan, B. (1977). Institutionalized organizations: Formal structure as myth and ceremony. *American Journal of Sociology, 83*(2), 340–363.
MFAN, M. O. (2018) *Climate change profile: Mozambique.* Ministry of Foreign Affairs of the Netherland.
MICOA—Ministry for Environment Coordinator. (2012). *Estratégia Nacional de Adaptação e Mitigação de Mudanças Climáticas 2013–2025.* ENAMMC.
ND-GAIN (Notre Dame-Global Adaptation Initiative). (2021). *"ND-GAIN country index, vulnerability and readiness."* Retrieved July 15, 2023, https://gain.nd.edu/our-work/country-index/rankings/.
Ndibuuza, F., & Langa, P. (2020). The tale of academic practice in a rising knowledge society: Focus on a university in South Africa. *Tertiary Education and Management, 26*, 117–130. https://doi.org/10.1007/s11233-019-09036-x
Prudente, M. S., Aguja, S. E., & Anito, J. (2015). Exploring climate change and attitudes: Drawing implications for a framework on environmental literacy. *Advanced Science Letters, 21*(7), 2413–2418.
Snelson-Powell, A., Grosvold, J., & Millington, A. (2016). Business school legitimacy and the challenge of sustainability: A fuzzy set analysis of institutional decoupling. *Academy of Management Learning and Education., 15*(4), 703–723.
Tunji-Olayeni, P. F., Adegboye, F., Oluwatobi, A., Adeyemi, G., Olagunju, O., Okoro, A., & Osabuohien, E. S. (2021). Accelerating progress on sustainable development goals: Assessing secondary schools students' knowledge of climate change actions. In *International conference on energy and sustainable environment. IOP Conference Series: Earth and Environmental Science* (Vol. 665, pp. 1–12). https://doi.org/10.1088/1755-1315/665/1/012041.
Uamusse, M. M., Tussupova, K., & Persson, K. M. (2020). Climate change effects on hydropower in Mozambique. *Applied Sciences, 10*(4842), 1–12. https://doi.org/10.3390/app10144824
UN (2015) *Transforming our world: The 2030 agenda for sustainable resolution adopted by the general assembly.* Agenda items 15 and 116. A/RES/70/1.
UNESCO. (2014). *Roadmap for implementing the global action programme on education for sustainable development*. UNESCO.
UNHCR. (2022). *Mozambique: Factsheet on climate related displacement.* Protection Cluster. Maputo. https://reliefweb.int/report/mozambique/mozambique-protection-cluster-factsheet-climate-related-displacement-april-2022
Velasco, A. (2009). Epistemología de la representación del cambio social. *Episteme NS, 29*(2), 115–144.
VNR. (2020). *Voluntary national review of agenda 2030 for sustainble development.* Maputo: VNR2020 Mozambique.
VNR. (2021). *Voluntary national review of agenda 2030 for sustainable development.* VNR2020 Mozambique.
Wall, T., Hindley, A., Mburayi, L., Cregan, K., & Evans, V. (2022). Integrating sustainability in business schools: The possibility of harmonic response across heterogenic landscapes? In *Research in management learning and education (RMLE) un–Conference proceedings* (pp. 21–22). Research in Management Learning and Education (RMLE).
Wals, A. E. (2015). *Beyond unreasonable doubt: Education and Learning for socio-ecological sustainability in the anthropocene.* Wageningen University.
Westphal, J. D., & Zajac, E. J. (2001). Decoupling policy from practice: The case of stock repurchases programs. *Administrative Science Quarterly., 46*, 202–228.
World Bank Group. (2023). *Country climate and development report: Mozambique.* The World Bank Group.
Wright, L. T., Moutinho, L., Stone, M., & Bagozzi, R. P. (2021). *The Routledge companion to marketing research.* Routledge.

Chapter 3
Is Climate Change Policy Fit for Purpose? Beyond Green Capitalism and Liberal Environmentalism

Nick J. Fox and Pam Alldred

Abstract Though international policy-makers have addressed environmental degradation and climate changes for many years, it is questionable whether current policies are scientifically, socially, politically or economically adequate to resolve the existential climate crisis now facing Earth and its human and non-human inhabitants. This chapter analyses policies as more-than-human assemblages. These assemblages are analysed in terms of their comprehensiveness, measured against the breadth of current scientific and social scientific knowledge of anthropogenic climate change. Two policy positions on climate change are assessed using this methodology: 'liberal environmentalism' and 'green capitalism'. Neither is found to be adequate as a policy to successfully counter the threats to the climate produced by human activity since the industrial revolution. In their stead, the chapter offers a way to develop a scientifically and politically adequate climate change policy, and what this may entail.

Keywords Climate change · Policy assemblage · Sustainability · Green capitalism · Posthumanism

Introduction

Since the end of the last century, sustainable development policy discussions have gained significance due to recognised environmental degradation and human-induced climate change (IPCC, 2014; United Nations, 2016). International treaties aim to curb greenhouse gas emissions, and efforts to combat various forms of pollution, from nitrous oxide to plastic waste, are integral to 21st-century policy (Chasek & Downie, 2020; Gunningham, 2019).

N. J. Fox (✉)
University of Huddersfield, Huddersfield, UK
e-mail: n.j.fox@hud.ac.uk

P. Alldred
Nottingham Trent University, Nottingham, UK
e-mail: pam.alldred@ntu.ac.uk

W. Leal Filho et al. (eds.), *Climate Change Mitigation and Adaptation in Practice*, Climate Change Management, https://doi.org/10.1007/978-3-031-85217-6_3

This chapter offers a critical assessment of such policies on sustainability and sustainable development. The starting point is a 'posthuman' (Braidotti, 2013: 5–6) or more-than-human ontology that aims to re-privilege the interests of non-human animate and inanimate matter, while acknowledging the part humans play in the Earth's ecosystem. Humans, from this perspective, 'are *part of* the environment, not separate from or in opposition to it' (Fox & Alldred, 2020b: 123). This ontology also supplies a materialist conceptual framework that enables us to develop a 'micropolitical' analysis of policy in terms of the capacities produced by a 'policy-assemblage' (Ureta, 2014).

Having established this framework for inquiry, the chapter explores in detail two of the sustainability policy assemblages that have dominated discussions of climate change in recent international fora: 'liberal environmentalism' and 'green capitalism' (also known as 'climate capitalism'). It analyses micropolitically what each policy aims to achieve, but also discloses what is excluded or ignored in each perspective. This analysis supplies a critical assessment of how this policy will impact on sustainable development. These critiques provide the basis for the chapter's conclusion, which will set out a more-than-human assessment of what is required socially, economically and politically for environmental sustainability.

Social Science, Posthumanism and the Environment

Social scientists have explored environmental and ecological issues from different angles over the years. Initially, they viewed the environment as a backdrop for human activities (Dunlap & Catton, 1979; Walker, 2005: 80), and sought ways to manipulate it for human benefit (Hoehner et al., 2003; Swinburn et al., 1999). Next, they delved into how the physical environment, such as climate and geography, influenced human existence (Urry, 2009), or studied the psychological and social impacts of the built environment (Halpern, 2013). More recently, they have examined the impact of human activity on the environment (Dunlap & Catton, 1994: 24), during the anthropocene era (Steffen et al., 2007), acknowledging that humans are part of the environment, and still governed by the same physical laws (Stevens, 2012: 580).

These approaches have typically maintained a distinction between humans and the environment, resulting in an anthropocentric perspective (Stevens, 2012: 8) that places society conceptually above the environment (Walker, 2005: 80). Some scholars have attempted to challenge this anthropocentrism in environmental sociology (Benton, 1991; Shove et al., 2012; Stevens, 2012; Walker, 2005). However, in this chapter it is the 'new' materialism (Coole & Frost, 2010; Fox & Alldred, 2017; Latour, 2005; Thrift, 2008; van der Tuin & Dolphijn, 2010) that most decisively transgresses the distinction between humans and 'environment', and provide a basis from which to establish a post-anthropocentric and more-than-human policy to address anthropogenic climate change.

One entry point into this materialist and more-than-human perspective is Rosi Braidotti's work, which argues that human interests are intertwined with those of other living beings and the Earth itself. Braidotti advocates for a materialist feminism that is relational, embodied, and embedded (Braidotti, 2011: 132–136), as well as a posthumanism that moves beyond both traditional humanism and anti-humanism (Braidotti, 2013: 37). While humanism prioritised secular human reason over religious authority (Carroll, 1993: 117), anti-humanism criticised this focus for its exclusionary and exploitative tendencies (Braidotti, 2011: 82, 88–89; 2013: 23). Braidotti proposes a posthuman project that recognises the vital capacities of all matter, animate and inanimate, for self-organisation and becoming (Braidotti, 2011: 16; cf. Bennett, 2010). In this view, matter is in a constant state of being affected and affecting other elements. Braidotti's work provides an eco-philosophical foundation for a 'new' materialist ontology centred on relationality and becoming (Braidotti, 2011: 41).

This new materialist perspective involves two key shifts: firstly, it moves away from essentialism towards an emphasis on relationality, and secondly, it recognises the capacity of non-human entities, including things, organisations and ideas, to exert agency traditionally attributed only to humans. In this framework, new materialism rejects the idea of pre-existing, fixed entities like bodies, animals, and even governmental structures. Instead, it views them as relational materialities that take shape and continuity through interactions with other materialities (Delanda, 2006: 3; Deleuze, 1988: 125). Events and interactions are understood as *assemblages* (Bennett, 2005: 445) or arrangements/orderings (Buchanan, 2017: 465) of *relations* among bodies, things, social institutions, and constructs. These relations are fluid and in constant flux, emphasising the dynamic nature of the world and challenging static and essentialist perspectives (Deleuze, 1988: 128; Lemke, 2015).

New materialism challenges the traditional notion of 'agency' by emphasising that all elements within an assemblage, including humans, possess the ability to influence or be influenced by other interconnected components (Deleuze, 1988: 101). In this perspective, humans no longer hold exclusive agency. New materialist scholars extend this concept to include expressive elements like beliefs, desires, feelings, values, ideologies, and discourses, which can materially impact other parts of the assemblage (Coole & Frost, 2010: 28; DeLanda, 2006: 12–13). The assemblage's collective 'economy' of *affects* (Clough, 2004: 15), as it combines both human and non-human elements, dictates its capabilities. These *capacities* for action, interaction, or emotion do not stem from inherent qualities but arise from the interactions between these elements (Barad, 2001: 96; DeLanda, 2006: 10–11). What any human or non-human, animate or inanimate relation can do depends entirely upon the extent of its affects with other matter.

This conceptual toolkit of relations, assemblages, affects and capacities supplies a monistic ontology of nature/culture (van der Tuin & Dolphijn, 2010). This consequently elides conventional differentiation between humans and their 'environment' (Latour, 1993), with human bodies, cultures and societies part of the environment. This more-than-human, materialist ontology of environment can be applied both to theorise sustainability (Braidotti, 2013: 5; Fox & Alldred, 2020b), but also to assess critically different policy propositions addressing sustainability and climate change (McCann, 2011).

Policy and the Policy Assemblage

Policy can be understood as material-semiotic engagements with a social or natural issue or event such as environmental pollution or climate change (Shore & Wright, 1997: 30–31), that—in some way—aims to materially affect that event (Taylor Webb and Gulson, 2012: 87–88). Unlike analyses of policy development and implementation explored at a macro-level of stakeholders (Burstein & Linton, 2002) or institutions (Wiktorowicz, 2003: 618), a materialist perspective affords a means to explore policy, policy-making and policy implementation *micropolitically,* addressing the affective movements in policy-making or policy implementation (DeLanda, 2006: 87; Patton, 2000: 68; Widder, 2012: 125).

Cultural geography scholars have explored policy-making in terms of a 'policy assemblage' (McCann, 2011; McCann & Ward, 2012; Prince, 2010; Ureta, 2014), and this suggests a basis for such a micropolitical approach. According to this perspective, a policy may be understood as an unstable and dynamic assemblage, comprising social actors, networks and institutions (Prince, 2010: 173; Ureta, 2014: 305). However, while these advocates of policy-assemblages have cited materialist authors such as Deleuze and Guattari and actor-network (ANT) theorists, this understanding of a 'policy assemblage' is under-theorised (McCann & Ward, 2012: 43). A more thoroughly new materialist grounding has subsequently supplied the concept with more analytical gravitas, allowing exploration of '*how* policies assemble micropolitically, and what they can do' (Fox & Alldred, 2020a: 273). This approach assesses the interactions between a more-than-human *event* such as global warming and the *policy* developed to somehow influence or change this event. So, for instance, anthropogenic climate change can be analysed as a more-than-human 'event-*assemblage*'. Using current natural and social science understandings of the climate, this assemblage comprises a range of human and non-human elements. The principal elements in such an arrangement can be depicted as follows (in alphabetical order):

atmosphere; fossil fuels (coal, oil); greenhouse gases; humans; industry; oceans; the Sun

This climate change event-assemblage is a consequence of the *affects* between these human and non-human elements. For example, the affect between fossil fuels and oxygen during industrial production generates both energy and greenhouse gases. These latter trap heat in the atmosphere, intensifying its 'greenhouse' effect. This 'affect economy' (Clough, 2004: 15) amplifies solar heat absorption by Earth's oceans, exacerbating climate change.

Policy-making aimed at addressing a complex event-assemblage—such as anthropogenic climate change by reducing greenhouse gas emissions, can also be seen as an event in its own right, and consequently also an assemblage. This 'policy-assemblage' includes various human and non-human elements, including scientists, scientific evidence, and diverse other stakeholders (Baer, 2012: 267; Dror, 2017; Yearley, 2014). Again, this assemblage can be depicted as an arrangement of various elements, including (in alphabetical order):

audience; evidence of climate change; experts; money and economics; policy documents; policy-makers; relevant natural and social science theories; stakeholders; social and political processes

Policy assemblages on topics such as anthropocentric climate change may also include consumers, energy producers, governments, the media, previous policies, as well as the political commitments, views and beliefs of all those concerned. This assemblage will be constituted by the affective interactions between these human and non-human elements. For instance, evidence of climate changes from scientific studies and experts will affect policy makers, as will politics, economics and their own commitments.

Unpacking these event- and policy-assemblages and the affects that constitute them allows an assessment of how a policy has emerged and what are its main foci, but also what it emphasises or ignores. As a consequence, this also supplies a means to assess what a specific policy can do, what it cannot do, and thus whether it is *effective* and *adequate* to address an event of concern such as anthropocentric climate change. The following Fig. 3.1 illustrates the interactions between an event and a policy to address it, according to this assemblage analysis.

Beginning with the event-assemblage (EA), the development of a successful policy (for instance, to address the negative impacts of human fossil fuel use on climate change) requires that a policy-assemblage (PA) is capable of identifying relevant affects in an EA (for instance, how industry's use of fossil fuels affects

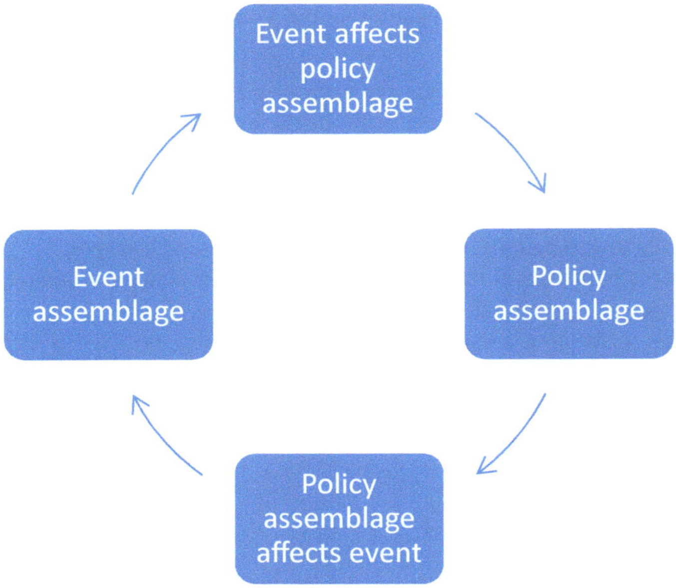

Fig. 3.1 How event- and policy-assemblages interact

levels of carbon dioxide in the atmosphere). Recognising these affects enables policies to comprehensively address an issue of concern. Thus, it has only been since scientists linked levels of 'greenhouse' gases in the atmosphere to industrialisation that this affect in a climate change assemblage could be identified by policymakers.[1]

However, Fig. 3.1 also indicates a second stage in the event/policy interaction. Once developed, a PA need to be *adequate* and *appropriate* to influence the EA it aims to address. If it does not have this capacity, then a policy will be ineffectual or even irrelevant (Dror, 2017: 34–35). For instance, while a global policy to use renewable energy sources rather than coal or oil for industry will reduce greenhouse gas emissions, a policy to raise tea prices to limit how often citizens boil water will be *inadequate* to reduce greenhouse gases significantly, and *ineffective* if they choose to drink coffee instead. The capacity of a policy to affect an EA will also be shaped by other human and non-human elements in the PA. For instance, if a policy to introduce renewable energy generation from solar or wind power is not adequately resourced, backed by fiscal policies or poorly promoted to stakeholders, it will not significantly impact energy use by industry or car-owners.

To summarise, this assemblage approach to policy addresses the affects that produce both events and policy. By analysing how EA and PA interact and affect each other, the adequacy of a policy can be assessed, and hence its likely success or failure to counter the effects of an event. This opportunity is the basis for the assessment of environmental policy assemblages in the remainder of this chapter, as it considers two popular policies—'liberal environmentalism' and 'green capitalism'—and whether either is fit for purpose to counter the anthropogenic climate crisis. For each of these positions, after a brief description, the analysis will apply the following steps:

- Using up-to-date and comprehensive natural and social scientific evidence, identify the component material elements in the event-assemblage and the affects between these elements. In the case of liberal environmentalism, the chosen EA is environmental pollution from the waste generated by a consumer society; the EA assessed in the case of green capitalism is global warming generated by using fossil fuels to generate energy for industry, housing and transport.
- Analyse the policy assemblage to identify which affects in the climate change EA it addresses, and those it does not address.
- From this, offer an assessment of whether the policy position is effective and adequate to address the EA in question, by assessing if what it can do (its capacities) is sufficient to actually resolve the challenge posed by the relevant EA (waste, climate change).

Liberal Environmentalism

Liberal environmentalism (Bernstein, 2002; Collard & Dempsey, 2022) is well-represented in many past and present 'good citizen' environmental initiatives. These aim to nudge or more radically change human behaviour, and have included anti-litter campaigns, current efforts to reduce plastic waste, and charities that seek to protect endangered species or habitats such as the world's rainforests (Yearley, 2014: 5). It was the starting point for United Nations policies on sustainable development (Bernstein, 2002: 3)[22] as well as the 'green capitalism' approach that is considered in the following section.

Bernstein (2002: 1) described liberal environmentalism as a compromise. On one hand, it formulates policies and actions to counter the impact of human activity and consumption upon the environment (Yearley, 2014: 98), while on the other it does not offer a critical analysis of how socioeconomic development/growth contribute to environmental depredations (Bernstein, 2002: 4; Collard & Dempsey, 2022: 1546; Talshir, 2012: 18; Whitehead, 2014). Sometimes, consumer sentiment is harnessed to encourage business into more environment-friendly practices, such as removing plastic micro-beads (harmful to marine life) from cosmetic and health products (Dauvergne, 2018). Waste recycling policy supplies a relevant illustration. Recent UK policies around waste recycling aim to maximise domestic and commercial efforts to reduce, reuse and recycle waste products (from industrial by-products to discarded or worn-out goods to product packaging) via incentives and penalties. Such policies may be promoted in terms of both reducing pollution and supplying economic benefits:

> A key priority for the Government is to boost growth in the economy whilst continuing to improve the environment. ... Moving towards a more resource efficient, circular economy offers scope for innovation, sustainable growth and saving money, as well as reducing the impact on the environment. (Department for Environment, Food and Rural Affairs, 2013: 5)

This policy from the UK government sets out a series of measures (from charging for plastic shopping bags to investing in waste prevention and reuse schemes), with the explicit programme aim to improve the environment and protect human health (ibid).

Micropolitically, this waste policy assemblage engages with affects in the waste EA in two ways. First, it embraces the positive aspects of the non-human natural and physical environment, while at the same time sustaining a privileged position for humans and their economic activity. Second, the policy assemblage focuses primarily on citizen behaviour, at the expense of any critical assessment of why human economic activity impacts negatively upon the environment. In the case of waste management, there are no criticisms of Western society's culture of rampant consumption, built-in obsolescence, or industry's use of plastics for manufacture and packaging to minimise overheads. Together, these two affective movements sustain a dualism between humans and the environment', congruent with a long-held Western perspective on the environment as a physical resource to be exploited for economic and social advantage. These priorities provide an analytical entry-point to assess how the liberal environmentalist policy-assemblage interacts with environmental event-assemblages (see Fig. 3.1), and to measure this interaction against a more comprehensive 'waste production/reduction event assemblage'. Continuing with the example of recycling and the DEFRA policy document quoted above, this more comprehensive EA comprises at least (in alphabetical order):

consumers; culture of consumption; economic activity; environment; government; industry; market economy; natural resources; producers; profit; waste

Such a critical approach understands 'waste' within a market economy as due to the affect driving producers in a capitalist economy to continually seek profit from the exploitation of environmental resources, and a consequent need for continued growth in both production and consumption (Pearce et al., 1989: 10). However, the liberal model of a waste-event *excludes* such key social, political and economic affects within the waste assemblage. From a critical perspective, liberal environmentalism's focus on recycling policies is only a minor part of the solution: a sticking plaster on the damage that industrial production wreaks on the environment. These gaps in the liberal environmentalist policy-assemblage challenge its adequacy and appropriateness as a viable intervention, whether to address waste or for global challenges such as halting anthropogenic climate change. By excluding from analysis the economic and political aspects of the dynamic interactions between human culture and the natural world, such polies are rendered inadequate as a means to address one of the core drivers of anthropogenic climate change: the capitalist mode of economic production.

Green Capitalism

The liberal environmentalist policy position has more recently morphed into policies that may be designated as 'green capitalism'. This latter has been a perspective favoured by policy-makers of multiple political hues, though most vociferously by politicians on the Right and entrepreneurs. Simply put, while liberal environmentalism ignored the effects of a market economy upon the environment and climate change, green capitalism considers a market economy as the means by which the world can be rescued from the current climate crisis (Tienhaara, 2014; Zysman & Huberty, 2014). It is reflected in the rapid shift of the car industry toward electric vehicle and battery manufacture, and the embrace by the political Left of renewable energy production as a new growth industry providing jobs and cheaper energy (Russon & Prescott, 2021).

Two versions of green capitalism have emerged. The first of these suggested that climate change was a failure in how markets should ideally operate (Stern, 2007: 1). This revisionist turn was best represented by the work of Nicholas Stern, former chief economist at the World Bank and a former senior UK Treasury civil servant. In Stern's (2007) assessment, capitalist markets failed due to the ability of greenhouse gas producers, mainly Northern nations, to limit global climate change consequences while impacting regions not responsible for emissions. He proposed market intervention via regulation, taxation, and international cooperation through three strategies: implementing efficient carbon pricing/trading, fostering low-carbon technology innovation, and employing incentives and disincentives to alter consumer and business behaviour (Stern, 2007: xviii–xxi). With these tweaks, a market economy can become the environment's saviour (see also Pearce et al., 1989: 153–171).

The second and more radical version of green capitalism transformed liberal environmentalism into a 'neoliberal environmentalism'. Proponents of this policy position argue that ingenuity, entrepreneurialism and capitalism's unending quest for profit together are the only means to reverse climate change (Prudham, 2009: 1596). These features, according to green capitalists, have created widespread prosperity through industrialisation, as technological innovation enhanced production and consequently capital accumulation. Green capitalism is distinctive in 'the increasing incorporation and internalisation of ecological conditions into the circuits of capital accumulation' (Prudham, 2009: 1596). Zysman and Huberty (2014: xiii) suggest synergy between a capitalist free-market and the environment, within a 'green spiral', that:

> reflects a process of mutually reinforcing feedback between climate policy and industrial interests, in which the development of new infrastructure and energy approaches creates new economic clienteles who then become advocates for further action. ... These green industrial interests help stabilise policies in place and push for new policies, offsetting opposition from interests tied to the pre-existing system.

These green spirals can fuel new technologies and methods of production, according to Zysman and Huberty. However, success will depend upon 'immediate material gains … that can sustain the search for broad growth opportunities capable of supporting our long-term goals for the economy, the environment, and the planet at large'. (Zysman & Huberty, 2014: xiv).

The continuities between these two green capitalist turns in environmental policy emerge when analysed micropolitically. In both analyses, capitalist economics *is* part of the climate change assemblage—unlike the liberal environmentalist perspective, which ignores this affective connection. At the same time, green capitalist policies de-couple environment sustainability from economic development and growth: the latter do not really feature in these green capitalist policy assemblages. In addition, both versions of green capitalism recognise that markets can and should alter how they operate, so that they enhance the environment rather than destroy it. While Stern (2007) saw governmental regulation and interventions as key to re-direct the flows of capital away from production of greenhouse gases, neo-liberal environmentalists consider the quest for profit within a free market as having the in-built affective capacity to move economies towards low-carbon energy technologies.

Once again, the interaction between event and policy assemblages illustrated in Fig. 3.1 can be applied to this policy position. In a green capitalist conception, the climate change event-assemblage (EA) comprises at least (in alphabetical order):

> capital; climate; consumers; developing and developed nations and governments; economic growth; energy; entrepreneurs; existing energy technologies; greenhouse gases; industry; innovative green technologies; material resources ('the environment'); market economy; means of production; profit; the Sun;

As can be seen, the green capitalist version of the climate change EA incorporates many elements excluded from the liberal environmentalist event-assemblage considered earlier. The principal affect in the green capitalist policy-assemblage that interacts with this EA links entrepreneurial efforts to transform raw materials such as aluminium and lithium into 'green' innovations such as solar panels, electric vehicles and wind turbines. Such products can then compete and eventually replace technologies generating greenhouse gases, and thereby halt global warming (Fox, 2022).

However, this policy-assemblage (PA) excludes from attention some other key affects that a comprehensive analysis of anthropogenic climate change identifies. By reconstituting the environment as a resource for capitalist production, and humans simply as workers or consumers, this event-assemblage is 'capitalocentric'. Concerns with environmental sustainability are replaced with an overarching concern for economic development, which is treated as the basis for worthwhile human existence: any benefits for the environment are an added bonus. Inherent problems deriving from the market competition baked into a capitalist mode of production, including an incessant need for growth (Bosquet, 1977: 166) and consequent wastefulness (Yearley, 2014; 106) are not acknowledged as fundamental causes of environmental degradation and greenhouse gas emissions, while national and global inequalities in wealth and well-being linked to a capitalist mode of production are also un-addressed.

The development of electric vehicles (EVs) supplies a good example of this capitalocentric affect-economy. While EVs can play a major part in reducing the production of greenhouse gases by drivers, this green capitalist approach ignores the environmental costs of extracting resources, in particular lithium and other rare elements for EV battery production, mostly in global South developing countries (Sovacool et al., 2019: 213). The shift to mass production of EVs (while supplying car manufacturers with a welcome opportunity for innovation and hence profit) will also result in substantial increases in waste: not only when EV batteries (which are only 50% recyclable) reach the end of their useful life in approximately 10 years (Richter, 2022), but also the wastage of billions of redundant fossil-fuelled vehicles and all the vehicle production plants that EV manufacturing will replace over the next twenty years. Furthermore, the EV industry depends upon rapid energy-intensive market growth, while cut-throat competition between major vehicle manufacturers will lead to further wastage as uncompetitive corporations fail or are taken over (Fox, 2022). These economic aspects of the shift to EVs challenge the capacity of this kind of technology-driven initiative to drive down greenhouse gas production.

This analysis of the green capitalist policy-assemblage supports the views of many scholars, who have argued that it is highly questionable whether green capitalism is adequate and/or appropriate to achieve anything approaching net zero carbon use (Blakeley, 2021; Harris, 2014; Keen, 2021; Prudham, 2009; Smith, 2016), and is indeed part of the problem underpinning anthropogenic climate change.

Discussion: Towards a More-Than-Human Climate Change Policy

A policy-assemblage (PA) approach has provided a micropolitical analysis of two of the principal policy positions on environmental sustainability and sustainable development in current political discourse. This novel methodological approach has allowed comparison between the affect-economies of liberal environmentalist and green capitalist PAs and the event-assemblage (EA) that they aim to address: anthropogenic climate change. This assessment revealed that neither liberal environmentalism nor green capitalism PAs fully engaged with the complex natural and social event assemblages of anthropogenic climate change and other environmental impacts. Neither, it may thus reasonably be argued, is consequently adequate or appropriate as a climate change policy.

However, this micropolitical perspective to policy analysis offers further opportunities not available in approaches such as a discourse analysis of policy statements (Gasper & Apthorpe, 1996). Not only does it enable evaluation of policies, but also the means to devise an adequate and appropriate policy assemblage. This concluding section develops a more-than-human policy assemblage that can supply a feasible way forward to address environmental sustainability and the climate crisis, acknowledging humans as an integral though not privileged *part* of environment (Fox & Alldred, 2020b).

Micropolitical analysis indicates that both liberal environmentalism and green capitalism PAs retain an anthropocentric bias, despite their claims to address the needs of the non-human 'environment'. In the former, the environmental damage caused by human actions are not subjected to critical assessment and the focus is upon further human action to ameliorate this damage. In the latter, human ingenuity, as demonstrated by both technological advances *and* the emergence of the capitalist economic mode of production, is hailed as the saviour of a climate unwittingly damaged by previous human technologies and the same economic model.

Earlier, this chapter offered a critique of anthropocentric ontologies of environment, and developed a more-than-human alternative that dissolves the dichotomy of human/environment. From a more-than-human standpoint, humans and their culture are integral to the environment, deserving equal consideration without privilege or exclusion (Fox & Alldred, 2020b). This perspective does not prioritise current or future human needs but acknowledges humans' distinctive abilities within the environment, such as attributing meaning to events, practicing altruism, innovating technologies, and employing reason for theorizing and foreseeing future events (Fox & Alldred, 2021; Schmidt, 2013: 189–190). Sustaining these positive unique capacities will be part of the task of any successful climate change policy.

This more-than-human ontology supplies a foundation upon which to design a climate change PA that has the capacity to overcome the anthropocentrism of existing policy positions. Such a comprehensive climate change EA comprises (in alphabetical order):

animals; atmosphere; capitalist economic model; competition; consumption; Earth; global North; global South; governments; greenhouse gases; growth; humans; ideologies; industry; market; nations; natural resources; oceans; plants; profit; Sun; waste; wealth and health inequalities

To be similarly comprehensive, a more-than-human PA needs to engage with the complex affect-economy in this EA, and devise actions that effectively address these affects, while also aligning itself with a broader perspective on environmental sustainability. While the development of such a policy would require extensive discussions among governments, experts, and stakeholders at both global and national levels, certain overarching affects crucial to this policy framework can be swiftly identified and integrated.

First: this PA should recognise the challenges currently facing many life-forms and habitats from human development and population expansion. Policy needs to respond swiftly and comprehensively to the negative environmental impacts of the industrialisation of production, the unending extraction of natural resources, and the pollution caused by waste products. However, this needs to go hand-in-hand with local and global income redistribution, to assure basic living standards for all humans, regardless of social contexts or nationality.

Second: it should acknowledge the part that human altruism, ingenuity and technology can play in replacing means of production that are generating greenhouse gases and otherwise damaging the environment and non-human species (Fox & Alldred, 2020b). But as a necessary qualification to this, a further acknowledgment is required: that it has been the capitalist market economy—with its objective of capital accumulation and consequent need for incessant growth—that drove the industrialisation that in turn resulted in environmental degradation and the current climate crisis, and continues so to do (Baer, 2012; Bosquet, 1977: 166). Human ingenuity and innovation needs to be de-coupled from capitalism and its pursuit of competition, economic growth and profit, with concomitant social, economic and political transformations required.

These affective movements supply a basis for a comprehensive policy approach involving incremental actions, spanning from local to global levels, to tackle the multifaceted impacts of climate change across natural, biological, social, economic and political realms. These actions are not standalone choices but are intricately interconnected, forming a coherent strategy to effectively address climate change complexities. Importantly, this is a long-term, worldwide initiative that hinges on political determination, effective leadership, and collaboration among diverse stakeholder groups to achieve success. Such a policy assemblage to counter anthropogenic climate change—assuring both environmental protection and sustainable development, encompasses several interconnected threads.

A first strand emphasises the use of fiscal and regulatory mechanisms to incorporate the true environmental costs of goods and services into their pricing. This approach aims to minimise the extraction of new environmental resources and promote recycling and reuse. The focus is on strengthening protection for non-human resources, acknowledging the interconnections between environmental and human well-being. Second, the programme highlights the importance of supporting environmentally-friendly technologies by offering tax incentives and promoting the global sharing of intellectual property. Key infrastructure sectors like transportation, food production and energy generation and distribution should be managed independently of market forces to facilitate more rational resource consumption. The expertise and technology required for efficient recycling and the replacement of non-recyclable materials, such as plastics, should be shared globally.

However, the programme must also acknowledge the need for intervention in capitalist markets through legislation, regulation, and taxation, reintroducing 'market distortions' that neoliberal advocates have sought to eliminate over the past 200 years (Fox, 2023). It aims to eliminate wasteful production processes and encourage the production of long-lasting and sustainable products. The proposal also advocates for regional trading zones, like the European Union and ASEAN, as alternatives to global free trade that often favours wealthier nations.

Alongside these economic and political intervention, a final strand of the programme would ensure human social and economic security by addressing wealth inequalities. It suggests the implementation of universal basic income schemes as a means to replace existing welfare and means-tested benefits systems. This would be financed through higher rates of taxation on personal income, corporate profits, and penalties for environmental pollution. Additionally, population growth management and migration policies aligned with global environmental objectives are considered essential.

Implementing this ambitious programme will necessitate extensive collaboration and alliances, both domestically and internationally. Depoliticising climate change policy and empowering organisations such as the United Nations and the IPCC with authority and resources could facilitate its implementation. Moreover, the cooperation and support of countries in the global South are seen as crucial for the success of these challenging policies.

In summary, such a comprehensive climate change policy-assemblage represents a comprehensive approach to address environmental challenges, wealth inequalities, and sustainable development. It recognises the need for significant changes in taxation, regulation, and international cooperation to transition toward a more sustainable and equitable global future. While challenging, this policy perspective is urgently required to make global and national climate change policies fit for purpose, and to reveal liberal environmentalism and green capitalism as at best sticking plasters and at worst insidious ideologies that will carry the world toward inevitable climate catastrophe.

Conclusion

The chapter has questioned the effectiveness of two of the principal current international policies to address the urgent climate crisis. These policies were explored from a holistic perspective, considering the comprehensive scientific and social understandings of anthropogenic climate change. Methodologically, it applied a 'policy assemblage' analysis, and used this to evaluate the liberal environmentalist and 'green capitalist' policy perspectives. Both were found inadequate as approaches to combating climate threats resulting from human activity. Instead, the chapter has suggested a more comprehensive, and scientifically and politically sound climate change policy as an effective approach to address the pressing environmental challenges faced by both human and non-human inhabitants of planet Earth.

Notes

1. Knowledge of a topic such as anthropogenic climate change will never be perfect, due to the highly complex physical, biological and social affect-economies in event assemblages such as these, as yet not fully understood by scientists and social scientists.
2. Liberal environmentalism was a foundational principle in UN sustainability policies since the 1970s. The *5th Assessment Report of the Intergovernmental Panel on Climate Change* (IPCC, 2014) reaffirmed the dual objectives of socioeconomic development and environmental sustainability. While non-human life relies solely on environmental sustainability, human well-being depends on both elements (IPCC, 2014: 137). The United Nations' (2016) *Agenda for Sustainable Development* reiterated these goals and outlined 17 sustainable development objectives. Surprisingly, 13 of these objectives focused primarily upon improving human life quality, including poverty eradication, gender equality, and access to clean water and energy. Only three goals were directed toward environmental concerns. Notably, the strategy relied heavily on economic growth rather than wealth redistribution between the global South and North. However, the assumed positive link between economic development and environmental protection has been widely questioned (Rees, 2003) and sometimes found to be contradictory (Wallis, 2010; Yearley, 2014: 104).

References

Baer, H. (2012). *Global capitalism and climate change*. AltaMira Press.

Barad, K. (2001). (Re)configuring space, time and matter. In M. Dekoven (Ed.), *Feminist locations* (pp. 75–109). Rutgers University Press.

Bennett, J. (2005). The agency of assemblages and the North American blackout. *Public Culture, 17*(3), 445–465.

Bennett, J. (2010). *Vibrant matter*. Duke University Press.

Benton, T. (1991). Biology and social science: Why the return of the repressed should be given a (cautious) welcome. *Sociology, 25*(1), 1–29.

Bernstein, S. (2002). Liberal environmentalism and global environmental governance. *Global Environmental Politics, 2*(3), 1–16.

Blakeley, G. (2021). Green capitalism is not enough. *Tribune* (20 Jan 2021). https://tribunemag.co.uk/2021/01/green-capitalism-is-not-enough

Bosquet, M. (1977). *Capitalism in crisis and everyday life*. Harvester Press.

Braidotti, R. (2011). *Nomadic theory*. Columbia University Press.

Braidotti, R. (2013). *The Posthuman*. Polity.

Buchanan, I. (1997). The problem of the body in Deleuze and Guattari, or, what can a body do? *Body & Society, 3*, 73–91.

Burstein, P., & Linton, A. (2002). The impact of political parties, interest groups, and social movement organizations on public policy: Some recent evidence and theoretical concerns. *Social Forces, 81*(2), 380–408.
Carroll, J. (1993). *Humanism. The wreck of modern culture.* Fontana.
Chasek, P. S., & Downie, D. L. (2020). *Global environmental politics.* Routledge.
Clough, P. (2004). Future matters: Technoscience, global politics, and cultural criticism. *Social Text, 22*(3), 1–23.
Collard, R. C., & Dempsey, J. (2022). Future eco-perfect: Temporal fixes of liberal environmentalism. *Antipode, 54*(5), 1545–1565.
Coole, D. H., & Frost, S. (2010). Introducing the new materialisms. In D. H. Coole, & S. Frost (Eds.), *New materialisms. Ontology, agency, and politics* (pp. 1–43). Duke University Press.
Dauvergne, P. (2018). The power of environmental norms: Marine plastic pollution and the politics of microbeads. *Environmental Politics, 27*(4), 579–597.
DeLanda, M. (2006). *A new philosophy of society.* Continuum.
Deleuze, G. (1988). *Spinoza: Practical philosophy.* City Lights.
Department for Environment, Food & Rural Affairs. (2013). *Prevention is better than cure. The role of waste prevention in moving to a more resource efficient economy.* DEFRA.
Dror, Y. (2017). *Public policymaking reexamined.* Routledge.
Dunlap, R. E., & Catton, W. R. (1979). Environmental sociology. *Annual Review of Sociology, 5*, 243–273.
Dunlap, R. E., & Catton, W. R. (1994). Struggling with human exemptionalism: The rise, decline and revitalization of environmental sociology. *The American Sociologist, 25*(1), 5–30.
Fox, N. J. (2022). Green capitalism, climate change and the technological fix: A more-than-human assessment. *The Sociological Review, 71*(5), 1115–1134.
Fox, N. J. (2023). Capitalism and the state: a new materialist perspective. *European Journal of Social Theory*, 28(1): 126-148.
Fox, N. J., & Alldred, P. (2017). *Sociology and the new materialism.* Sage.
Fox, N. J., & Alldred, P. (2020a). Re-assembling climate change policy: Materialism, micropolitics and the policy assemblage. *British Journal of Sociology, 71*(2), 269–283.
Fox, N. J., & Alldred, P. (2020b). Sustainable development, posthumanism and the unusual capacities of humans. *Environmental Sociology, 6*(2), 121–131.
Fox, N. J., & Alldred, P. (2021). Economics, the climate change policy-assemblage and the new materialisms: Towards a comprehensive policy. *Globalizations, 18*(7), 1248–1258.
Gasper, D., & Apthorpe, R. (1996). Introduction: Discourse analysis and policy discourse. *European Journal of Development Research, 8*(1), 1–15.
Gunningham, N. (2019). Averting climate catastrophe: Environmental activism, extinction rebellion and coalitions of influence. *King's Law Journal, 30*(2), 194–202.
Halpern, D. (2013). *Mental health and the built environment: More than bricks and mortar?* Routledge.
Harris, J. (2014). Can green capitalism build a sustainable society? *Perspectives on Global Development and Technology, 13*(1–2), 43–60.
Hoehner, C. M., Brennan, L. K., & Brownson, R. C. (2003). Opportunities for integrating public health and urban planning approaches to promote active community environments. *American Journal of Health Promotion, 18*(1), 14–20.
IPCC. (2014). *Climate change 2014: Mitigation of climate change. Working group III contribution to the fifth assessment report of the intergovernmental panel on climate change.* Cambridge University Press.
Keen, S. (2021). The appallingly bad neoclassical economics of climate change. *Globalizations, 18*(7), 1149–1177.
Latour, B. (1993). *We have never been modern.* Harvard University Press.
Latour, B. (2005). *Reassembling the social. An introduction to actor network theory.* Oxford University Press.

Lemke, T. (2015). New materialisms: Foucault and the 'government of things.' *Theory, Culture & Society, 32*(4), 3–25.
McCann, E. (2011). Veritable inventions: Cities, policies and assemblage. *Area, 43*(2), 143–147.
McCann, E., & Ward, K. (2012). Assembling urbanism: Following policies and 'studying through' the sites and situations of policy making. *Environment and Planning A, 44*(1), 42–51.
Patton, P. (2000). *Deleuze and the political*. Routledge.
Pearce, D., Markandya, A., & Barbier, E. B. (1989). *Blueprint for a green economy*. Earthscan.
Prince, R. (2010). Policy transfer as policy assemblage: Making policy for the creative industries in New Zealand. *Environment and Planning A, 42*(1), 169–186.
Prudham, S. (2009). Pimping climate change: Richard Branson, global warming, and the performance of green capitalism. *Environment and Planning A, 41*(7), 1594–1613.
Rees, W. E. (2003). Economic development and environmental protection: An ecological economics perspective. *Environmental Monitoring and Assessment, 86*(1), 29–45.
Richter, J. L. (2022). A circular economy approach is needed for electric vehicles. *Nature Electronics, 5*(1), 5–7.
Russon, M.-A., & Prescott, A. (2021). TUC: Jobs at risk if UK fails to hit carbon emissions target. *BBC News* (12 Sept 2021). https://www.bbc.co.uk/news/business-58519996
Schmidt, J. (2013). The empirical falsity of the human subject: New materialism, climate change and the shared critique of artifice. *Resilience, 1*(3), 174–192.
Shore, C., & Wright, S. (1997). Policy: A new field of anthropology. In C. Shore & S. Wright (Eds.), *Anthropology of policy* (pp. 3–36). Routledge.
Shove, E., Pantzar, M., & Watson, M. (2012). *The dynamics of social practice: Everyday life and how it changes*. Sage.
Smith, R. (2016). *Green capitalism: The god that failed*. WEA/College Publications.
Sovacool, B. K., Kester, J., Noel, L., & de Rubens, G. Z. (2019). Energy injustice and Nordic electric mobility: Inequality, elitism, and externalities in the electrification of vehicle-to-grid (V2G) transport. *Ecological Economics, 157*, 205–217.
Steffen, W., Crutzen, P. J., & McNeill, J. R. (2007). The Anthropocene: Are humans now overwhelming the great forces of nature? *AMBIO: A Journal of the Human Environment, 36*(8), 614–621.
Stern, N. (2007). *The economics of climate change: The Stern report*. Cambridge University Press.
Stevens, P. (2012). Towards an ecosociology. *Sociology, 46*(4), 579–595.
Swinburn, B., Egger, G., & Raza, F. (1999). Dissecting obesogenic environments: The development and application of a framework for identifying and prioritizing environmental interventions for obesity. *Preventive Medicine, 29*(6), 563–570.
Talshir, G. (2012). The role of environmentalism. In Y. Levy, & M. Wissenburg (Eds.), *Liberal democracy and environmentalism. The end of environmentalism?* (pp. 22–43). Routledge.
Taylor Webb, P., & Gulson, K. N. (2012). Policy prolepsis in education: Encounters, becomings, and phantasms. *Discourse: Studies in the Cultural Politics of Education, 33*(1), 87–99.
Thrift, N. (2008). *Non-representational theory: Space, politics, affect*. Routledge.
Tienhaara, K. (2014). Varieties of green capitalism: Economy and environment in the wake of the global financial crisis. *Environmental Politics, 23*(2), 187–204.
United Nations. (2016). *Sustainable development goals*. United Nations. http://www.un.org/sustainabledevelopment/
Ureta, S. (2014). Policy assemblages: Proposing an alternative conceptual framework to study public action. *Policy Studies, 35*(3), 303–318.
Urry, J. (2009). Sociology and climate change. *The Sociological Review, 57*(s2), 84–100.
van der Tuin, I., & Dolphijn, R. (2010). The transversality of new materialism. *Women: A Cultural Review, 21*(2), 153–171.
Walker, G. (2005). Sociological theory and the natural environment. *History of the Human Sciences, 18*(1), 77–106.
Wallis, V. (2010). Beyond 'green capitalism'. *Monthly Review, 61*(9): 32–48.

Whitehead, M. (2014). Sustainability. In C. Death (Ed.), *Critical environmental politics* (pp. 257–266). Routledge.
Widder, N. (2012). *Political theory after deleuze*. Continuum.
Wiktorowicz, M. (2003). Emergent patterns in the regulation of pharmaceuticals: Institutions and interests in the United States, Canada, Britain, and France. *Journal of Health Politics, Policy and Law, 28*(4), 616–658.
Yearley, S. (2014). *The green case: A sociology of environmental issues, arguments and politics*. Routledge.
Zysman, J., & Huberty, M. (2014). *Can green sustain growth?* Stanford University Press.

Nick J. Fox is professor of sociology at the University of Huddersfield and is also adjunct professor at Torrens University Adelaide. Recent work using new materialist and more-than-human approaches includes studies of sustainable development, social disadvantage and health, political sociology, and materialist research methodology. He is the author of many papers and five books, including (with Pam Alldred) *Sociology and the New Materialism* (Sage, 2017).

Pam Alldred is professor in the Social Work, Social Care and Communities department, Nottingham Trent University. She researches sexualities, parenting and sex education and has led two large international projects on gender-related violence and then on sexual violence, with European Union co-funding. Pam recently published *Sociology and the New Materialism* (with Nick J. Fox, Sage, 2017) and is co-editor of the *Handbook of Youth Work Practice* (Sage, 2018).

Chapter 4
Policy Interventions for Inclusive Sustainable Development and Climate Change Management

Jacqueline Njambi Kamau, Justine Magutu, and Henry Amadi

Abstract Climate change poses significant challenges to Kenya with erratic weather patterns, including prolonged droughts and increased flooding. To adjust to these significant variances, science, and technology were integrated to support policy-making. The Climate Change Act, 2016, National Climate Change Action Plan 2018–2022, and Sessional Paper No. 3 National Climate Change Framework Policy are among Kenya's government's policies for climate change mitigation and adaptation. The objective of the study was to examine whether gender-based inequality is promoted through policy gaps and policy limitations. The study used desk research and database analysis on the available literature and policy papers. Ecofeminism theoretical framework was applied while data analysis used qualitative content analysis. The results show that while climate change legislation has supported resilience, there are no interventions for gender-based inequality. The policies emphasize agricultural production to be responsive to climate change using procedures that enhance standards and productivity levels. For instance, helping a corn farmer to boost productivity per acre. The policies are insensitive to gender prejudices, and there are no provisions to enhance the working conditions for female laborers. The climate change policies have targeted the right actors; however, integrating science and technology will enhance the success of climate-smart agriculture.

Keywords Science · Technology · Sustainable development · Social inclusion · Just transition

J. N. Kamau (✉) · J. Magutu · H. Amadi
University of Nairobi, Nairobi, Kenya
e-mail: jacqueline.kamau@gmail.com

J. Magutu
e-mail: magutu@uonbi.ac.ke

H. Amadi
e-mail: henry.odongo@uonbi.ac.ke

W. Leal Filho et al. (eds.), *Climate Change Mitigation and Adaptation in Practice*,
Climate Change Management, https://doi.org/10.1007/978-3-031-85217-6_4

Introduction

Policy interventions are pivotal instruments in the contemporary pursuit of harmonizing progress, sustainability, and climate action. Rooted in the Constitution of Kenya under the right to a safe and secure environment, these interventions hold the power to shape nations towards inclusive sustainable development and effective climate change management (Timu & Moses, 2022). The intricate interplay of science, technology, and socio-economic considerations necessitates a strategic approach that safeguards the environment and ensures social inclusion, particularly within the framework of a just transition.

The Constitution of Kenya's commitment to environmental sustainability is enshrined in Article 42 of the Bill of Rights. It espouses the entitlement of every individual to a clean and salubrious environment, emphasizing its preservation for current and future generations' welfare (Kogo et al., 2022). Moreover, Chapter Five of the Constitution expounds upon the critical realm of land and environmental concerns, amplifying their importance as keystones for sustainable progress. Aiming higher, Kenya's Vision 2030, a sweeping national development blueprint, prominently prioritizes combatting climate change as a cornerstone of its agenda (Gebre et al., 2023).

Kenya's governance mechanisms demonstrate a pronounced commitment to steering policy implementations and reforms towards gender-responsive actions, with a particular focus on Climate Smart Agriculture (CSA). This commitment has propelled intensified conservation endeavours across both rural hinterlands and urban landscapes (Bedeke, 2023). Imbued with multifaceted aspirations, one of the central tenets of Kenya's emphasis on CSA is the assurance of robust food security. This strategic imperative underscores the imperative of sustainable food production within the broader climate agenda.

The discourse on CSA is closely interwoven with women's roles in this paradigm. As highlighted by Beal et al. (2021), women play a significant labor force role in CSA. Consequently, any entrepreneurial endeavour in this sphere demands the prioritization of equitable remuneration and social security, establishing a foundation for greener and more climate-resilient agri-food systems. However, a salient challenge persists: prevailing policies lack provisions that empower young women to engage in CSA practices. This void in policy formulation has exacerbated gender-based disparities and the marginalization of vulnerable groups, imperilling the attainment of globally agreed-upon development goals, such as the Sustainable Development Goals (SDGs) and the Paris Agreement (Friedman et al., 2022).

A litany of policy instruments has been fashioned to address the difficulties of climate change in Kenya, exemplifying the country's conscientious approach to environmental stewardship (Kogo et al., 2021). The Environmental and Management Coordination Act of 1999, the Climate Change Act of 2016 (No. 11 of 2016), and the National Climate Change Action Plan (NCCAP), spanning 2018 to 2022, form the bedrock of Kenya's policy apparatus. Supplementary frameworks, including Sessional Paper No. 3 of 2016, the National Climate Change Response Strategy

2010 (NCCRS), and the National Adaptation Plan 2015–2030, intricately choreograph Kenya's climate action strategy. Comprehensively addressing the economic facet, the National Climate Finance Policy of 2018 and the Green Economy Strategy and Implementation Plan of 2017 underscore the pragmatic integration of fiscal mechanisms (Mairura et al., 2021).

A panoramic view of Kenya's climate-conscious policies also encompasses agriculture. The Kenya Climate Smart Agriculture Implementation Framework 2018–2027 (KCSAIF) and the Kenya Climate Smart Agriculture Strategy 2017–2026 (KCSAS) manifest the nation's commitment to fostering sustainable agricultural practices amidst evolving climatic dynamics. To synergize the trajectory, even the Energy Act of 2019 orchestrates policies within the energy sector, bridging the chasm between sustainable energy and climate resilience.

The policy interventions in Kenya's pursuit of inclusive sustainable development and climate change management are a testament to the nation's unwavering commitment. Anchored in the constitutional bedrock and fortified by multifaceted policy instruments, these interventions echo the intricate interplay between science, technology, and socio-economic dimensions. As Kenya surges forward with its policies for a greener and more equitable future, these interventions hold the promise of shaping the national landscape and influencing the global discourse on sustainable development and climate resilience.

Background

The agricultural structures in Kenya are very susceptible to climate change impacts since the systems are rain-fed and highly underdeveloped. Thus, women involved are very vulnerable since they are exposed to shocks such as drought and floods caused by changes in weather conditions (Derenoncourt, 2022). There is a need for a comprehensive policy approach to promote gender-sensitive Climate Change Policies.

Kenyans are grappling with substantial shifts in their climatic conditions, prompting critical concerns. By January 2023, 40 of the country's 47 counties are ensnared in severe drought and famine (Howlett & Ramesh, 2023). This dire situation is rooted in erratic rainfall patterns and the escalating trajectory of temperatures. The conventional lengthy rainy season has shrunk into a parched period, while the abbreviated rainy interval has transformed into an extended and deluge-laden stretch (Howlett & Ramesh, 2023). The oscillation between drought and floods persistently plagues the country, accentuating its climate vulnerabilities.

According to Bedeke (2023), amidst the catastrophic backdrop of drought and famine, the plight of women and girls takes on a particularly distressing dimension. These vulnerable groups are not only subjected to gender-based violence. Still, they are also relegated to the fringes of food availability, often receiving what is left after boys and men have consumed their share (Friedman et al., 2022). The staggering levels of malnutrition among women in comparison to men reflect deeply entrenched disparities. These inequities are closely tied to illiteracy, lack of awareness, limited progress opportunities, and pervasive poverty (Howlett & Ramesh, 2023). Women experience malnutrition at a significantly higher rate of 25.2%, compared to men at 20.2% (Watson et al., 2022). Uduji and Okolo-Obasi (2023), underscore the necessity for a comprehensive spectrum of development strategies to enhance women's food security and nutrition amidst these adversities. Consequently, climate actions are increasingly being adopted as measures of adaptation to navigate these profound deviations.

Kenya's economic backbone heavily relies on tourism and rain-fed agriculture, rendering these sectors acutely susceptible to the repercussions of climate change and the ensuing extreme weather events (Kim et al., 2022). Escalating temperatures and recurring droughts translate into catastrophic losses in crops and livestock, imperiling human health and well-being. The resulting adversities span from starvation to displacement, unraveling the very fabric of existence (Howlett & Ramesh, 2023). Such climate-induced perturbations have considerably exacerbated the vulnerability of marginalized communities across various regions of Kenya. The fallout is manifested through amplified job losses, heightened food insecurity, intensified competition over scant resources, amplified human migration, and the exacerbation of political and economic instability (Seddon, 2022).

According to Howlett and Ramesh (2023), just like other societal concerns, gender has a divisive impact on how well people can adjust to climatic challenges. It has been found that a combination of factors, including power conceptions, social norms, asset control, and decision-making processes, affects how vulnerable people, both men and women, improve their capacity to respond to climate change impacts (Bryan et al., 2024).

The government has made efforts, including developing climate change policies. However, the policies have not been effective (Seddon, 2022). Policy principles enshrined in EPSK may have contributed since climate change policy interventions are still gender-responsive. Despite various efforts by the government, gender parity, access to resources, and social inclusion remain significant issues, leaving women exposed to discrimination and social disorder (Bryan et al., 2024).

Problem Statement

Climate change management and pursuing inclusive, sustainable development represent a critical global imperative. In the Kenyan context, the intersection of these two imperatives underscores the necessity for effective policy interventions that harmonize environmental protection, economic growth, and social equity (Kamau, 2018). While policy frameworks are in place, a persistent problem arises from the insufficient integration of these policies into a coherent strategy that holistically addresses the multifaceted challenges posed by climate change while ensuring equitable benefits for all segments of society (Babugura, 2021).

The existing policy landscape in Kenya reveals a comprehensive assortment of climate change management policies, ranging from the Climate Change Act of 2016 to the National Climate Smart Agriculture Strategy of 2017–2026 (Kenya Climate Change Working Group, 2019). Despite the breadth of these initiatives, their effectiveness is hindered by the fragmentation of efforts, resulting in suboptimal coordination and collaboration among diverse stakeholders (Kenya Climate Change Working Group, 2019). This deficiency undermines the overarching objective of sustainable development that integrates economic growth, social inclusivity, and environmental stewardship (Babugura, 2021).

Furthermore, the engagement of marginalized and vulnerable groups, such as women and rural communities, remains inadequate within the current policy landscape (Bedeke, 2023). Gender-responsive policy implementation in Climate Smart Agriculture, for instance, is hindered by the absence of tailored policies that empower young women and ensure their equitable participation in sustainable agricultural practices (Beal et al., 2021). Consequently, this absence perpetuates gender-based inequalities and constrains the realization of internationally agreed-upon development goals, such as the Sustainable Development Goals and the Paris Agreement (Friedman et al., 2022).

Addressing these policy implementation gaps and achieving inclusive, sustainable development in the face of climate change in Kenya is a pressing challenge. A comprehensive analysis of the existing policies, their interconnections, and the integration of gender-responsive strategies is imperative (Kogo et al., 2022). Moreover, bridging the gap between policy design and actual practice requires a cohesive approach that fosters collaboration among government agencies, non-governmental organizations, private sectors, and local communities (Kenya Climate Change Working Group, 2019). By aligning policies with equity, inclusivity, and environmental integrity, Kenya can navigate the complex terrain of climate change while ensuring that sustainable development benefits are accessible to all strata of society.

Kenyans continue to experience drastic climatic changes. In January 2023, 40 counties out of the country's forty-seven counties are experiencing drought and famine (Rajabov, 2022). The drought is a result of poor rainfall patterns and rising temperatures. The long rainy season has become short and dry, while a short rainy season is long and wet (Seddon, 2022). The country continues to experience drought and floods in its seasons. In the wake of drought and famine, women and girls have been exposed to gender-based violence, and they are the last to eat what is left by boys and men (Rajabov, 2022).

Kim et al. (2022) argue that levels of malnutrition among women are high compared to men. Malnutrition in women is linked to illiteracy, lack of awareness, lack of progress, and poverty. According to the most recent poll, women are 25.2% more malnourished than men (20.2%). Based on Bedeke (2023), a wide range of development measures are necessary to improve women's nutrition and food security. Climate Actions are applied to adapt to these extreme deviations. Tourism and rain-fed agriculture, vulnerable to climate change and extreme weather events, are the two main drivers of Kenya's economy. Rising temperatures and frequent droughts cause catastrophic crop and livestock losses that risk human health and well-being by causing starvation, relocation, and other problems (Rajabov, 2022).

Climate change has already exacerbated the insecurity of vulnerable populations in numerous regions of Kenya by increasing job loss, food insecurity, competition for scarce resources, human movement, and political and economic instability (Uduji & Okolo-Obasi, 2023). Despite applying climate actions and gender equality interventions in public policies, the challenges continue to grow (Rajabov, 2022). Thus, this study sought to evaluate the policy interventions implemented to promote inclusive governance regimes and reduce gender-based inequalities.

Study Objectives

General and Specific Objectives

The study's general objective was to evaluate the policy interventions for inclusive climate governance regimes towards reducing gender-based inequalities. The study's specific objectives include examining policy gaps and barriers that promote gender-based inequalities and social exclusion and analyzing the promotion of decent green jobs. Further, the study seeks to evaluate gender responsiveness in reporting on climate change.

Literature Review

Just and Inclusive Adaptation

The public Policies in climate change mitigation and adaptation promote just Transition. In pursuing inclusive, sustainable development and effective climate change management in Kenya, policy interventions are pivotal in shaping adaptive strategies that embrace technological advancements and social inclusion (Uduji and Okolo-Obasi, (2023). As climate change poses significant challenges to societies worldwide, the differential impacts on genders become increasingly evident (Kamau, 2018). Disparities in roles, access to resources, and power dynamics can undermine the adaptive capacities of women and men in the face of climate risks. Thus, integrating just and inclusive adaptation measures within public policies becomes imperative.

The climate change sector in Kenya recognizes the multifaceted nature of adaptation and emphasizes integrating technology and social inclusion to foster resilient communities. The policies enacted strive to bridge the gender gap in adaptation strategies, acknowledging the diverse roles and responsibilities that men and women hold within society. However, the existing disparities in resource allocation, decision-making, and access to opportunities often leave women disproportionately vulnerable to climate-induced risks (Githinji & Nyangena, 2021).

Gender-sensitive climate policies have gained prominence due to the recognition that climate change impacts are not gender-neutral. It is widely acknowledged that climatic shifts disproportionately affect women due to their distinct socio-economic roles and limited access to resources. Women are more likely to be engaged in sectors highly exposed to climate vulnerabilities, such as agriculture, and their responsibilities often encompass tasks directly influenced by climate variations (Nyika, 2022). This intertwining of gender dynamics and climate impacts necessitates a comprehensive policy formulation—one sensitive to the diverse experiences and needs of men and women.

The significance of incorporating social inclusion within climate adaptation policies lies in its potential to empower marginalized groups, including women, youths, and persons with disabilities. Policies can create an enabling environment for just and inclusive adaptation by recognizing and addressing the structural inequalities that amplify vulnerabilities. Inclusive adaptation should prioritize the well being of all the population (Heath, 2025). However, while policies may articulate these principles, effective implementation remains challenging due to various factors, including traditional gender roles, limited access to education, and decision-making processes that often marginalize women (FAO, 2016).

Further, recent studies highlight the importance of context-specific gender analyses in climate adaptation policies. Research by Ahmed et al. (2020) underscores the necessity of addressing cultural norms and social hierarchies that intersect with climate vulnerabilities. Similarly, exploring the experiences of women farmers in Kenya, Githinji and Nyangena (2021) emphasize the need for policy interventions that consider the gendered dimensions of climate impacts and adaptive strategies.

Just and inclusive adaptation within policy interventions for climate change management in Kenya is very important. The differential impacts of climate change on genders necessitate policies that acknowledge these disparities and actively address them. While policy frameworks aspire to integrate technology and social inclusion, translating these intentions into tangible outcomes requires persistent efforts to dismantle gender-based barriers. By fostering an environment where all members of society can effectively contribute to and benefit from adaptive strategies, Kenya can move closer to achieving inclusive sustainable development and resilient climate change management.

Climate Smart Agriculture

CSA emerges as a pivotal approach in addressing the multifaceted challenges of climate change, striving to realign agricultural development with the demands of a changing climate (Beal et al., 2021). At its core, CSA seeks to achieve food and nutrition security, generate employment opportunities, and enhance agricultural productivity while fostering environmental sustainability (Derenoncourt, 2022). This strategic approach recognizes the intricate interplay between agricultural practices and climate dynamics, underpinning the need for policies that promote adaptability and inclusivity (World Bank, 2021).

In Kenya, the implementation of CSA is substantiated by a series of policies that delineate its scope and aspirations. For instance, The Kenya Climate Smart Agriculture Implementation Framework 2018–2027 (KCSAIF) and Kenya Climate Smart Agriculture Strategy 2017–2026 (KCSAS) underscore the nation's commitment to climate-resilient agriculture (Phiri et al., 2022a, 2022b). These policies are anchored in the ambition of elevating agricultural standards and productivity levels, thereby ensuring the well-being of farming communities and the nation. Kenya's CSA policies endeavour to translate theoretical objectives into tangible outcomes that bolster food security and economic prosperity by facilitating strategies that augment crop yields, such as aiding maize farmers in increasing production per acre.

An exemplary facet of Kenya's CSA policies is their emphasis on enhancing agricultural resilience. As depicted by Babugura (2021), these policies target increased productivity and encompass measures that prepare farmers for diverse climate stresses. For instance, the policies acknowledge the variable rainfall patterns that punctuate Kenya's climatic profile and consequently advocate for cultivating maize varieties adept at thriving under low and high rainfall levels. This approach exemplifies a synergy between sustainable agricultural practices and climate adaptation, reiterating the holistic nature of climate-smart agriculture that encompasses increased productivity and the fortification of agricultural systems against the uncertainties of a changing climate.

Climate Smart Agriculture stands as a pragmatic response to the challenges posed by climate change, harmonizing agricultural development with climate adaptation imperatives. Incorporating policies like KCSAIF and KCSAS underscores Kenya's commitment to redefining its agricultural landscape to foster sustainability, inclusivity, and resilience. The ongoing trajectory of CSA in Kenya holds the potential to amplify agricultural productivity and economic growth and the resilience of farming communities in the face of evolving climatic conditions.

Social Inclusion in Climate Management

Integrating marginalized and vulnerable groups, such as women, youths, and persons with disabilities, is pivotal to fostering resilient communities. The literature underscores the necessity to bridge the gap between policy formulations and these groups' lived experiences to achieve equitable climate outcomes (Beal et al., 2021). The Constitution of Kenya 2010 enshrines the right to a clean and healthy environment for all its citizens (Babugura, 2021). This constitutional provision is a foundation for policy interventions, ensuring that environmental protection and climate change management are intrinsically linked with social equity (Ngara, 2017).

Engaging these marginalized communities in decision-making processes is vital for developing policies that truly address their unique vulnerabilities and amplify their adaptive capacities. Based on the views of Derenoncourt (2022), gender parity can be attributed to the equal contribution of males and females in all dimensions of life, including having equal opportunities. Gender parity entails equality when the proportions of men and women are calculated (Friedman et al., 2022).

Beal et al. (2021) delve into making CSA gender-smart in West Africa. Their study examines the influence of a focus on gender equity and empowerment on the adoption and productivity of CSA practices. Their findings emphasize the importance of gender-sensitive interventions in agricultural practices, where women play a significant role. They also highlight the need to ensure that policy interventions align with the socio-cultural context to promote inclusivity and sustainability effectively (World Bank, 2021).

Furthermore, Ahmed et al. (2020) provide a case study from coastal Bangladesh that underscores the interplay between gender and climate change adaptation. The study sheds light on how gender-specific vulnerabilities intersect with climate risks, making women particularly susceptible to the impacts of climatic shifts. This research accentuates the urgency of integrating gender-responsive measures within climate policies to create adaptive strategies that genuinely resonate with the needs and capacities of diverse communities.

Examining the gender dimension in climate adaptation policies is central to understanding the differential impacts of climate change on genders. Nyika (2022), demonstrates, through their study of communities dependent on livestock and forests in northern Mali, that adaptation strategies can inadvertently perpetuate gender inequalities. As such, policy interventions must consider the roles and responsibilities that genders undertake within society and develop measures that empower and include women as active participants in adaptation efforts (Githinji & Nyangena, 2021). On the other hand, Social Inclusion entails establishing systems and structures to improve how a person and groups of people can participate in a community's life and activities. Babugura (2021) describes Social Inclusion as a process that enables and creates opportunities for specific people or groups of people to live a decent and dignified life rather than being disadvantaged or excluded based on their identity. In CSA, the groups can include young women or women who are poor and deprived. Further, the socially excluded people include persons with disabilities and HIV/AIDS affected, among other groups (World Bank, 2021). Beal et al. (2021) believe that access to resources denotes the capacity to use and profit from a particular resource. On the other hand, control of resources can be described as having the power to decide on the use and benefit of a resource. Effective CSA can only happen when both genders have access to and control resources (Babugura, 2021). Gender-responsive policies refer to laws, regulations, and guidelines that fulfill gender relations, roles, and norms (Kim et al., 2022). These policies promote gender parity, social inclusion inclusion, and access to resources. According to Bryan et al. (2024), the Social Inclusion agenda contributes significantly towards equality and gender-based empowerment in the practice of CSA. The policy review reveals a considerable gap in the drafting and implementation of the policies. For example, The National Climate Change Action Plan (NCCAP), 2018–2022, does not recognize the adaptive capabilities of men and women. The policies do not consider the use of technology and best practices that enhance an enabling environment for women to thrive in CSA.

Climate Smart Solutions to support gender equality have led to the development of a new approach, Gender Smart Agriculture. The approach promotes aspects such as improving the control of vulnerable persons over resources and other agricultural productive assets. CSA requires capacity development for policy actors (Bryan et al., 2024). The capacity development will help to integrate gender equality in the country-based and global climate change policies. The policymaking process for The National Climate Change Action Plan (NCCAP), 2022–2027, will require gender experts to make the **clauses gender sensitive** and to promote social inclusion.

Theoretical Framework

The present study adopts the ecofeminism theoretical framework to comprehensively analyze the patriarchal systems that underlie policy interventions within the climate change sector (Hosio & Aquarista, 2023). Ecofeminism, a multifaceted and interdisciplinary theory, provides a lens through which the dynamics of gender, environment, and power can be critically examined. According to Schmitt (2023), this framework facilitates exploring how patriarchal ideologies and structures shape policy decisions and perpetuate inequalities, ultimately influencing the effectiveness of inclusive sustainable development and climate change management efforts (Echavarren, 2023).

According to Hosio and Aquarista (2023), ecofeminist thought is the notion that patriarchal systems reinforce a hierarchical and exploitative relationship between gender and the environment. In the context of climate sector policy interventions, this framework helps unveil how patriarchal dominance influences the allocation of resources, decision-making processes, and the prioritization of certain groups' needs over others (Schmitt, 2023).

Moreover, ecofeminism employs key feminist concepts such as gender equality, evaluating non-patriarchal or nonlinear systems, and endorsing a worldview that appreciates organic processes, holistic connections, intuition, and collaboration (Echavarren, 2023). Applying this tenet to climate change policy interventions in Kenya allows for a more comprehensive understanding of how gender disparities intersect with environmental vulnerability, enabling the identification of entry points for transformative change.

The ecofeminism theoretical framework is a potent tool for dissecting the patriarchal underpinnings of policy interventions within the climate change sector (Schmitt, 2023). By elucidating the interplay between gender, power, and environmental exploitation, this framework enables a comprehensive assessment of policy effectiveness and the identification of avenues to promote inclusive, sustainable development (Hosio & Aquarista, 2023). Through its synthesis of feminist principles and environmental concerns, ecofeminism lays the groundwork for dismantling patriarchal systems and nurturing a more equitable and resilient response to climate challenges.

Methodology

Research Design

The paper uses desk-based reviews and database confirmations as research methods to collect both qualitative and quantitative interviews. The study reviewed specialized policies on the gendered effects of climate change. The research is based on an analysis of regional and international public policies on the effects of climate on women in these areas. The study analyzed the contents of the policies in the climate change sector. There was a focus on the priority areas of each policy and the target population. The study also evaluated the beneficiaries of the Climate change policies. The study employed a qualitative content analysis of the climate change policies in Kenya.

Desk-Based Review

The analyzed data was obtained from secondary public sources of information that were already available covering between 2012 and 2022. These sources included acts of parliament, climate change policies, strategy documents, and United Nations conventions, among others, from global organizations with subject-matter knowledge. The primary sources consulted for this study are mostly government publications.

Policy Analysis and Discussions

Policy Interventions

Examining existing policy interventions within Kenya's climate change management sector reveals a complex landscape with both promising elements and significant challenges. While some frameworks have been developed to promote gender-responsive CSA, the predominant issue lies in executing these policies. A comprehensive review underscores substantial gaps, particularly the absence of specific gender considerations within these policies (Nyika, 2022). This lack of gender sensitivity within the frameworks hinders the effective integration of women, who play a pivotal role in agricultural practices, into the adaptation processes. This observation resonates with the findings of Beal et al. (2021), emphasizing the importance of gender equity in agricultural interventions for enhanced productivity and resilience.

A notable concern identified within the policy landscape is the multiplication and duplication of policies, leading to a convoluted and fragmented framework. This fragmentation is further exacerbated by unclear policymaking processes that hinder the streamlined development and implementation of effective strategies (Githinji & Nyangena, 2021). The gender bias embedded within these policies also emerges as a substantial challenge, reflecting the broader patriarchal structures that tend to marginalize women's voices and contributions (Nyika, 2022). This lack of gender sensitivity in policy planning perpetuates gender-based inequalities and restricts the potential benefits of inclusive adaptation.

Furthermore, the study highlights the inadequacy of policy planning, evidenced by poor coordination between agricultural institutions and the minimal involvement of key policy actors (Leal Filho et al., 2022). The importance of involving diverse stakeholders, including women and marginalized communities, in policy formulation and implementation cannot be understated. The absence of such involvement hampers policies' effectiveness and perpetuates existing disparities and vulnerabilities (FAO, 2016). The lack of monitoring and evaluation structures compounds the challenges, inhibiting the tracking of policy outcomes and hindering opportunities for iterative improvements.

Additionally, the dearth of research focusing on women's and men's roles in CSA is a significant limitation that affects the development of targeted and evidence-based policy interventions. Kim et al. (2022), are of the view that this kind of knowledge gap perpetuates gender-blind policies that do not adequately address the differentiated impacts and contributions of genders in climate adaptation. Thus, the policy landscape demands a comprehensive reevaluation and restructuring to address these challenges and ensure that policy interventions are well-formulated and effectively executed, yielding tangible benefits for all segments of society.

Gender Responses in the Climate Change Act 2016

Under the Climate Change Act 2016 Art 6 (d), NCCC shall approve gender and intergenerational responsive public education awareness strategy and implementation program. The strategy is still pending. The government has developed NCCAP 2013–2017 and NCCAP 2018–2022. Thus, there is a need to **develop the gender strategy** along with NCCAP 2023–2027. According to Mehar and Prasad (2022), gender mainstreaming strategies are more effective in implementation based on action plans that specify the actual roles and timelines.

Article 7(6) The president shall, in appointing members, ensure compliance with **two third-gender principles.** The article emphasizes Art 81 of the Constitution of Kenya 2010. There is a need for **strict Adherence and compliance** with the two-third gender principle. Further, Article 8(c) states that The Cabinet Secretary has to **formulate a** National Gender and Intergeneration-responsive public education and awareness strategy. The strategy is yet to be formulated. The finding is similar to that of Jangbrand (2022), who analyzed national Action plans, which require formulation of strategies, and the strategies have not been developed yet.

Under Article 21, educational institutions must collaborate with the Kenya Institute of Curriculum Development to integrate climate change into various disciplines. There is an appeal for climate change to be **integrated with gender studies.** There has been progress since, in September 2022, the **Kenya School of Government** launched a county climate change fund (CCCF) Mechanism Curriculum and facilitators guide (Nyika, 2022).

Gender-Responsive Reporting on Climate Change

Kenya's Nationally Determined Contribution (NDC) in July 2015 reported on adaptation and mitigation contributions. The study analyzed this report along with Kenya's Updated National Determined Contribution in December 2020. Section 4.3 of the report on Adaptation, Loss, and Damages aims to bridge the implementation gaps, especially in productive sectors of the economy however, there is **no highlight of women**. Despite the underrepresentation, as indicated in Table 2 of the report on Prioritized Adaptation programs which mentions gender, youth, and other vulnerable groups on page 38 of the report. There is a requirement to develop social safety net structures for women and other vulnerable groups and on page 41 Consolidate successful technologies and develop **transfer strategy** for women. The studies finding indicate that there is minimal reporting on women and gender-disaggregated data is not available.

The Sessional Paper No. 3 of 2016 on the National Climate Change Framework Policy seeks to promote **gender transformative approaches** and interventions for gender parity and social inclusion (Kamau, 2018). The other intelligent policy intervention is to lower emissions and encourage forestation to absorb carbon from the atmosphere. The Kenya Climate Change Act 2016 Art 17(C) requires the National Environmental Authority to regulate, enforce, and monitor compliance with levels of greenhouse gas emissions. The regulations and roles allocated to environmental authorities to control levels of greenhouse emissions enhance compliance, leading to decreased emissions (Beal et al., 2021).

Gender-Sensitive Climate Smart Agriculture

The policy analysis establishes significant gaps within the existing policy framework, particularly concerning gender sensitivity in CSA practices. These policies often overlook the distinct differences between men and women, failing to recognize the varied roles, knowledge, and experiences each gender brings to the agricultural landscape (Jangbrand, 2022). However, recognizing these gaps presents an opportunity to enhance the effectiveness of CSA strategies and policies through a gender-sensitive lens, ultimately fostering inclusive and equitable adaptation processes (Mehar & Prasad, 2022).

CSA practices and gender-sensitive policies can be effective when they harness gender-specific advantages. Acknowledging women's and men's unique talents, skills, and expertise can lead to more effective resource utilization and increased agricultural productivity. According to Githinji and Nyangena (2021), by nurturing these talents and skills through targeted capacity-building initiatives, CSA can promote economic empowerment for both genders, fostering self-reliance and contributing to overall household income.

Gender-sensitive CSA practices prioritize resilience to climate variability by incorporating strategies that account for the distinct challenges women and men face. The strategies could entail promoting diversified working seasons that align with women's roles and responsibilities, ensuring they have the flexibility to engage in agricultural activities alongside their other duties (Jangbrand, 2022). Furthermore, by recognizing the differentiated impacts of climate change on genders, CSA policies can drive the development of adaptation strategies that enhance the capacity of both men and women to cope with climatic shifts.

Mitigating agriculture's contribution to climate change necessitates the active involvement of key actors, including women, youths, young men, mothers, and fathers. Gender-sensitive CSA practices can spotlight these actors' roles in reducing greenhouse gas emissions, promoting sustainable resource management, and encouraging the adoption of climate-friendly agricultural techniques. By acknowledging and involving diverse stakeholders, CSA policies can tap into a broader range of innovative ideas and approaches to combat climate change (Babugura, 2021).

Promoting gender pay equity and ensuring equal representation of genders in the agricultural workforce and leadership positions are central components of gender-sensitive CSA (Phiri et al., 2022a, 2022b). Creating an environment that values gender equality can contribute to developing adaptable and self-reliant men and women who actively participate in and benefit from CSA initiatives. Self-reliant men and women can catalyze transformative change, breaking traditional gender norms and fostering a more inclusive and equitable agricultural sector (Jangbrand, 2022).

Recommendations and Conclusion

Recommendations

The study recommends a multifaceted approach to enhance gender inclusivity and effectiveness in CSA. Firstly, there is a critical need to bolster the capacities of both women and men through robust support for technical training and educational systems. Capacity building will pave the way for more active and informed participation in CSA initiatives. Moreover, a pivotal step entails the consolidation and revision of existing legislative frameworks related to gender intelligence and CSA. This process should culminate in the development of legislation that is not only comprehensive but also accessible and user-friendly. Furthermore, it is imperative to ensure that the underrepresented gender is integrated into leadership structures within the CSA sector.

Legislative advancements require the establishment of equitable labour practices. Equitable labour laws can be achieved through a thorough review of regulations aimed at eradicating precarious employment and guaranteeing gender-equitable and secure working conditions. Additionally, a robust monitoring system must be instituted to rigorously assess the impact of policies on both gender equity and overall productivity. Lastly, a strategic focus on research and data collection is indispensable. Data collection will provide an evidence-based foundation for fostering inclusivity, thereby informing policy formulation and effectively addressing existing gender disparities within the CSA domain.

Conclusion

Inclusive, sustainable development and effective climate change management in Kenya hinges on the principle of inclusion, particularly in the context of CSA. CSA's strength lies in its potential to encompass and empower vulnerable groups, including women, youths, and persons with disabilities. The database analysis on various reports such as World Bank shows that the prevailing policy interventions in climate change and CSA have fallen short of embracing gender sensitivity. Through the ecofeminism theoretical framework, the study indicates that there is an urgent need to enhance these policies by infusing them with inclusive and gender-responsive measures to unlock the full potential of CSA and mitigate the disproportionate impacts of climate change. This enhancement is pivotal for fostering equitable opportunities in CSA for both men and women.

Seamless collaboration and coordination of all stakeholders involved in CSA activities will improve the effectiveness of these interventions. More productive and resilient agricultural practices can be developed by fostering the active participation of farmers, policymakers, researchers, and civil society. The interconnectedness of these actors contributes to a holistic approach that amplifies agricultural productivity and resilience and prioritizes inclusivity, an indispensable pillar of CSA.

The journey towards inclusive, sustainable development and climate change management in Kenya demands a paradigm shift in policy formulation and implementation. The principles of equity, inclusivity, and gender responsiveness must guide the trajectory. This transformation is crucial for fostering more sustainable agricultural practices and nurturing a society that thrives in the face of climatic challenges while ensuring that all population segments reap the benefits of progress.

References

Ahmed, F., Farnaz, N., Al-Amin, A. Q., & Islam, M. S. (2020). Gender and climate change adaptation: A case study from coastal Bangladesh. *Sustainability, 12*(9), 3780. https://www.mdpi.com/2225-1154/10/11/172

Babugura, A. A. (2021). Kenya's constitution of 2010 and environmental protection. In *Environmental law in Kenya* (pp. 55–79). Springer.https://www.saflii.org/za/journals/AHRLJ/2014/25.html

Beal, T., Kuhl, L., Dieng, I. V., & Gitter, S. (2021). Making climate smart agriculture gender smart in West Africa: Does a focus on gender equity and empowerment affect adoption and productivity? *World Development, 146*, 105571. https://www.sciencedirect.com/science/article/pii/S2949911923000199

Bedeke, S. B. (2023). Climate change vulnerability and adaptation of crop producers in sub-Saharan Africa: A review on concepts, approaches, and methods. *Environment, Development and Sustainability, 25*(2), 1017–1051. https://doi.org/10.1007/s10668-022-02118-8

Bryan, E., Alvi, M., Huyer, S., & Ringler, C. (2024). Addressing gender inequalities and strengthening women's agency to create more climate-resilient and sustainable food systems. *Global Food Security, 40*, 100731. https://www.sciencedirect.com/science/article/pii/S2211912423000615

Derenoncourt, M. E. (2022). Gender-smart investing to scale climate-smart agriculture in the Ghanaian context. https://cgspace.cgiar.org/handle/10568/119771

Echavarren, J. M. (2023). The gender gap in environmental concern: Support for an ecofeminist perspective and the role of gender egalitarian attitudes. *Sex Roles,* 1–14. https://doi.org/10.1007/s11199-023-01397-3

FAO. (2016). *Climate change and gender.* Food and Agriculture Organization of the United Nations. https://www.fao.org/3/i5188e/I5188E.pdf

Friedman, E. A., Way, J. A., Carney, T. S., & Hamrick, F. A. (2022). Climate change, gender, and vulnerability: How effective climate policy can address the needs of vulnerable women. *Journal of International Affairs, 76*(1), 65–87. https://www.un.org/womenwatch/feature/climate_change/downloads/Women_and_Climate_Change_Factsheet.pdf

Heath, S. C. (2025). Navigating psychosocial dimensions: Understanding the intersections of adaptation strategies and well-being outcomes in the context of climate change. *Current Opinion in Environmental Sustainability, 72*, 101493. https://www.sciencedirect.com/science/article/pii/S1877343524000800

Gebre, G. G., Amekawa, Y., & Fikadu, A. A. (2023). Farmers' use of climate change adaptation strategies and their impacts on food security in Kenya. *Climate Risk Management*, *40*, 100495. https://www.sciencedirect.com/science/article/pii/S2212096323000219

Githinji, M., & Nyangena, W. (2021). Gendered vulnerability and adaptation to climate change among smallholder farmers in Kenya. *Climate and Development*, 1–17. https://www.nmwe.kssl.org/pmc/articles/PMC9558303/

Hosio, Y. F., & Aquarista, A. (2023). Ecological impact of mangrove forest damage on women's economic resources based on ecofeminism theory in Youtefa Bay, Jayapura City. *Jambu Air: Journal of Accounting Management Business And International Research, 1*(2), 86–97. http://www.rayyanjurnal.com/index.php/jambu/article/view/73

Howlett, M., & Ramesh, M. (2023). Designing for adaptation: static and dynamic robustness in policymaking. *Public Administration, 101*(1), 23–35. https://doi.org/10.1126/science.abn9668

Jangbrand, A. (2022). Gender-responsive peacebuilding in a changing climate: A qualitative content analysis of strengths and weaknesses in National Action Plans. https://www.diva-portal.org/smash/record.jsf?pid=diva2%3A1699705&dswid=-365

Kamau, P. (2018). Gender mainstreaming in climate change adaptation: A Kenyan case study. In *Mainstreaming gender in climate change adaptation* (pp. 97–115). Springer. https://doi.org/10.1007/s10584-019-02447-0

Kenya Climate Change Working Group. (2019). Climate change policy brief: The need for a comprehensive climate finance strategy for Kenya. https://www.kenyaclimatechange.org/policy-briefs/

Kim, J. H., Le, T. M., & Glenday, G. (2022). *Stocktaking survey on climate-smart public investment management in Africa: What did we learn?* https://documents1.worldbank.org/curated/en/099840105112221101/pdf/IDU0170f7893095a5049cb0b0410768b868a94a0.pdf

Kogo, B. K., Kumar, L., & Koech, R. (2021). Climate change and variability in Kenya: a review of impacts on agriculture and food security. *Environment, Development and Sustainability, 23*, 23–43. https://doi.org/10.1007/s10668-020-00589-1

Kogo, B. K., Kumar, L., Koech, R., & Hasan, M. K. (2022). Response to climate change in a rain-fed crop production system: insights from maize farmers of western Kenya. *Mitigation and Adaptation Strategies for Global Change, 27*(8), 50. https://doi.org/10.1007/s11027-022-10023-8

Leal Filho, W., Barbir, J., Gwenzi, J., Ayal, D., Simpson, N. P., Adeleke, L., Tilahun, B., Chirisa, I., Gbedemah, S. F., Nzengya, D. M., & Yaffa, S. (2022). The role of indigenous knowledge in climate change adaptation in Africa. *Environmental Science & Policy, 136*, 250–260. https://www.sciencedirect.com/science/article/abs/pii/S1462901122001915

Mairura, F. S., Musafiri, C. M., Kiboi, M. N., Macharia, J. M., Ng'etich, O. K., Shisanya, C. A., Okeyo, J. M., Mugendi, D. N., Okwuosa, E. A., & Ngetich, F. K. (2021). Determinants of farmers' perceptions of climate variability, mitigation, and adaptation strategies in the central highlands of Kenya. *Weather and Climate Extremes, 34*, 100374. https://www.sciencedirect.com/science/article/pii/S2212094721000645?via%3Dihub

Mehar, M., & Prasad, N. (Eds.). (2022). Climate change, adaptation and gender: Policy, practice and methodological underpinnings. *CABI*. https://blog.cabi.org/2023/10/24/fostering-food-safety

Ngara, T. (Ed.) (2017). Climate-smart agriculture manual for agriculture education in Zimbabwe. UNEP DTU. Partnership. https://backend.orbit.dtu.dk/ws/files/159520287/Climate_smart_Agriculture_Manual_FINAL_2_2_sent_in_Sierra_leones.pdf

Nyika, J. M. (2022). Climate change situation in Kenya and measures towards adaptive management in the water sector. In *Research anthology on environmental and societal impacts of climate change* (pp. 1857–1872). IGI Global. https://www.igi-global.com/chapter/climate-change-situation-in-kenya-and-measures-towards-adaptive-management-in-the-water-sector/293987

Phiri, A. T., Charimbu, M., Edewor, S. E., & Gaveta, E. (2022). Sustainable scaling of climate-smart agricultural technologies and practices in Sub-Saharan Africa: The case of Kenya, Malawi, and Nigeria. *Sustainability, 14*(22), 14709. https://www.mdpi.com/2071-1050/14/22/14709?trk=public_post_main-feed-card_reshare_feed-article-content

Phiri, A. T., Toure, H. M., Kipkogei, O., Traore, R., Afokpe, P. M., & Lamore, A. A. (2022b). A review of gender inclusivity in agriculture and natural resources management under the changing climate in sub-Saharan Africa. *Cogent Social Sciences, 8*(1), 2024674.

Rajabov, N. (2022). Zonal policy concept: Prospects for effective investment climate management. *Economics and Education, 23*(Maxsus_son), 212–219. https://cedr.tsue.uz/index.php/journal/article/view/658

Schmitt, M. (2023). Women farmers and the influence of ecofeminism on the greening of German agriculture. In *Gender and rurality* (pp. 102–116). Routledge. https://doi.org/10.4324/9781003395386-7/women-farmers-influence-ecofeminism-greening-german-agriculture-mathilde-schmitt

Seddon, N. (2022). Harnessing the potential of nature-based solutions for mitigating and adapting to climate change. *Science, 376*(6600), 1410–1416. https://doi.org/10.1126/science.abn9668

Timu, L. J., & Moses, A. L. (2022). Evaluation of drought management strategies for enhanced food security in west Pokot county, Kenya. *African Journal of Climate Change and Resource Sustainability, 1*(1), 76–89. https://journals.eanso.org/index.php/ajccrs/article/view/1008

Uduji, J. I., & Okolo-Obasi, E. N. (2023). Gender inequality in nutrition and food security in Sub-Saharan Africa. In *Gender inequality and the potential for change in Sub-Saharan Africa*. https://doi.org/10.1080/23311886.2021.2024674?scroll=top&needAccess=true

Watson, R., Kundzewicz, Z. W., & Borrell-Damián, L. (2022). Covid-19, and the climate change and biodiversity emergencies. *Science of the Total Environment, 844*, 157188. https://www.sciencedirect.com/science/article/pii/S0048969722042863
World Bank. (2021). Climate-smart agriculture in Kenya. CSA country profiles for Africa, Asia, and Latin America and the Caribbean series. The World Bank Group. https://hdl.handle.net/10568/6956875

Jacqueline Njambi Kamau is a certified Public Policy analyst and Researcher based at University of Nairobi Department of Political Science and Public Administration. She has speciality in Labour, social Protection, Gender, Climate change and Governance. She has worked with ILO, KIPPRA, AWSC WEEHUB, PASGR among others. She is a Board Member of Public Policy Research Center.

Dr. Justine Magutu is a senior lecturer at the University of Nairobi. She teaches in the Department of Political Science and Public Administration. Her speciality includes Political Science, Gender and public policies.

Dr. Henry Amadi is a Senior Lecturer at University of Nairobi. He serves in the Department of Political Science and Public administration. He is an expert in Public policies and political Analysis. He has also participated in a variety of workshops on democracy, electoral systems, conflict management and development in Africa. He attained his Ph.D. Candidate at Leipzig University, Germany. His current research interests are in security governance and public policies.

Chapter 5
Enhancing Climate Change Resilience by Changing the Default Energy Policy Approach

Abdul-Mumin Abdulai

Abstract Climate change poses daunting development and existential challenges to the world at large. The impacts of climate change are far-reaching. For this reason, concerted global efforts are required to tackle the menace of climate change by seriously tackling greenhouse gas emissions. This chapter seeks to contribute to the needed global efforts by critically investigating the extent to which climate change resilience can be enhanced by policy making that seriously promotes green or renewable energy sources as the new default energy options, i.e. green energy default policy approach (GEDPA); thereby replacing fossil fuels as the current default energy options. By using largely secondary data and some expert opinions, this chapter argues that GEDPA can greatly facilitate public and private investments in science and technology to create many green energy options and manufacture affordable green energy-based appliances or gadgets. And to instill climate change sensitive public behaviour through readily available, accessible and affordable green energy options.

Keywords Fossil fuels · Green energy · Default policy · Appliances · Resilience · Climate change

Introduction

Climate change poses daunting development and existential challenges to the world at large. The impacts of climate change are far-reaching. Due to varying adaptive capacities among countries, communities and people to climate change, its impacts are felt differently globally. This challenging situation posed by climate change calls

A.-M. Abdulai (✉)
Department of Development Management and Policy Studies (Primary), University for Development Studies, Dungu, Tamale, Ghana
e-mail: amabdulai@uds.edu.gh

Department of Environment and Sustainability Sciences (Secondary), University for Development Studies, Nyankplala, Tamale, Ghana

W. Leal Filho et al. (eds.), *Climate Change Mitigation and Adaptation in Practice*, Climate Change Management, https://doi.org/10.1007/978-3-031-85217-6_5

for concerted global efforts to tackle the menace of climate change by tackling greenhouse gas emissions head-on. In response to the need for concerted efforts resulted in the United Nations Organization (UN) taking a leading role to adopt transdisciplinary and multi-stakeholder approach to address climate change.

Many UN's conferences and initiatives (e.g. 1972 Conference on Human Environment in Stockholm, 1987 World Commission on Environment and Development, the Earth Summit in 1992 which produced Agenda 21 document and the millennium developments goals (MDGs), the sustainable development goals (SDGs) in 2015, etc.) aim at addressing development and environmental challenges by working with member countries to improve quality of lives in those countries.

Scholars or the academia are among the diverse stakeholders (e.g., environmental non-governmental organizations-NGOs, community-based organizations-CBOs, traditional authorities, the clergy, environmental activists, companies, etc.) engaged by the UN to address the anthropogenic greenhouse gas emissions. For instance, the inter-governmental panel on climate change (IPCC), which is the scientific mouthpiece of the UN on climate change, is composed of renowned scholars whose scientific evidences on climate change inform policy direction as to which measures to adopt to address greenhouse gas emissions. Therefore, green energy transition, which underscores the need for environmentally friendly practices and policies, seeks to reverse rising global temperature to pre-industrial level by phasing out fossil fuels (e.g., petroleum, coal and natural gas) as the default energy sources.

Specifically, the conference of parties (COP) is an initiative by the UN to create an avenue for both Party A (comprising industrialized and heavy greenhouse gas emitting countries) and Party B (comprising non-industrialized and less greenhouse gas emitting countries) to deliberate on the best ways to tackle the rising global greenhouse gas emissions. At the end of the 28th UN Conference of Parties (COP28) held in Dubai, UAE from November 30 to December 13, 2023, a unanimous agreement was adopted which aimed to end the dominance of fossil fuels as the major or default sources of global energy. Reacting to this historic agreement on global greenhouse gas emissions, the UN Climate Change Executive Secretary Simon Stiell remarked that "whilst we didn't turn the page on the fossil fuel era in Dubai, this outcome is the beginning of the end" (UN, 2023a). The Dubai Climate Change Agreement is deemed historic because disagreements as to how to curb global greenhouse gas emissions and maintain global temperature increase at 1.5 °C bedeviled the international community and world leaders for nearly three decades, i.e. 27 out of the 28 COPs ended acrimoniously on how to curb global greenhouse gas emissions. Again, COP28 affirmed the scientific recognition that global greenhouse gas emissions must be reduced 43% by 2030, down from 2019 levels, to maintain global warming at 1.5 °C. However, red flags were raised regarding the likely failure of many Parties to meet the goals enshrined in the Paris Agreement by 2030 (UN, 2023a). This fear expressed by the UN is more likely to change when GEDPA proposed in this chapter is widely adopted. This is due to the fact that GEDPA is capable of creating readily available, accessible and affordable green energy options to bolster resilience of many vulnerable countries to climate change. Green energy, which is also known as renewable energy, is generated solely from natural sources such as sunlight, wind, water, etc. Which means these natural sources rejuvenate or replenish rapidly more than the rate at which they are consumed. Energy generated from these sources is called 'green' because it does not produce gases that are harmful to the environment, e.g. carbon dioxide. In other words, green energy does not pollute the environment like fossil fuels. For this reason, green energy is also referred to as clean energy. It is, again, referred to as sustainable energy largely because it is readily available for long-term use. The following are some of the advantages and disadvantages of green energy:

Solar Energy

This green energy type is the most available for human use. Solar energy is 10, 000 times more readily available than human consumption rate. Solar technologies are currently more advanced. The lifespan of solar panels is almost 30 years (UN, 2023b). Solar energy is regarded as the only green energy which is on track to meet the Net Zero Emissions (NZE) goal by 2050 (IEA, 2023a). The remaining green energy sources require massive expansion to get back on track to meet the NZE goal (IEA, 2023a). However, the disadvantage of solar energy is the fact that not all countries are endowed with sunlight equally. Nevertheless, countries with excess solar energy can export to less endowed countries.

Hydro Energy

This green energy type is generated from rivers and reservoirs. Generally, it is generated from water running from higher gradient to lower gradient. Hydropower is the largest source of global green or renewable electricity. However, its disadvantage emanates from its heavy reliance on rainfall. Drought or unreliable rainfall patterns can affect hydro energy generation. Again, hydro energy infrastructure greatly affects biodiversity (UN, 2023b). Furthermore, occasional spillage from hydro dams causes massive damage to social and economic resources of communities.

Geothermal Energy

It is generated from accessible heat in the interior of earth called geothermal reservoirs. UN (2023b) classified geothermal reservoirs into: (i) hydrothermal reservoirs, which are naturally hot enough, and easily accessible for electricity generation; and (ii) enhanced geothermal systems, which are hot but are supported with hydraulic stimulation. Although geothermal energy technology is advanced and in use for nearly a hundred years, it is not easily affordable in many developing countries.

Wind Energy

This type of green energy is generated from energy of moving wind or air called kinetic energy. Huge turbines are either sited on land surfaces (called onshore) or in vast water bodies (called offshore) to generate wind energy. Despite the vast global wind energy potential for electricity generation, its disadvantage is highlighted by UN (2023b) that not all countries or locations have the same average wind speed necessary for wind energy generation.

Bioenergy or Biofuel

This type of green energy is generated from biomass, i.e., organic materials such as animal droppings (e.g., cow dung), wood, certain cultivated crops and tree species for biofuel. UN (2023b) underscored the significant role of bioenergy in meeting the energy needs of poor rural communities. IEA (2023a) noted that bioenergy is rapidly developing into one of the largest sources of renewable energy globally. However, the downside of bioenergy is that burning biomass emits greenhouse gases though not at the scale of burning fossil fuels (UN, 2023b). Ocean energy is another type of green energy, which is generated from thermal and kinetic energy from ocean currents or waves. However, ocean energy technology is at its infancy, and not fully developed yet (UN, 2023b).

In tracking global green or clean energy, IEA (2023a) observed that the introduction of new green energy policies in 2022 by the world's leading economies such as China, EU, USA and India will produce positive impact on global green energy use. Therefore, IEA (2023a) strongly recommended among others (i) the integration of "variable renewables" into national energy policies, (ii) effectively tackling technology challenges to unleash the full potentials of many of the green energy sources examined above. Therefore, pragmatic policy interventions at global, national, and local levels are required to reverse the reigning default fossil fuel regime with its associated adverse impacts on climate change resilience in many countries. Adopting GEDPA will create a congenial environment for many countries to meet the goals of the Paris Agreement.

Conceptualizing the Global Default Energy Policy Approach and the Need for Change

The default energy policy approach is conceptualized in this chapter as the prevailing dominance of fossil fuels (i.e., oil, natural gas and coal) in global energy policies. The DEPA is the recognition that global energy policies overwhelmingly focus on fossil fuels as the main or default energy source to be bolstered to meet global energy demand. In the same vein, the strong global recognition or acceptance that fossil fuels constitute the major source of greenhouse gases (UN, 2023a) underscores the urgent need for policymakers to move away from the current reigning default fossil fuel regime to new default approach called green energy default policy approach (GEDPA).

It is the strong conviction of this chapter that climate change resilience in the majority of vulnerable countries and communities can be enhanced greatly by policy making processes that are heavily skewed toward green or renewable energy sources as the new default energy options. Such a shift in energy policy will curb global emissions of greenhouse gases greatly; thereby reducing climate change risks confronting many countries globally. Especially the vulnerable countries. Adopting GEDPA will require deliberate efforts by policymakers to facilitate public and private support and investments in science and technology to produce many green energy options and manufacture affordable green energy-based appliances or gadgets to meet public demand.

Negative human behaviour (i.e., unfriendly environmental attitudes, e.g., anthropogenic global carbon emissions) coupled with largely inadequate efforts by the international community has been identified by UN Secretary-General (Mr. António Guterres) as the major contributor to the increasing global greenhouse gas emissions, which weakens the adaptive capacity and resilience of many countries and communities to climate change ravages (see Abdulai, 2022a, 2022b). Incidentally, adopting GEDPA will boost global efforts to curb global emissions of greenhouse gases. It can also boost the resilience of many vulnerable countries to climate change.

Studies and Expert Opinions on the Need to Change the Current Default Energy Policy Approach

Many studies (e.g., Cassingham, 2006; Heinrichs et al., 2016; Hiroshi et al., 2011; Pichert & Katsikopoulos, 2008; Rempel & Gupta, 2022) have demonstrated the need to change the current default energy policy approach. Hiroshi et al. (2011) argued that the international goal to cut greenhouse gas emissions by almost 80% by 2050 resulted in serious restrictions on fossil fuels. However, such restrictions can be counterproductive if the right approaches are not applied.

For this reason, Rempel and Gupta (2022) explored 28 approaches and found only 12 as being environmentally appropriate to tackle fossil fuels. However, only 3 out of the 12 environmentally appropriate approaches were found to be cost-effective. Examples include: (i) financial capital regulation for fossil fuel projects; (ii) fossil fuel subsidies removal and (iii) moratoria and bans. These measures, Rempel and Gupta (2022) argued, would enable the international community to achieve its overarching goal of "leaving fossil fuels underground".

Pichert and Katsikopoulos (2008) studied home electricity consumption choices in USA, UK, and EU. The study found that approximately 50–90% of the respondents preferred renewable or green energy sources. However, only 0.4 to 1.0% of the respondents were actually using renewable or green energy sources. Pichert and Katsikopoulos (2008) further probed the respondents to get the reason for the wide difference between the huge preference for green energy and the small percentage of actual patronage by asking the following question: "what happens to green electricity when it is the default option?" By using two field studies and two laboratory experiments Pichert and Katsikopoulos (2008) found that the percentage of customers willing to buy "green electricity" increased tremendously due to switching the default energy to green. However, it is not very clear from the finding by Pichert and Katsikopoulos (2008) whether inadequate green electricity sources to meet domestic demand contributed to the small percentage of green energy usage in these countries, or inadequate public awareness on the associated benefits of using green energy sources. Therefore, successful GEDPA should focus on green energy availability, accessibility, affordability and awareness creation.

With reference to the social proof theory, Cassingham (2006) argued that the overall potential water savings in USA is about180 million gallons of water monthly. This is because USA has over 4.5 million hotel rooms. Randomly selecting 150 rooms showed that about 6,000 gallons of water monthly can be conserved when towels and linens are reused. This projection underscores the significance of the idea to occasionally change prevailing default policy approaches or change the business-as-usual attitudes and practices that promote unsustainable consumption of environmental or natural resources, which increases global greenhouse gas emissions. In other words, default energy policies that are counter-productive to the fight against climate change.

The proponents of social proof theory believe strongly "that the [positive] behavior of others is positive reinforcement for our behaviors" (Harlow et al., 2016, p. 296). For this reason, if countries are incentivized to move away from using dirty and polluting energy sources and adopt GEDPA, sustainable consumption can be enhanced to achieve the global greenhouse gas emission reduction target by 2050.

Furthermore, climate finance, which is referred to as the "great enabler of climate action", received massive boost at COP28 in Dubai. For instance, the Green Climate Fund has increased to USD 12.8 billion from the contributions of 31 countries. Whereas more financial pledges are expected to be fulfilled (UN, 2023a). By this development, massive investments including subsidies in green or renewable energy sources should be made a climate action priority. This will easily translate into making green or renewable energies the new default global energy sources to curb global greenhouse gas emissions.

Enablers of Fossil Fuels as the Prevailing Default Energy Sources

The primary enabler of fossil fuels as the prevailing default energy sources is the massive subsidies that many governments or countries are allocating to fossil fuels. Subsidies are, in turn, enabled by the intention of governments to support living conditions of their citizens by maintaining lower prices. This policy direction, as IMF (2023a) argued, has unpleasant fiscal consequences that include: (i) inefficient resource allocation that distorts growth; (ii) increasing pollution levels leading to climate change with its numerous ramifications such as deaths due to air pollution, etc.; (iii) high cost of borrowing or higher taxes or lower social spending; (iv) the poor may not be properly targeted leading to the sole benefit of rich households or individuals; (v) and many other consequences such as smuggling of highly subsidized fossil fuels to places where prices are higher; thereby contributing to supply volatility. Fossil fuel subsidies make the adoption of green or cleaner renewable energy policies extremely difficult in many countries (IEA, 2023b).

IMF (2023a) classified subsidies into two main types, i.e., (i) explicit subsidies which refer to retail price being lower than the supply cost–which comprises the domestic costs of production and distribution to consumers for non-tradable product like electricity. However, tradable product like oil the supply cost refers to the opportunity cost of utilizing the product domestically instead of exporting it, and the cost of distribution to consumers; and (ii) implicit subsidies refer to when external costs such as emission of gases that pollute the air (i.e., particulates that affect public health and increase mortality rates) and increased temperatures (i.e. climate change) etc., are not captured fully in the retail price of fossil fuels. This chapter argues that implicit subsidies constitute the main driver of the prevailing default fossil fuel energy policy approach. For this reason, it is a fervent belief of this chapter that when countries begin to fully internalize the external costs of fossil fuel subsidies, they will realize it will be cheaper or cost effective to invest in green or renewable energy sources.

However, recognition of the dangers associated with fossil fuel subsidies and the numerous benefits accompanying green or renewable energy sources moved world leaders, beginning with G20 Leaders in 2009, to call for the phasing out of fossil fuel subsidies to create the enabling environment to completely phase out fossil fuels as the default global energy sources. This call was reaffirmed in 2012 (IMF, 2023a). Subsequently at 26 and 27 conference of parties (i.e., COP26 and COP27) Climate Summits in 2021 and 2022 in Glasgow, UK and Sham El Sheik, Egypt respectively, governments again called for rapid response to achieve the goals of removing counter-productive subsidies as enablers of the default fossil fuel regime.

Unfortunately, fossil fuel subsidies increased to 7 trillion US Dollars in 2022 from 5 trillion in 2020, which represents 7.1% of GDP (IMF, 2023a). The reasons for such increases are attributed to (i) attempts by governments to support their people against the adverse impacts of increasing energy prices (IMF, 2023a), and (ii) in the case of EU Governments, the conflict between Russian and Ukraine pushed EU Governments to increase support for fossil fuels following serious concerns for energy security (McFarlane, 2023). It is highly possible that the urgent need for the majority of countries to recover from the ravages of Covid-19, this chapter believes, has also contributed to the amount of fossil fuel subsidies witnessed in 2022.

Countries with Highest Fossil Fuel Subsidies as of 2022

As indicated in Table 5.1 in a descending order, China has the highest fossil fuel subsidies with more than two trillion US dollars. USA, Russia, India and Japan are countries with the next highest fossil fuel subsidies. As this chapter argues in the preceding paragraphs, Table 5.1 has shown clearly that implicit subsidies constitute greater percentage of total fossil fuel subsidies in these countries with highest fossil fuel subsidies.

Unfortunately, many of these countries failed to honour the urgent call for global efforts to devise measures to safeguard or protect the general populace against the ravages of negative externalities or impacts from climate change. And to realize the likely adverse impact that their failure to internalize implicit fossil fuel subsidies and double their efforts to phase out fossil fuels will have on global efforts to transition to green or renewable energy sources, i.e., the new default approach called green energy default policy approach (GEDPA) in this chapter.

The importance of data captured in Tables 5.1 and 5.2 is to identify the most polluting countries and regions. This will indicate which countries and regions should heed the humanitarian plea or call by international bodies like UN and the private sector comprising non-governmental organizations (NGOs), civil society organizations (CSOs) and community-based organizations (CBOs) for countries and/or regions to adopt green or renewable energy policy approaches to enhance global human prosperity. Therefore, this chapter posits that Tables 5.1, 5.2 and 5.3 constitute a wake-up call. That is, countries and regions that may not want to appear or remain in Tables 5.1 and 5.2 for being identified as the heavy polluting countries must work extremely hard to appear in Table 5.3 to be identified as countries leading the global transition to clean or green energy. And countries that appear in Table 5.3 must work extremely hard to maintain their enviable records in the energy transition or migration from the prevailing default fossil fuel energy policy approach to the new green energy default policy approach.

Regions with Highest Fossil Fuel Subsidies as of 2022

Table 5.2 shows that regions housing countries with highest fossil fuel subsidies invariably reflect similar trend or trajectory in global fossil fuel subsidies. Specifically, East Asia & Pacific, housing China as a country with highest global fossil fuel subsidies, is the Region with highest global fossil fuel subsidies in both 2019 and 2022. However, East Asia & Pacific experienced reduction in total fossil fuel

Table 5.1 Countries with highest fossil fuel subsidies in 2022 in descending order

Countries	Explicit subsidies (US$ billion)	% of GDP	Implicit subsidies (US$ billion)	% of GDP	Total (US$ billion)	% of GDP total
China	270	1.5	1966	11.0	2.235	12.5
USA	3	0.0	754	3.2	757	3.2
Russia	71	4.0	351	19.6	421	23.6
India	32	1.0	314	9.6	346	10.6
Japan	34	0.6	276	5.2	310	5.8
Saudi Arabia	129	13.8	124	13.2	253	27
Indonesia	78	6.2	116	9.2	194	15.4
Iran	63	10.5	100	16.7	163	27.2
Korea	65	3.2	97	4.8	162	8.0
Turkey	59	5.9	93	9.3	152	15.2
Germany	43	1.0	86	2.0	129	3.0
Mexico	15	1.1	83	6.5	98	7.6
United Kingdom	19	0.6	55	1.7	74	2.3
Brazil	2	0.1	67	3.1	68	3.2
France	18	0.6	46	1.5	64	2.1
Italy	10	0.4	54	2.4	64	2.8
South Africa	5	1.2	56	12.8	61	13.9
Vietnam	7	1.7	50	12.6	57	14.3

This is not exhaustive of the countries used by IMF in its Report on fossil fuel subsidies. The author selected these countries based on their total amount of fossil fuel subsidies in descending order (see total subsidies column)
Source Author's selection from IMF's Report (2023b)

subsidies in 2023 i.e., USD 3039 trillion down from USD 3227 trillion in 2022. This is a direct impact of China changing its energy trajectory from the prevailing default fossil fuel policy approach to green energy default policy approach (GEDPA) by massively increasing subsidies for green or renewable energy sources in 2022. By maintaining its investment momentum, China is poised to reduce fossil fuel subsidies substantially to pave way for smooth green energy transition. China's energy scenario may suggest its preparedness to shift its energy policy from fossil fuels. The Chinese case can influence energy transition efforts of many countries due to the fact that a heavy polluting country is showing serious commitment in green energy generation. Table 5.3 presents examples of drastic shifts in public policies that constitute best practices in the energy transition.

Table 5.2 Trend of fossil fuel subsidies by regions in 2019 and 2022

Regions	2019	2022	2019	2022	2019	2022
	Explicit subsidies-E.S (US$ billion)	Explicit subsidies-E.S (US$ billion)	Implicit subsidies-I.S (US$ billion)	Implicit subsidies-I.S (US$ billion)	Total ES + IS (US$ billion)	Total ES + IS (US$ billion)
East Asia and Pacific	172	502	2248	2713	2432	3227
Europe and Central Asia	98	299	840	984	974	1304
North America	0	0	906	791	912	795
Middle East and North Africa	147	336	371	439	518	776
South Asia	39	66	287	365	326	431
Latin America and Caribbean	31	43	263	302	302	359
Sub-Saharan Africa	19	29	95	113	114	143
Global	505	1275	5010	5708	5577	7035

The author ranked these regions in descending order based on the total amount of fossil fuel subsidies of countries selected from each region in Table 5.2 (see total subsidies columns)
Source Author's extraction from IMF's Report (2023b)

In a similar vein, North America comprising only USA and Canada posed third highest fossil fuel subsidies nearing fossil fuel subsidies in Europe & Central Asia comprising 39 countries. The trend in North America is occasioned by USA being the second highest country in fossil fuel subsidies after China in 2022. Another significant information revealed by the regional analysis is about trend in Sub-Saharan Africa where fossil fuel subsidies are negligible. Sub-Saharan Africa, as the least polluting region in terms of fossil fuel subsidies, coupled with the untapped and unparalleled green or renewable energy resources is a region that holds the key to unlock global green energy transition when the right investments are undertaken in this region.

Sub-Saharan Africa's energy scenario constitutes a wake-up call for leaders in this region to rise up to the occasion of working for sustainable regional development by capitalizing on the Region's massive green or renewable energy resources coupled with the rising global demand for green energy. Sub-Saharan Africa is the region of paradoxes. For example, this is a region with nearly 60% of global best solar resources (Climate Council, 2022), yet Sub-Saharan Africa has the least electricity coverage per capita in the world.

As a citizen of Sub-Saharan Africa, the author of this chapter has first-hand experience of some of these paradoxes in the region. In a region of plenty freshwater

Table 5.3 Countries leading the way in green or renewable energy generation

Countries	Renewable energy type	Renewable energy generation efforts
China	Wind and solar	As the biggest global investor in green or renewable energy, China is on track to double its power or energy capacity from solar and wind energy sources. It is believed that China is on track to break its national power target of generating 1200 GW from green or renewable sources by 2030. That is, by five years earlier than 2030
Morocco	Solar	With its vast and intense sun light, Morocco is now a global leader in solar power. Morocco has the largest solar farm in the world, which is the Noor-Ouarzazate Complex located in the Sahara Desert. This farm generates energy capacity of 580 MW, which can power twice the size of Marrakesh in Morocco
United Kingdom	Wind	UK's efforts in wind energy paid off by making the country the global leader in offshore wind energy generation. UK has installed the most capacity globally. Power generated from offshore wind provides power to more than 7.5 million homes. UK's national target is to increase its offshore wind power capacity fourfold by 2030. This target is to help the country decarbonise by 2035
Germany	Renewables not specified	Germany in 2022, increased consumption of green or renewable energy to 46.9%, which went up from 42% in 2021. Germany has committed to achieving a target of 80% green or renewable power by 2030 and nearly 100% by 2035
New Zealand	Renewables not specified	New Zealand generates 84% of its electricity currently from green or renewable energy sources but planned to achieve 100% renewable electricity by 2030. Again, the country strives to achieve a target of 50% of total national energy consumption from green or renewable energy by 2035
Norway	Hydro and wind	Norway generated 98% of electricity from renewables in 2016 mainly using hydropower. Norway began harnessing power from rivers and waterfalls over a century ago. The country has recently added wind power to its national consumption
Kenya	Wind	With Lake Turkana Wind Power Project, Kenya hosts the largest wind farm in Africa. It generates more than 310 MW energy to power more than one million households This project attracted US$ 650 Million investment, which is the largest private investment in the history of Kenya

(continued)

Table 5.3 (continued)

Countries	Renewable energy type	Renewable energy generation efforts
Uruguay	Wind, solar, hydro and biofuels	Uruguay doubled its efforts in 2007 to meet its energy requirements from green or renewable energy sources. Uruguay, before embarking on its green energy revolution in 2007, generated more than three quarters of its energy from fossil fuels. The bad news then was that these fossil fuels were imported The good news now is that Uruguay in 2021 generated 98% of its electricity from renewable sources, which went down slightly to 91% in 2022. Uruguay do export excess energy to some of its neighbouring countries
Iceland	Hydropower and geothermal	Iceland combined hydropower and geothermal power to generate nearly 100% of its electricity requirements. Iceland heats nearly 9 out of 10 households from geothermal power, which makes the country one of the leading ten global producers of geothermal power. The UN is hopeful that Iceland's transition could serve as best practice for many countries to emulate
Costa Rica	Hydro, wind, geothermal, solar, and biomass	Costa Rica combined hydro, wind, geothermal, solar, and biomass to generate 98% of its electricity from green or renewable energy sources for eight years from 2015. Costa Rica is the world record holder for using renewable energy for 300 days consecutively in 2018. The country exports excess power to its neighbours in Central America
Sweden	Solar, wind, bioenergy	By combining wind, solar, and bioenergy, Sweden attained its 2020 target of generating 50% of its electricity from green or renewable sources in 2012. By this performance, the country is poised to achieve its 2040 target of generating 100% electricity from green or renewable sources

These are countries generating almost 90 of electricity from green sources. Climate Council (2022) are very likely to achieve 100% by 2030. There are eleven countries contained in Climate Council's document. They are all selected and tabulated in this chapter
Source Author's Sketch based on Data from Climate Council (2022)

resources, yet majority of its inhabitants are thirsty, which is occasioned by frequent and prolonged water scarcity due to poor investment in water supply infrastructure. In a region of abundant solar, wind and hydro resources, the majority of countries experience erratic electricity supply.

In some cases, there will be no electricity supply for a good number of days (e.g., 4 days). Yet electricity tariffs keep on rising in the midst of such poor electricity supply. In the case of Ghana, this erratic electricity supply is referred to as '*Dum Sor*'—meaning 'off–on' in the local Twi dialect. Such erratic electricity supply affected completing this chapter on time, as the author sometimes had to wait close to 15 h to have electricity on again. Poor electricity supply is cited as one of the major development challenges of Sub-Saharan Africa because effective electricity supply will support businesses to flourish; thereby boosting economic production and job creation for the teeming youthful population in the region. Addressing such paradoxes requires governments or state actors, in partnership with non-state actors or the private sector in Sub-Saharan Africa, to eschew corrupt practices and rent-seeking. This will pave the way for development resources to reach their intended destinations and yield the expected results or outcomes.

Countries Leading the Way in Green or Renewable Energy Generation

Table 5.3 presents countries that are recognized as doing extremely well in the global energy transition pact. These countries are steadily changing the prevailing default fossil fuel policy approach to the new default green energy policy approach. The efforts of these countries are duly applauded, which serve as examples of best practice for many other countries to emulate. In fact, it is a wake-up call for many countries to change their polluting energy policy approaches. A classic example is the case of USA, which was the second highest Country in fossil fuel subsidies in 2022 has opted to increase renewable energy subsidies to US$ 15.6 billion in 2022 up from US$ 7.4 billion in 2016 (Reuters 2023). Another example is the seemingly paradoxical case of China. As the country with the highest fossil fuel subsidies in 2022, China has metamorphosized into a country with the world highest renewable energy subsidies in 2022 amounting to US$ 546 billion (Climate Council, 2022).

Areas in Which Science, Technology and Policy Can Collaborate to Enhance the Needed Energy Transition

This section highlights possible areas for effective collaboration among science, technology and public policy to advance the course of rapid, but smooth energy transition. These areas include: (i) manufacturing, (ii) transportation and (iii) empowerment (i.e., financial and social components). These three areas are largely informed by the availability, accessibility and affordability (i.e., the 3As) Framework conceptualized in this chapter.

Availability

This requires policy makers, especially governments, to ensure that all gadgets or appliances that are needed to facilitate the use of green or renewable energies are readily available. For instance, solar panels for solar energy must be available for people to use. The availability component will play a crucial role in enhancing the energy transition because a study undertaken by Pichert and Katsikopoulos (2008) on home electricity consumption choices in USA, UK, and EU showed how unavailability of renewable or green energy sources (let alone the gadgets) contributed to a huge gap between 50 and 90% of preference for renewable or green energy sources and only 0.4 to 1.0% of actual usage of renewable or green energy sources. For this reason, governments (and for that matter public policies) have to collaborate with science and technology by ensuring that sufficient funds are allocated to science and technology to support research and the manufacturing of green or renewable energies. Undoubtedly, scientific knowledge on green or renewable energies abounds in institutions of higher learning across the globe.

For instance, University for Development, Tamale-Ghana has Energy Technology Center that explores innovative technologies to boost green or cleaner energy production to reduce global carbon emissions (Abdulai, 2022b). Incidentally, Keen (2002) argued that complaints are growing in developed countries patterning to sick building syndrome which is due to fossil fuel linked chemicals found in materials for buildings and ventilation. This syndrome affects nearly 30% of new buildings in the developed countries. Allocating sufficient financial resources to support green or clean energy technology will avert the spread of this sick building syndrome. Unfortunately, what is clearly missing is the lack of political will on the part of governments to adequately support the application of existing scientific knowledge which is usually left to gather dust in the shelves of many institutions of higher learning. Therefore, public sector subsidies for fossil fuel should be redirected to support the 3As Framework.

Accessibility

Accessibility simply seeks to draw the attention of policy makers to the reality that (i) lack of (or inadequate) information as to where and how to access green or renewable energies and/or the gadgets and (ii) lack of (or inadequate) logistics and/or good transportation network can pose serious challenges to efforts by individual countries and the world at large to ensure smooth energy transition. From the study conducted by Pichert and Katsikopoulos (2008), it is highly likely that lack of (or inadequate) information regarding where and how to access green or renewable energies led to the small number of citizens actually utilizing green or renewable energies.

Therefore, lack of (or inadequate) public awareness can restraint wider acceptance and usage of green or renewable energies. Adequate logistics and good transportation network will facilitate scaling-up green or renewable energy coverage in the majority of countries to attain the countless benefits associated with adopting green or renewable energy policy approach. Once again, the need for policy makers to redirect public sector subsidies for fossil fuel to support measures to increase people's accessibility to green or renewable energy sources and/or gadgets has been underscored by the 3As-Framework in this chapter.

Affordability

Affordability draws the attention of world leaders or policy makers to the numerous socio- cultural, economic, political and environmental benefits accompanying people's ability to pay for using green or renewable energies. Attaining these benefits is possible only when countries make the right subsidy reforms by accounting for externalities captured in their implicit fossil fuel subsidies (IMF, 2023b). Affordability will ensure sustainability in providing green or renewable energies because strong purchasing power of people will create vibrant market to support green energy production.

Furthermore, affordability will enable wider coverage of these energy sources, especially in poor and low-income countries. The need to scale-up affordability for their people largely informs public empowerment (i.e., financial and social) policies and/or strategies in the majority of countries in the world. Empowerment strategies are capable of promoting active participation of people or citizens in economic and social activities of their countries.

Taking a critical look at affordability suggests that availability and accessibility alone cannot produce the needed green or renewable energy outcomes unless affordability complements. In other words, if sufficient green or renewable energy is generated or produced (i.e. availability), and easy to find (i.e., accessibility), the impact will not be profound if majority of the people are unable to purchase or pay for green or renewable energy (i.e., affordability). For this reason, pragmatic policy reforms are needed to support green or clean energy technologies.

IMF (2023b) identified some of major benefits that countries will acquire from fossil fuel reforms to include the following: (i) improvements in environmental conditions by significantly reducing greenhouse gas emissions; (ii) substantial reduction in air pollution-related mortalities; (iii) increase in revenue generation by reducing the costs of environmental remediation and (iv) improving general welfare of the people.

Conclusion

This chapter has explored the urgent need for the majority of governments in the world to swiftly alter their predominantly fossil fuel-oriented energy policies to a new default policy approach referred to as green energy default policy approach (GEDPA). The green energy transition pact is not meant to truncate economic growth of any country. Rather it is simply a call for world economies to infuse environmentally friendly practices in their economic activities by using green or clean energies which are less injurious to the environment.

A critical review of energy cum climate change discourse revealed countless benefits (e.g. improvements in environmental conditions; reduction in air pollution-related deaths; saving funds from environmental remediation; people's welfare improvement, etc.) which are associated with fast-tracking the green energy transition pact signed by world leaders. This chapter aims to facilitate pragmatic policy changes in favour of green or renewable energy sources by reviewing data on countries having the highest fossil fuel subsidies, highlighting implicit fossil fuel subsidies as bane to efforts to the green energy transition, and juxtaposing with leading countries in the green or renewable energy transition initiative.

Heart-warming finding in this chapter is the fact that citizens of countries with highest fossil fuel subsidies are not too happy with the lackadaisical or knee-jerking approaches of their governments to the energy transition pact. For instance, Climate Council (2022), an Australian Organization called on Australian Authorities to beef-up national efforts to become one of the countries which are leading the green energy transition charge. Climate Council (2022) argued that Australia has abundant wind, sun and vast land. By harnessing these abundant resources, Australia could become a global player in green or renewable energy and demonstrate to the world that Australia is seriously committed to climate action.

This chapter strongly supports the call by Climate Council (2022) for Australia to get serious about the energy transition pact. This is necessary because Australia and many similar countries have directly experienced the ravages or adverse impacts of climate change like extreme heat leading to intense bush fires affecting businesses, properties and loss of lives in such countries. The question is: what more climate calamities do Australia, and such similar countries want to experience before getting serious about changing their default fossil fuel policies to green energy default policy approach (GEDPA)?

This chapter puts forward the following recommendations to boost climate change resilience through GEDPA:

1. World leaders must show serious commitments and fulfil their pledges to rapidly phase out fossil fuels as the default global energy sources.
2. The green energy default policy approach (GEDPA) should be endorsed and embraced to enable countries, especially the most vulnerable countries, to reap the full benefits of green or renewable energy sources.
3. Collaboration between academia and policymakers should be strengthened to make green energies readily available, accessible and affordable, and finally
4. Governments in Sub-Saharan Africa should seize the golden opportunities accompanying the transition to green or renewable energies by harnessing the vast green or renewable energy resources to support sustainable development of the region, especially the SDGs.

Limitations of the Study

The use of largely secondary data for this study will not fulfill the needs of readers who prefer empirical studies on climate change. For instance, empirical data for this study could have been gathered from diplomatic missions or consulates of all the countries used in this study, which are identified as countries with heavy fossil fuels subsidies (i.e., polluting countries) in IMF's (2023a) Report and those leading the global transition to clean or green energy sources. The author's inability to gather such empirical data is largely due to financial constraints.

Future Prospects and Further Research

Green energy sources will rapidly become a global game changer in terms of promoting economic growth, social progress and environmental health. For instance, many countries (see Table 5.2) are currently generating nearly 98% of their electricity from green energy initiatives. This is a motivation for many other countries, especially natural resource rich developing countries in Asia, Africa and South America, to scale-down their fossil fuel subsidies in favour of green energy sources. It will yield many benefits in their respective countries if they fast-track the green energy transition. This is because green or renewable energy is the future of sustainable energy sources. IEA (2023a) noted that global green energy use in 2022 increased astronomically. However, investment is required to enhance green technologies. Also, further research is needed to explore empirically the economic, social, environment and related advantages or benefits that countries leading the green energy transition are reaping from their efforts following the astronomical rise in global green energy use.

References

Abdulai, A. M. (2022a). Factors that underscore the strength of religious organisations to boost the fight against climate change. *African Journal of Climate Change and Resource Sustainability, 1*(1), 37–48.

Abdulai, A. M. (2022b). How development oriented is the University for Development Studies Tamale-Ghana? From the lens of sustainable development goals (SDGs). In *Proceedings of ICERI2022 conference 7th-9th November*. ISBN: 978-84-09-45476-1.

Cassingham, K. (2006). Towel/sheet reuse program savings. *Laundry Today*, Sunday, 03 December 2006. www.laundrytoday.com. Accessed April 2, 2024

Climate Council. (2022). 11 Countries leading the charge on renewable energy. https://www.climatecouncil.org.au/11-countries-leading-the-charge-on-renewable-energy/. Accessed April 1, 2024.

Harlow, J., et al. (2016). Consumption and sustainability. In H. Heinrichs et al. (Eds.), *Sustainability science: An introduction*. Springer.

Heinrichs, H., et al. (Eds). (2016). *Sustainability science: An introduction*. Springer.

Hiroshi, K., et al. (Eds.). (2011). *Sustainability science: A multidisciplinary approach*. UN.

International Energy Agency (IEA). (2023a). Renewables. https://www.iea.org/energy-system/renewables. Accessed August 24, 2024.

International Energy Agency (IEA). (2023b). Energy subsidies: Tracking the impact of fossil fuel subsidies. https://www.iea.org/topics/energy-subsidies. Accessed April 1, 2024.

International Monetary Fund. (2023a). Climate change: Fossil fuel subsidies. https://www.imf.org/en/Topics/climate-change/energy-subsidies. Accessed April 1, 2024.

International Monetary Fund. (2023b). IMF fossil fuel subsidies data: 2023 update. Working paper No.23/169.

Keen M (2002) Urban ecology. In J. Birkeland (Ed.), *Design for sustainability: A source book of integrated ecological solutions*. Earthscan.

McFarlane, S. (2023). Explainer: global fossil fuel subsidies on the rise despite calls for phase-out. *Reuters*. https://www.reuters.com/business/environment/global-fossil-fuel-subsidies-rise-despite-calls-phase-out. Accessed April 1, 2024.

Pichert, D., & Katsikopoulos, K. V. (2008). Green defaults: Information presentation and pro-environmental behaviour. *Journal of Environmental Psychology, 28*(1), 63–73.

Rempel, A., & Gupta, J. (2022). Equitable, effective, and feasible approaches for a prospective fossil fuel transition. *Wires Climate Change, 13*(2), 756.

Reuters. (August 2, 2023). US doubles renewable subsidies to $15.6 billion in last seven years, EIA says. https://www.reuters.com/business/energy/us-doubles-renewable-subsidies-156-billion-last-seven-years-eia-2023-08-02/. Accessed 1 April 2024.

UN. (2023a). COP28 agreement signals "beginning of the end" of the fossil fuel era. Available at: https://unfccc.int/news/cop28-agreement-signals-beginning-of-the-end-o. Accessed April 1, 2024.

UN. (2023b). What is renewable energy? https://www.un.org/en/climatechange/what-is-renewable-energy. Accessed August 24, 2024.

Abdul-Mumin Abdulai is a Senior Lecturer at the Faculty of Sustainable Development Studies at Dungu CampusTamale and Faculty of Natural Resources and Environment at Nyankpala Campus both Faculties are in University for Development Studies, Tamale-Ghana. He is a sustainability scientist whose training spans geography and resource development, sociology and anthropology and environment and development at Bachelor's, Master's and Ph.D. levels respectively. He supervised a significant number of undergraduate and postgraduate theses and has an appreciable number of publications comprising articles in peer-reviewed journals, books, book chapters, and conference papers.

Chapter 6
Legal and Policy Challenges in the Development of Artificial Intelligence for Adaptation and Mitigation Policies in Nigeria

Erimma Gloria Orie, Orubebe Bibobra Bello, and King James Nkum

Abstract Artificial Intelligence (AI) has gained global popularity as a technological innovation in many spheres that recreates a process or situation in an automated mode of intelligence like humans. For Climate Change (CC), AI can support applications in adaptation and mitigation covering energy, agriculture, economy, etc. Using AI in constantly evolving environmental scenario enables access to climate insights informing adaptation and mitigation efforts. AI can be leveraged to enumerate physically uncountable people and other assets endangered by CC, causing natural disasters as climate crisis exacerbates. It can also be used for wildfire estimated at annual global cost of about $50 billion for prediction, prevention and optimal resource allocation for sustainable forest management. In Nigeria, though CC policies are a vital component for implementation of climate actions, including the implementation of SDG13—regulatory agencies are reluctant to adopt AI applications. Using the doctrinal methodology, this paper finds that, although Nigeria has adopted diverse approaches to climate change governance, many are inadequate due to the inability to embrace AI technology and other innovations. The challenge is that legislative arrangements in Nigeria do not regulate robotics. Thus, the paper argues for the application of AI to enhance adaptation and mitigation policies. It concludes that legal and institutional arrangements in Nigeria should integrate AI for sustainable development efforts to anticipate future CC developments and provide adequate solutions.

E. G. Orie (✉)
FCArb, Faculty of Law, National Open University of Nigeria, Abuja, Nigeria
e-mail: eorie@noun.edu.ng

O. B. Bello
Faculty of Law Oko, Thomas Adewumi University, Oko Irese, Kwara, Nigeria

K. J. Nkum
Private and Property Law Department, Faculty of Law, Taraba State University, Jalingo, Nigeria

Triune Biblical University (TBU), New York, USA

W. Leal Filho et al. (eds.), *Climate Change Mitigation and Adaptation in Practice*,
Climate Change Management, https://doi.org/10.1007/978-3-031-85217-6_6

Keywords Artificial intelligence · Climate change · Adaptation · Mitigation · Sustainable development

Introduction

Artificial Intelligence has gained global popularity as a technological innovation in many spheres including Climate Change (CC) which recreates a process or situation in an automated mode of intelligence like humans. This ingenious technology has the capacity to evaluate, foretell and diminish the possibility of CC using statistics, and modelled precepts by developing accurate data, meteorological conditions, prognostication, and studies become accurate. AI gives understanding of the repercussions of CC in a particular area. It elucidates baroscopic details with which to develop improved scenarios that depict extreme weather events, and related socio-economic and cultural effects of CC. This enables effects such as rising sea level, loss of traditional occupations, flash floods, etc. Another reality of CC is that it is mainly caused by human activities resulting in climate emergency, chiefly attributable to greenhouse gases (GHG) emission arising from fossil fuels consumption, rapid urbanization amongst others. While there are adaptation and mitigation measures that could influence CC, most of them could be feeble as the deleterious effects become catastrophic. A fact that makes more reliable solutions such as AI inevitable (Xin et al., 2023).

Nigeria, a nation known for computer deficiency, using constantly evolving environmental scenarios, offers useful and climate insightful information that could sustain resilient adaptation and mitigation efforts. AI can be leveraged to map out endemic population centres at high risk of CC that cause natural disasters as climate crisis exacerbates cases such as wildfire which is estimated at annual global cost of about $50 billion for prediction, prevention and optimal resource allocation for sustainable forest management (World Economic Forum, 2023). Furthermore, extraordinary aerial occurrences coupled with CC effects could lead to unforeseen consequences. This is because baroscopic records of adaptation methodologies quickly become impotent due to the dynamics associated with fast changing data on record, particularly in the areas of CC modelling, impact, and mitigation. The various applications of AI are already the norm in many developed economies and could be deployed in Africa including Nigeria.

Nigeria is inexplicably an indicator in the global climate discourse because its huge population of over 200 million people cannot be ignored (Worldometer, 2024). Sadly, increasing human activities and natural disasters pose existential risk to people's livelihood in Nigeria mainly due to a combination of geographical and socio-economic factors such as pollution, desertification, flooding, sea level rise, deforestation, and temperature increase from gas flaring, etc. These, amongst others explain why the country is recognised as susceptible to CC repercussions (World Bank Group, 2021). Interestingly, Nigeria is a signatory to several UN conventions and has enacted laws to drive its CC adaptation and mitigation initiatives including the Nigeria's Climate Act 2021 with the ultimate aim to mainstream CC actions into national plans and programmes including the implementation of SDG 13. However, regulatory agencies are reluctant to adopt AI applications, perhaps due to their lack of use. The challenge is that legislative arrangements in Nigeria do not regulate robotics. Other impediments as regards the deployment of AI technologies include skill shortages, fragmented data, ecosystem, ethics, inadequate infrastructure, uncertainty and conservatism of user attitude.

The above presupposes that Nigeria needs a comprehensive AI regulatory framework drawn from international best practices in AI governance. Therefore, using the doctrinal methodology, the paper canvases for the application of AI to enhance CC adaptation and mitigation whereby legal and institutional arrangements in Nigeria would be integrated into AI for sustainable development efforts to anticipate for future CC developments and provide adequate climate solutions. The paper covers four sections. Section 'Conceptual Clarifications of Key Terms' deals with the conceptual clarifications of key terms. Section 'The Legal, Policy, and Institutional Framework for AI Deployment Required for CC Mitigation in Nigeria' espouses the legal, policy and institutional framework for AI deployment in CC adaptation and mitigation in Nigeria. Section 'The Imperative of Artificial Intelligence in Nigeria's Climate Change Adaptation' documents the state of climate adaptation and mitigation in Nigeria. Section 'Conclusion and Recommendations' is conclusion and recommendations.

Conceptual Clarifications of Key Terms

There is need to define some key terms to situate the arguments in proper perspective. The perspective and meaning of key terms such as climate crisis (CC), assuefaction and extenuation expressed in this treatise are within the tenure of the national legislation (Climate Change Act, Nigeria, 2021).

Climate Crisis

This is a prolonged change or switch in temperature and weather pattern ascribed straight or circumlocutory to anthropogenic activity or spontaneous meteorological mutability analogous yet modifies the composition of the overall aerosphere patterns noticed within similar periods (Climate Change Act, Nigeria, 2021). However, there exists a euro-centric CC literature that specifically acknowledges that shifts in weather could be automatic, due to several factors, including the sun's activity or volcanic eruptions (Cole-Dai et al., 2009). Scientists further identify anthropogenic movements as the real driver of CC, due to the heating of substances like naphtha, anthracite, rock oil, and effluvium, which generate GHG (Eurostat, 2010). These emissions trap the surface of the Earth, thus, causing the sun's heat and raising temperatures (United Nations, 2021). In Nigeria, power, commercial enterprise, freightage, constructions, cultivation and the utilization of land are the major sectors responsible for GHG (NDC, 2021).

Adaptation

This has to do with the course of adjustments to confirmed or anticipated negative impacts thereof (Sect. 35, Climate Change Act, Nigeria, 2021). It is the action carried out to succour groups and biocoensis enduring atmospheric transformation. Comparatively, the United Nations and, indeed, global CC literature often espouse a little wider scope and meaning of adaptation to include anticipated adverse effects of CC and contextualized steps to bar or reduce the wreckage in action or exploit the chances that may arise. Instances of refitting course of actions are: extensive basic facilities replacement such as constructing shore-wall barricade to safeguard coastal inundation and change in behavior, such that individuals waste less food or consumption patterns. Fundamentally, acclimatization is the act of conforming to the present and eventual condition of CC (Climate Change Act, Nigeria, 2021).

Mitigation

Mitigation is the summation of effects to forestall or decrease the upsurge of atmospheric GHG focus by curbing ongoing or impending discharges and bettering possible sinks for GHG. Several UN documents emphasis volatility in the average current state and the impacts of CC, less severe mitigation measures aimed at the capriciousness or diminishing discharge of gasses that trap heat in the atmosphere (Climate Change Act, Nigeria, 2021). In other words, amelioration is obtained by curbing the origin of these gases through deliberate increase of the share of geothermal power or setting up a refined movement pattern or by boosting the repository effluvium by advancing the volume of woodland. Under the global context, alleviation is a mankind contribution which steps down the origin of GHG issues or upgrade the sinks. The Nigerian statute includes a complete prevention, that is, total stop of further production of GHG.

Artificial Intelligence (AI)

The Nigeria's Climate Change Act, 2021 did not define AI, but resort is often made to the United Nations and other international documented formulations. The general perception is drawn from the definition proffered by both the OECD (OECD, 2016) and UNCTAD (UNCTAD, 2017) which view the technology as cognitive "capacity of machines and systems to acquire and apply knowledge, to carry out intelligent behaviour". Also, some others adopt the Asia- Pacific perspective. It views AI as a set of "procedures that assume human cognition and reasoning using devices with computational power and large-scale data sources including Machine learning" (Cheong et al., 2022).

Sustainable Development

Green development is a worldwide enterprise introduced at the instance of the foremost international organisation and known to be "development that meets the needs of the present without compromising the ability of future generations to meet their own needs." (United Nations, 1987). Sustainable Development (SD) is a "transformative vision emphasised by the pursuit of the UN 2030 Agenda for Sustainable Development" that registers diverse targets, "including decarbonisation" (Nerini et al., 2018), climate action, (Nerini et al., 2019a, 2019b) among other SD aspects. Presently, green development aspirations like the ones in focus, "address global challenges such as poverty, inequality, CC, environmental degradation, peace and justice."

It should be noted that Sustainable Development Goal (SDG) Target 13.1 encourages all countries to "step-up CC knowledge-based" response. It requires the inclusion of "measures into national policies, strategies and planning" (UN SDG 13). It improves knowledge, consciousness, individual, and organised competency on CC amelioration and "forewarning". With regards to Nigeria and in the context of this discussion, it would mean that the more the adaptation and mitigation policies are strengthened, the better the strengthening of "resilience and adaptive capacity to climate-related hazards and natural disasters" (UN SDG 13). One major way to achieve the SDG 13 target is by the use of AI.

The Legal, Policy, and Institutional Framework for AI Deployment Required for CC Mitigation in Nigeria

Legal Framework

The Constitution of the Federal Republic of Nigeria, 1999, (as Amended)

The Constitution is the 'grundnorm' and provided a federal structure of government in which power is shared between "three (3) tiers of government—the federal, 36 states and 774 LGAs with a FCT" (Constitution of The Federal Republic of Nigeria, 1999). Essentially, under the Constitution, "the state shall protect and improve the environment and safeguard the air, land, forest and wildlife of Nigeria (Sect. 20 of the

Constitution of Nigeria, 1999)". There is a consensus that this clear and broad proviso was enough guarantee for effective response and, in particular, good governance of the environment including CC. However, this is not the result in Nigeria because there are legal and policy encumbrances. For example, the same Constitution provides that "Sect. 20 (Sect. 20 of the Constitution of Nigeria, 1999) and "chapter II of the Fundamental Objectives and Directive Principles of State" (FODPSP) "are not justiciable and enforceable in Nigerian courts" (Orie, 2014). In addition to the above, non-justiciable cum non-enforceable clause, the "ground norm" also entrenched what has become the popular "open ended derogatory clause "The judicial powers vested in the judiciary shall not extend to any issue or question as to whether any act or omission by any authority or person as to whether any judicial decision is in conformity with the fundamental objectives and directive principles of state policy" (Sect. (6)(c) of the Constitution of Nigeria, 1999).

The Climate Change Act, Nigeria (2021)

Nigeria is already experiencing the negative impacts of CC. It is a party to the United Nations Framework Convention on Climate Change (UNFCCC) (UNFCCC, 1992) and other treaties such as Kyoto Protocol. The country's goals were "to keep the 1.5degrees target within reach; protect communities and natural habitats; mobilise finance to deliver on the first two goals; and to work together with the rest of the world to deliver on the various CC targets", committed to achieve net-zero by 2060. The legislation in the main "seeks to provide a framework for achieving low GHG emissions with the ultimate aim to mainstream CC actions into national plans and programmes".

The Climate Change Act established the "National Council on Climate Change (NCCC)" with the statutory authority". The legislation enshrined that "the NCCC collaborates with the Federal Inland Revenue Service (FIRS)" to coordinate the development and implementation of a coherent carbon tax regime throughout the country including emissions trading among other funding mechanisms to finance the "Climate Change Fund provided in the Act".

The National Climate Change Policy (NCCP), (Federal Ministry of Environment, 2021) sets out the sovereign CC policy goals that would make the country attain "a climate resilient economy", and nationwide sectoral adaptation and mitigation strategy to cushion the deleterious impacts of CC in Nigeria. The Act empowered the NCCC, "ministries in charge of environment, budget and national planning to be responsible for formulating Nigeria's National Climate Change Action Plan (Action Plan)". The CC legislation introduced Nigeria's carbon budget which determines "the approved quantity of GHG emission that is acceptable over a specified time and GHG emission cap for key sectors" of the Nigerian "economy; incentives for private and public entities that achieve their GHG emission reduction targets" as well as "the level of Nigeria's compliance with its international climate obligation".

The statute amongst other novel provisos enjoined "all ministries, departments and agencies (MDAs) of the Federal Government; and public and private" business entities throughout the country to draw up implementable CC mechanisms that will enable the country to achieve the desired green economy and a climate resilient society. Under the Act, private entities that employ 50 or more employees are also enjoined to comply with the annual carbon emission reduction targets to be set under the "Adaptation Action Plan". The agency created several positions including "Climate Change Officer" and "Environmental Sustainability Officer," for all purposes and intent, responsible for the development of "an annual report that shows the status of such entity's carbon emission reduction efforts" for a given person. The legislation also established the "Climate Change Fund" "to warehouse" duties and levies related to CC adaptation efforts nationwide. The law provides for monetary incentives to both public and private organisations which are expected to transparently run their affairs.

Policy Framework for Adaptation

Nigeria has a myriad of policy formulations. A taxonomy of some key related critical policies include the First National Communication (FNC, 2003); National Disaster Framework (NDF, 2010); National Adaptation Strategy and Plan of Action for Climate Change for Nigeria (NASPA-CCN, 2011); Second National Communication (SNC, 2014); Intended Nationally Determined Contribution (INDC, 2015); First Biennial Update Report (FBUR, 2018); National Agricultural Resilience Framework (NARF, 2015); Third National Communication (TNC, 2020); and Nigeria's First Nationally Determined Contribution (NDC, 2021).

Nigeria is a party to the Paris Agreement and, as required, the country must continue with its transparent reporting obligations which started in 2016. As a follow through to the country's commitment, Nigeria unveiled and submitted its First Biennial Update Report (FBUR, 2018). Nigeria's declared national goal in terms of approach are coordinated inter-sectorial policies and strategies aimed at mainstreaming environmental sustainability, efficient adaptation, and CC mitigation measures. Another bold deliverable lined out was improvements through national legislation "to strengthen the country's regulatory" frameworks, and "capacity development" aimed at a seamless transfer of technology key priority areas to boost local production of goods. In 2021, the country submitted its first updated NDC to the UNFCCC, even without the promised funding. Be that as it may, the quintessential document confirmed and indeed further articulated Nigeria's desire to reinforce its scope of national implementation of the earlier declared CC goals that will advance carbon neutrality as well as green industrial and economic growth. Such reinforcement is a panacea to youth unemployment and ultimately will also increase R&D in CC adaptation science including AI. It should be noted however that the substantive goal of this strategy was to increase the country's coordinated participation and benefit from global innovations, disruptive technological cooperation on CC, particularly AI.

The National Adaptation Strategy and Plan of Action on Climate Change

The above programme aimed to adapt and mitigate the deleterious effects of CC. The action plan is designed and aimed at fulfilling the country's developmental objectives of reducing GHG emission intensity of its economy. The plan is targeted and independently focused on keeping Nigeria's carbon emissions below the threshold of developed economies at any point in time.

National Council for Artificial Intelligence and Robotics (NCAIR)

This novel organisation has the responsibility to conduct research and deepen "understanding of the application and use of emerging technologies like AI, Deep Learning, Extended Reality (XR-VR/MR/AR), Robotics", "Drones, and the Internet of Things (IoT). (NCAIR) This infant organization ought to be strengthened to cover aspects of AI, robotics and machine learning applicable to CC as espoused in this paper and in line with AI governance models adopted by other countries.

Learning for instance from capacity building training programmes through international assistance by developed countries, the national AI governance initiative has been made a train the trainer initiative that now provides for coordinated AI R&D, national security, as "well as use of AI in the public and private" sectors. A few states have directed their science and technology ministries to liaise with the agency to prepare a more robust "present and future" Nigerian "workforce for the integration of AI systems across all sectors of the economy and society" following the US model (USA National Artificial Intelligence Initiative Act, 2020). Similarly, China is seen as one of the emerging AI Super Powers "leading the era of AI implementation". Its "development plan of the new generation AI was released in 2017" emphasizing the need for "the dual technical and social attributes of AI" to be "properly managed" to ensure that AI is trustable and reliable (National Governance Committee for the New Generation Artificial Intelligence, 2019). Also, the EU has established a new AI law to drive its various AI applications in all sectors" (EU Artificial Intelligence Act, 2024). These frameworks invariably will factor in the "various ethical and legal dilemmas associated with the use of AI". Also to be incorporated are such principles like safety and reliability, transparency, accountability, fairness of justice into the practicable AI operation for common good (Obianyo & Ater, 2022). Therefore, it is in the interest of Nigeria to leverage these AI initiatives gaining momentum globally to enhance its CC adaptation and mitigation efforts.

Institutional Framework for Climate Change Adaptation and Mitigation

Nigeria's approach towards addressing CC from the institutional viewpoint is "focused on ensuring that "strategies are consistent with national development priorities" and using the energy sector as a "key driver for high economic growth". In order to improve policy formulation and co-ordination in this area, "the Ministry of the Environment" created a Special Climate Change Unit which has been renamed the "Department of Climate Change (DCC)". This department has institutional policy mandate to initiate and implement all CC related activities and in particular, adaptation and mitigation. The country also has the quasi-institutional "Inter-Ministerial Committee on Climate Change facilitates" "multi-disciplinary and sectorial coordination at the federal level". This co-operation between ministries and other stakeholders culminated in the Federal Cabinet chaired by the President adopting "the Nigeria Climate Change Policy Response Strategy (NCCPRS)".

The country's CC strategies focus on several sectors. These include agriculture, food security, freshwater resources, coastal water resources and fisheries, forests. Others are biodiversity, health and sanitation, human settlements, housing, energy, transportation, communications, industry and commerce, disaster, migration, security, livelihoods, vulnerable groups, and education. Although there have been a couple of institutional and multi-disciplinary and sectorial successes, several issue specific policies and programmes including climate adaptation and mitigation measures were created under these strategies at the federal level but adequate capacity building and computer assets acquisition being the most pronounced.

Regardless of the laudable goals expressed in this and related national documentations, it is sad to note that the research found that national institutions and mechanisms (policy, legislative, economic and social) are at best still weak. The requisite institutional, policy and legal framework for CC governance are non-functional due to poor and inefficient implementation, corruption; and adaptation efforts remain uncoordinated between the federal, states, local government council areas, and critical private sector stakeholders. These coupled with Nigeria's inability to focus adaptation to the country's urgently needed and most vulnerable sectors: agriculture, ecosystem resources management, transportation, building/infrastructure, energy, forestry, water resources, health, and manufacturing, etc. have reduced Nigeria's CC adaption and mitigation resilience capabilities; thus, worsening the country's human security, social and economic vulnerabilities (Okon et al., 2021).

The State of CC in Nigeria

To attempt an enduring summary of CC case studies, particularly on adaptation and mitigation projects, is necessary. However, the point must be emphasised that documentation, data analysis and other methodologies used vary. The following three case studies summarise and buttress the observations and conclusion, and they can be shared:

(i) One of the earliest climate adaptation projects carried out in Nigeria is the Great Green Wall, which spanned across twenty-two countries. It was a multi-faceted adaptation project that cut across energy, forest system and water resources management. The over-reaching ambitious goal of the project was to grow a carbon sink of several thousand kilometers of degraded land and establish a green "sink forest across 100 million hectares of degraded land by 2030" (Bueb et al., 2021). The project was aimed at addressing the perennial effects of droughts, illegal migration, food insecurity and persistent natural resources conflict. The project was initiated by the African Union (AU) in partnership with the UN (UNCCD, FAO and GEF), and a host of other development partners.
(ii) Another case study is the Nigeria Erosion and Watershed Management Project (NEWMAP) (The World Bank, 2018). This project was intended to provide an integrated CC adaptation to the perennial removal of soil along drainage lines by surface water runoff and similar types of land degradation in the country. The project commenced with a huge expenditure outlay of approximately USD500 million by the World Bank. This climate adaptation project is also regarded as a multi-sectorial project (The World Bank, 2018).
(iii) Another incongruous adaptation project is the Building Nigeria's Response to Climate Change (BNRCC, 2020). This pilot project was estimated to last for five-years and was funded by the Canadian Government. The project was peculiar because it was premised on empirical vulnerability research, with gender mainstreaming components. The BNRCC costs USD 4.9 million and most Nigerian researchers in climate adaptation are of the view that this project culminated in the development of CC policies. This project NASPA—CCN was in collaboration with the government and embraced the strategic goal of increased local food production and aqua-culture development. It is also believed to have pioneered the fuel wood efficiency and diversification of human livelihood incomes initiative amongst under-privileged people. This project improved access to water sources and to some significant extent contributed to the rehabilitation through the strategic planting of trees in endangered ecosystems.

Nigeria's CC adaptation and mitigation despite the lofty targets/goals at first blush looks promising, however, upon an in-depth analysis, it simply looks least blissful. This can be inferred from a summary of Nigeria's natural disaster events related to CC, environmental sustainability index, disaster risk reduction efforts amongst others, despite the above summarised case studies.

Between 2012 and 2022, Nigeria experienced extreme flooding with unprecedented deaths and damages worth billions of Naira (Orie & Ugbejeh, 2021). In Nigeria there is expectation that CC related impacts will aggravate the force of overflow of rainwater. The north-eastern and north-central parts of the country could experience heightened atmospheric conditions with an expressive effect on human survival. Several credible reports point to the inescapable fact that the deleterious effects of CC have already occasioned hunger, resource disputes and extinction of species, and illegal migration in Nigeria (Orie, 2018). In fact, Nigeria has experienced major flooding events which resulted in inestimable economic damages (Federal Government of Nigeria, 2013). The Niger Delta of Nigeria is currently experiencing oil pollution coupled with environmental degradation through persistent flooding, rural–urban migration, hyperinflation, and outbreak of diseases like cholera, malnutrition, and intestinal cancers, etc.

Most of these research projects, including this, carried out by joint public and private universities clearly show that Nigeria is already experiencing phenomenal deluge of floods, drought, deleterious weather events that have astronomically precipitated high increase of food prices, and insecurity. Already, CC precipitated conflict has resulted in widespread Fulani headers and traditional famers infractions. This has exacerbated critical ethno-religious relations. This is a fact that has worsened Nigeria's security situation, and at times, leads to calls by some minority and affected ethnic nationalities for the outright disintegration of Nigeria (NDC, 2021).

It is against this backdrop that science and technology particularly, AI based adaptation and mitigation at the community, sectorial, and indeed, all the three tiers of Nigerian government are of crucial importance to enable Nigeria to develop practical climate solutions. Herein lies the imperative of AI in Nigeria for CC.

The Imperative of Artificial Intelligence for Nigeria's Climate Change Adaptation and Mitigation

Prospects of Artificial Intelligence in Nigeria

In general, Al means different things to different governments. At the moment, its applications and R&D activities are minimal in the country when compared to other industrialised nations. However, there is enthusiasm and desire for AI by private and public sectors' actors and the utilization level will surely increase in due cause (Rutenberg et al., 2021). Meanwhile, global AI companies are increasing their bid to capture existing AI market share and penetration into Nigeria.

Another remarkable aspect of this discourse is the effect of technology on food security. The importance of the foregoing cannot be over emphasized because majority of Nigerians are engaged in "subsistence agriculture". However, deforestation and overgrazing threaten Nigerian soil drastically. The application of AI in the CC "adaptation" will help address the challenges inherent in the current agro-economics. Furthermore, successful deployment of AI can significantly address vexed issues such as plant pests, diseases, stunted growth, and low yield amongst others (Rutenberg et al., 2021). Building on the momentum in AI development in Nigeria, which is experiencing similar climatic crises as other African countries, requires the deployment of AI applications to CC adaptations and mitigations.

Requirements for Artificial Intelligence in Nigeria's CC Adaptation and Mitigation

Nigeria is experiencing the deleterious effects of CC in the spread of diseases, drug resistant strains, irreversible alteration of critical ecosystems, etc. Under these dire situations, inadequate adaptation, and mitigation of the current impacts of CC would be profound. Besides, several recent reports (Sixth Assessment Report) such as "Climate Change 2022 Impacts, Adaptation and Vulnerability" and Nigeria: Climate Risk Country Profile (The World Bank Group, 2021) point to increased risks to humankind. These growing risks presuppose that urgent steps ought to be taken nationwide to avert climate catastrophe.

It is against this background that this research comes to terms with the improved and science-based climate adaptation as a necessary strategic approach to combat CC. Although the use of AI is gaining prominence the world over, it is rarely explored in Nigeria. This situation notwithstanding, Nigeria needs to improve its climate related evaluation and predict flood arising and rain fall patterns. This will also entail deploying enormous data to gather information on oil and gas pollution, droughts, locust and agriculture impacting insects amongst others. A1 assists in making accurate predictions about weather, regardless of variability, a fact that enables mitigation efforts to be deployed early to reduce CC "loss and damage".

Deployment of AI for Adaptation and Mitigation in Nigeria

In order to reach its adaptation goals, Nigeria needs to mainstream environmental sustainability and CC adaptation strategies in its national development goals (Orie, 2020). It must improve its legal, policy and institutional framework to attract adaptation, enhancing machine tools and digital equipment that can foster seamless transfer of AI technology (Oxford Insights, 2022). This technology requires a legal and policy framework that will facilitate human capacity building, flexible mainstreaming of the benefits, and adequately manage the risks occasioned by their use, knowing the level of literacy of the Nigerian population (Obianyo & Ater, 2022). In this regard, it is instructive that Nigeria invests in R&D to enable her to participate and benefit from current international scientific and technological cooperation on CC. It is imperative for the country to urgently increase its climate adaptation capabilities against catastrophic climate vulnerabilities. All these and many more could be achieved through targeted and deliberate acquisition and improvement of AI assets across the country. This is imperative because understanding the climate entails the ecosystem which is apposite to the fight against the impact of CC. One of the very complex aspects of the ecosystem is the ocean. The ocean is "prohibitively expensive" to deploy and maintain marine vessels with AI devices to "observe the ocean and collect needed data". Besides, it is expensive to develop Robots and other AI tools for "their autonomous capabilities" which the use of AI could improve.

In addition, there are several advantages derivable from the deployment of AI. The country could "enhance capabilities" on "CC data" at all levels of government through a decentralized institutional arrangement between "the Climate Change Department of Federal Ministry of Environment" and other MDAs in a coordinated networking arrangement to leverage "partnerships with NASRDA and NBS" to adopt disruptive technologies" that AI provides. This will amongst others employ "layered data" through "satellites", "drones", "geospatial and data analytics" (AI and machine learning inclusive).

Other imperatives would be to:

(a) foster "effective early warning system" in the control of "human diseases" intrinsic in CC.
(b) Create and "develop a national strategy" that allows "technology transfer" and "renewable energy know-how" (NDC, 2021).
(c) Undertake data driven "risk assessment", "risk reduction" measures to increase the "resilience of the transportation" and "communication sectors".
(d) Create organizational efficiency to supply appropriate anticipatory signaling mechanism for enhanced resolutions as well as being sensitive to periodic change for major farming regions.
(e) Strengthen management and implementation in order to protect forests and protected areas.
(f) Advance cross border flood and precarious widespread flood observation, prediction and design to create contingency planning and alertness; deepen awareness of threats to aquatic assets and subsurface water hazards to maximize long term operation and increase the supply of agricultural water needs and city planning (NDC, 2021).
(g) Initiate competent and effective media coverage by partnering with entertainment and advertisement organizations in order to take advantage of their expertise to create awareness among the population, as well as molding public perception to improve public involvement in the conversation of CC adaptation and alleviation.
(h) Institute effective groundwork to identify the connectivity of CC determinants in poverty, insurrections and dispute prompted adaptation and mitigation (NDC, 2021).
(i) Improve Gross Domestic Product (GDP). In sum, it has been submitted in some quarters that AI can improve GDP. While 20 to 25% net earnings can be figured out in the GDP of developed nations, developing nations such as Nigeria can put in 5 to 15%. In aid of their adaptation and mitigation efforts (NCAIR, 2020).

All these could be achieved and enhanced using the instrumentality of the newly established National Council for Artificial Intelligence and Robotics which is under the National Information Technology Development Agency. Thus, the use of AI in consideration of the ever-evolving components of CC enables accurate forecasts on environmental changes, wherefore mitigation efforts could be put out promptly. This method guarantees a high degree of certainty or confidence which is an essential feature of AI usage (Johns Hopkins University, 2023).

Contending Issues to Deployment of AI for CC in Nigeria

Notwithstanding the potentials of AI, it is important to note that the supplementary surveillance and forecasting which AI furnishes researchers is at a huge energy cost as AI uses computers, which in turn use electrical power to function. Fairly, every latest grade of processor performs additional computation for a lower degree of energy. However, AI requirements for computing are fuelling and robust endowment in computational energy. The present AI technology is partly tapping into the enormous computational power available (Johns Hopkins University, 2023). All these lend credence to the requirement for regulation of AI systems deployed for CC adaptation and mitigation. Furthermore, regulation will project the liability and accountability of "the AI possessors, creators, or manufacturers for the activities of their products (Joshi, 2019).

AI will undoubtedly improve the understanding of CC and enhance greater authentic forecasting standards. Invariably, it will improve Nigeria's capacity to withstand CC. Nonetheless, this does not in any way suggest that other scientific know-how will not be relevant. Certainly, technological innovations like the widespread connectivity of sensors, which follow the outcome of Internet of Things (IoT), tend to have a meaningful impact on the capability of Nigeria to accumulate valuable data. It is worthy of note that some innovations may have disordered or contradictory effects. This is not unconnected with the fact that any hike in the demand of power adversely impact on CC. These innovations will grow steadily; likewise, Nigeria's capacity to determine their maximum effect will advance. Although several kinds of AI are common and have been dealt with in many different contexts by private and public sector partners, this study strictly considers AI as a material that treats a significant volume of data as well as stepping up predictions from the available data.

Predictive principles in AI are accessible to business executives or investors by applying two major channels. Applying and adopting broadly accessible open medium is the primary principle. Several of the world automation firms discharge sequences science-based apparatus on free software certificates. The second channel is building, experimenting, and equipping innovative formula. Firms following the above channels, and as combinations, can be effortlessly established, especially in Lagos, which is a prominent technological nerve. This explains the complex nature of AI application in CC in a developing country like Nigeria.

Creating algorithms is merely an aspect of a broader source of synthesizing expert systems into upshots and the process of climate adaptation. In practice, efficient and adequate anticipative algorithms, using AI, depends on the development of three basic constituents: smart computers, large datasets, and abundance of cheap labour. In other words, there is the need for smart computers as reliable computing power is sufficiently available in Nigeria. But adaptation projects that integrate verified supercomputers, cheap well trained human labour is required to segment or tag data. Nigeria lacks the requisite computer trained labour force for AI technology.

This leaves three critical factors in the Nigerian context; namely, the availableness of text files, appropriateness of blueprints as well as the resourcefulness of the creators, investors, et al. making use of blueprint in CC. That said, in the establishment of novel projects, resources and results, these determinants are interwoven, and not easy to be examined alone. AI is a multifunctional apparatus. Nevertheless, the turnout of AI project or enterprise relies on the dataset of a particular code. In spite of the program, the capability of critical word processing files is imperative for AI to be of any substance to support CC adaptation and mitigation. This fact is further exacerbated because the chunk of the innovation takes place outside the shores of Nigeria. Another dominant variable is the fact that most foreign investors and their partners remain the key drivers of the technology in Nigeria. Regardless of the above, AI remains a tool that is capable of transforming Nigeria's climate adaptation process.

Even if it is because of inefficient dissemination or a fundamental lack of actionable data, its efficient collection, or as a result of other factors, the unavailability of data has serious consequences for the use of AI in adapting to CC. What is more? Without enough data, AI algorithms will be considerably less accurate and less useful in the Nigerian CC adaptation and mitigation process. Notwithstanding the risk of AI prejudices, the concern of job loss, discrimination, ethical and legal impasse, etc. all of which are the negative sides of A1, proper regulation will resolve the associated AI negatives as being displayed in leading AI technology climes.

Conclusion and Recommendations

Nigeria is currently experiencing the deleterious effects of CC. There is substantial evidence that even if significant reductions in the country's GHG emissions are achieved, a great majority of the effects of CC appears to be inevitable. Federal, state, local and community adaptation and mitigation with AI planning and regulation is urgently required to address how the country can adapt to the changes of climate variability to avoid the impeding climate catastrophe. This formed the central theme and purpose of this paper.

The paper presented a legal, policy and institutional synthesis of the vexed issue of adaptation planning and mitigation with Nigeria as a case study. The vignette examined the institutional and regulatory challenges and trade-offs that CC poses in three vulnerable areas: building/infrastructure, energy/electricity, and ecosystem resources. It discussed obstacles to Nigeria's adaptation and mitigation and analysed possible successes which the introduction of AI will impact in the country's effort.

Although Nigeria has adopted diverse approaches including legal, policy and institutional frameworks to CC governance, many are inadequate due partly to inability to deploy AI technology and other innovations. Nigeria needs to use AI adaptation methodologies that are climate data and science dependent adaptation and mitigation strategies to combat flooding, precipitated sea-level rise and wildfires when making decisions about future development, infrastructure investments, environmental protection, and disaster risk preparedness. However, the contending challenges in terms of lack of robust legal statute, poor legislative arrangements to regulate robotics, dearth of relevant data and manpower for effective deployment of AI to combat CC remain critical. Therefore, the paper concludes with recommendations that the federal government of Nigeria should:

(a) Establish a national law on AI
(b) Mainstream AI and related technology innovations in other Nigerian laws and policies of CC adaptation and mitigation programmes.
(c) Acquire the necessary equipment and training of high skilled manpower to give impetus to AI for sustainable development efforts to better anticipate future CC developments and provide adequate climate solutions.

These recommendations will ensure that AI governance framework together with proper regulation will make AI systems deployment safe, effective, and put the country on the same pedestal with her contemporaries.

References

Bueb, B., Tröltzsch, J., Reichwein, D., et al. (2021). Towards sustainable adaptation pathways: A concept for integrative actions to achieve the 2030 Agenda, Paris agreement and Sendai framework. https://inis.iaea.org/search/search.aspx?orig_q=RN:53029772. Accessed June 1, 2024.

Building Nigeria's Response to Climate Change (BNRCC). (2020). https://www.climatescorecard.org/2020/11/building-nigerias-response-to-climate-change-bnrcc/. Accessed June 1, 2024.

Cheong, S.-M., Sankaran, K., & Bastani, H. (2022). Artificial intelligence for climate change adaptation. *WIREs Data Mining Knowledge Discovery, 12*(5). https://doi.org/10.1002/widm.1459. Accessed May 2, 2024.

Climate Change Act, Nigeria. (2021). https://faolex.fao.org/docs/pdf/NIG208055.pdf. Accessed June 1, 2024.

Cole-Dai, J., Ferris, D., Lanciki, A., Savarino, J., Baroni, M., & Thiemens, M. (2009). Cold decade (AD 1810–1819) caused by Tambora (1815) and another (1809) stratospheric volcanic eruption. *Geophysical Research Letters., 36*, L22703. https://doi.org/10.1029/2009GL040882

Constitution of the Federal Republic of Nigeria. (1999). https://faolex.fao.org/docs/pdf/nig164561.pdf

EU Artificial Intelligence Act. (2024). https://www.artificial-intelligence-act.com. Accessed May 2, 2024.

Eurostat. (2010). Using official statistics to calculate greenhouse gas emissions: A statistical guide. https://ec.europa.eu/eurostat/web/products-statistical-books/-/ks-31-09-272. Accessed June 1, 2024.

FBUR (2018) Nigeria. First biennial update report. https://unfccc.int/sites/default/files/resource/Nigeria%20BUR1_Final%20%282%29.pdf. Accessed May 1, 2024.

Federal Government of Nigeria. (2013). Nigeria, post-disaster needs assessment 2012 Floods. https://www.gfdrr.org›default›files›publication. Accessed May 2, 2024.

Federal Ministry of Environment. (2021). The national climate change policy (NCCP). https://climatechange.gov.ng/wp-content/uploads/2020/09/national-climate-change-policy-1-1.pdf. Accessed June 1, 2024.

FNC. (2003). Nigeria's first national communication. https://unfccc.int/resource/docs/natc/nignc1.pdf. Accessed May 2, 2024.
Intended Nationally Determined Contribution (INDC). (2015). https://www.fao.org › details › LEX-FAOC187295. Accessed August 26, 2024.
Johns Hopkins University. (2023). How AI can help combat climate change. *JHU Hub*. https://hub.jhu.edu. Accessed May 5, 2024.
Joshi, N. (2019). Why governments need to regulate AI. *Allerin*. https://www.allerin.com/blog/why-governments-need-to-regulate-ai. Accessed February 1, 2024.
NARF. (2015). National agricultural resilience framework.
NASPA-CCN. (2011). National adaptation strategy and plan of action on climate change for Nigeria. https://csdevnet.org/wp-content/uploads/NATIONAL-ADAPTATION-STRATEGY-AND-PLAN-OF-ACTION.pdf. Accessed May 5, 2024.
National Centre for Artificial Intelligence and Robotic (NCAIR). (2020). https://prnigeria.com › 2020/11/13 › nitda-artificial-in. Accessed June 7, 2023.
National Disaster Framework (NDF). (2010). https://disasterlaw.ifrc.org›media. Accessed April 1, 2024.
National Governance Committee for the New Generation Artificial Intelligence. (2019). Governance principles for the new generation artificial intelligence—Developing responsible artificial intelligence. *China Daily*. https://www.chinadaily.com.cn/a/201906/17/WS5d07486ba3103dbf14328ab7.html?from=timeline&isappinstalled=0. Accessed August 2, 2023.
NDC. (2021). Nigeria's first nationally determined contribution. Updated Submission. https://unfccc.int/documents/497790. Accessed May 4, 2024.
Nerini, F., Slob, A., Engström, R., et al. (2019a). A research and innovation agenda for zero-emission European cities. *Sustainability, 11*(6), 1692. https://doi.org/10.3390/su11061692
Nerini, F., Sovacool, B., Hughes, N., et al. (2019b). Connecting climate action with other sustainable development goals. *Nature Sustainability, 2*, 674–680. https://doi.org/10.1038/s41893-019-0334-y
Nerini, F., Tomei, J., To, L., et al. (2018). Mapping synergies and trade-offs between energy and the sustainable development goals. *Nature Energy, 3*(1), 10–15.
Obianyo, C., & Ater, S. V. (2022). A critical appraisal of the legal framework of artificial intelligence governance in Nigeria. *Chukwuemeka Odumegwu Ojukwu University Journal of Private and Public Law (COOUJPPL), 4*(1).
OECD. (2016). OECD science, technology and innovation outlook. https://doi.org/10.1787/sti_in_outlook-2016-en
Okon, E. M., Falama, B. M., et al. (2021). Systematic review of climate change impact research in Nigeria: Implication for sustainable development. *Heliyon, 7*(9), e07941. https://doi.org/10.1016/j.heliyon.2021.e07941
Orie, E. G. (2014). Environmental protection and fundamental human right to life: A review of the Nigerian constitutional provision and the judicial posture. *NOUN Current Issues in Nigerian Law, 4*, 148–196.
Orie, E. G. (2018). Regulating climate—Induced migration for attainment of sustainable development goals in Nigeria. *International Journal of Current Research in the Humanities, 22*, 293–315.
Orie, E. G. (2020). Climate change adaptation mechanism for sustainable development goal 1 in Nigeria: Legal imperative. In W. Leal Filho, N. Oguge, D. Ayal, et al. (Eds.), *African handbook of climate change adaptation* (pp. 1897–1918).
Orie, E. G., & Ugbejeh, E. (2021). Re-thinking the legal framework for flood management in Nigeria in the light of climate change and sustainable development goals (SDGs). *International Journal of Current Research in the Humanities, 25*, 269–290.
Oxford Insights. (2022). Government AI readiness index 2021. https://oxfordinsights.com/wp-content/uploads/2023/11/Government_AI_Readiness_21.pdf. Accessed May 10, 2024.

Rutenberg, I., Gwagwa, A., & Omino, M. (2021). Use and impact of artificial intelligence on climate change adaptation in Africa. In W. Leal Filho, N. Oguge, D. Ayal, et al. (Eds.), *African handbook of climate change adaptation* (pp. 1107–1126).
SNC. (2014). Nigeria's second national communication. https://unfccc.int/resource/docs/natc/nganc2.pdf. Accessed May 2, 2024.
The World Bank. (2018). Nigeria erosion and watershed management project (NEWMAP). https://www.worldbank.org/en/news/loans-credits/2018/06/27/nigeria-erosion-and-watershed-management-project-newmap-additional-financing. Accessed June 1, 2024.
The World Bank Group. (2021). Climate risk country profile: Nigeria. https://climateknowledgeportal.worldbank.org/sites/default/files/2021-07/15918-WB_Nigeria%20Country%20Profile-WEB.pdf. Accessed June 3, 2024.
TNC. (2020). Nigeria. Third national communication under the UNFCCC. https://unfccc.int/documents/226453
UN Sustainable Development Goal (SDG) 13. https://nigeria.un.org/en/sdgs/13. Accessed June 1, 2024.
UNCTAD. (2017). Information economy report 2017. http://unctad.org/en/pages/PublicationWebflyer.aspx?publicationid=1872. Accessed January 10, 2024.
UNFCCC (1992) United Nations Framework Convention on Climate Change (UNFCCC). May 9, 1992, S. Treaty Doc No. 102–38, 1771 U.N.T.S. 107.
United Nations. (1987). Our common future: Report of the UN world commission on environment and development. https://sustainabledevelopment.un.org/content/documents/5987our-common-future.pdf. Accessed June 1, 2024.
United Nations. (2021). What is climate change? https://www.un.org/en/climatechange/what-is-climate-change. Accessed June 4, 2024.
USA National Artificial Intelligence Initiative Act of 2020. Division E, Sec. 5001. (2020). https://www.congress.gov/bill/116th-congress/house-bill/6216. Accessed June 1, 2024.
World Economic Forum. (2023). Globally, wildfires cost $50 bn every year; AI can help fight them. https://economictimes.indiatimes.com/news/international/world-news/globally-wildfires-cost-50-bn-every-year-ai-can-help-fight-them-wef/articleshow/97035706.cms?utm_source=contentofinterest&utm_medium=text&utm_campaign=cppst. Accessed June 1, 2024.
Worldometer. (2024). Nigeria population. https://www.worldometers.info/world-population/nicaragua-population/. Accessed June 1, 2024.
Xin, Q., et al. (2023). Artificial intelligence for climate change risk prediction, adaptation, & mitigation. Ecological Processes. https://www.springeropen.com/collections/AICC. Accessed May 2, 2024.

Chapter 7
Navigating Uncertainty: Legal Considerations for Marine Carbon Dioxide Removal as a Climate Solution

Tony Cabus

Abstract According to the latest IPCC reports, the world will likely need carbon dioxide removal (CDR) to limit the warming of the planet to 1.5 °C by the end of the century. Marine CDR (mCDR) technologies have therefore emerged as a potential solution to mitigate climate change. While mCDR can be deployed unilaterally, its maximum efficiency requires international coordination due to potentially extensive transboundary effects. However, the current international legal framework, including the UNFCCC and the Paris Agreement lack a dedicated regime for mCDR, resulting in the application of ill-suited regulations. In this context, principles of international law, which could provide guidance on the permissibility of mCDR, instead add to the confusion. This chapter addresses the unrealistic expectation of current legal frameworks to overly rely on unequivocal instructions from natural sciences to eliminate the risks, and provides two avenues to contend with risks. One comes to encouraging the use of best science through transparency and standardisation. The other circumvents risks by restraining overuse of mCDR while improving liability mechanisms. Contrary to the prevailing notion of the law's impotency in the face of scientific uncertainty, it seeks to oppose this scepticism by offering legal remedies that, if not eliminating the risk entirely, aim to mitigate and frame it effectively.

Keywords Climate change · Carbon dioxide removal · International law · Uncertainty

Introduction: The Time Crunch and the Dilemma of Climate Policy

It is now well known that limiting the warming of the planet to 1.5 °C by the end of the century will require some degree of carbon dioxide removal (CDR) (IPCC, 2018, 2022). Whether it is for neutralising residual emissions from fields that involve emissions 'hard-to-abate', or to remove the excess of previously emitted CO_2 in the

T. Cabus (✉)
Walther Schücking Institute for International Law, Kiel, Germany

W. Leal Filho et al. (eds.), *Climate Change Mitigation and Adaptation in Practice*, Climate Change Management, https://doi.org/10.1007/978-3-031-85217-6_7

atmosphere keeping the global temperature above 1.5 °C, CDR will be needed to reach the target of the Paris Agreement. Indeed, every ton of CO_2 exceeding our remaining carbon budget (i.e., the amount of carbon the world can still emit before warming the global temperature over 1.5 °C) will hamper the 1.5 °C target and will need to be removed. On its current trajectory, the world will soon exhaust its carbon budget. Area business-as-usual scenario would lead to a 2.6–2.9 °C raise of temperature from pre-industrial times and it will further increase in the next century (MCC, 2021; CAT, 2022). Regrettably, even the pledges issued by States are expected to be insufficient for reaching the 1.5 °C scenario and would lead instead to a warming of 2.5 °C (UNFCCC, 2022). Considering that pledges typically represent the optimistic version of State policies, this failure to meet the global target, even under optimistic assumptions, poignantly reflects the difficulty to reduce our emissions. For this reason, CDR seems more than ever relevant, and not only will we probably need to use it to compensate residual emissions, but also to remove the excess of CO_2 resulting from insufficient and delayed mitigation (IPCC, 2018).

While most climate strategies currently envisage land-based CDR, marine CDR (mCDR) also has promising potential. The ocean constitutes the largest long-term sink for anthropogenic CO_2 and absorbs roughly 25% of all CO_2 emissions (Gattuso & Hansson, 2017). Several methods have been proposed to influence the carbon cycle of the ocean to ultimately reinforce its role as a carbon sink and/or reservoir. Ocean fertilisation hopes to trigger a phytoplankton bloom increasing uptake of atmospheric CO_2 through photosynthesis by releasing nutrient such as iron in the ocean. Ocean weathering (or ocean alkalinity enhancement) focuses on the chemical reactions taking place in the ocean and aims to influence the pH of the water and thereafter the solubility of atmospheric CO_2 with the ocean by releasing alkaline substances into it. Upwelling attempts to transfer deep-sea minerals to the surface to feed plankton and improve the CO_2 biological pump of the ocean while downwelling tries to artificially sink CO_2 rich waters to the depths where it will be store for thousands of years. Finally, blue carbon solutions aim at capturing atmospheric CO_2 through plant cultivation (e.g., macroalgae) that will subsequently be stored or used as biomass feed products (e.g., energy or fertiliser) (McGee et al., 2017).

Unlike land-based options, these methods are prone to transboundary effects since oceanic phenomena do not confine themselves to artificially drawn maritime boundaries (Henocque, 2017). Consequently, international governance and regulations should set benchmarks for a common use of the ocean in their regard. Regrettably, mCDR activities lack a dedicated legal framework and predominantly fall within the scope of indirectly related international instruments (Webb et al., 2021). This is either due to the generality of such instruments, which encompass a wide array of environmental threats, or because they address specific aspects of mCDR activities, such as their location or the materials used. Falling under the umbrella term "‘geoengineering’", mCDR has however been addressed by a major international forum, the Conference of the Parties of the Convention on Biological Diversity (CBD COP) gathering virtually all UN States except the United States. In a series of (non-binding) resolution, the conference called for a moratorium on the deployment of climate-related geoengineering (CBD, 2008, 2010, 2012, 2016). Based on a traditional understanding of the precautionary principle, this decision rests on the scientific uncertainties associated with mCDR activities. While precaution is usually laudable, it should also be recognised that the preservation of the status quo and the delay of mitigating action bears a long-term cost even if the price to pay today (both in financial terms and by ecosystems) appears clearer than the long-term benefits of mitigation (La Vina, 1993). As greenhouse gases (GhGs) accumulate in the atmosphere, mitigation measures taken today are more influential than those taken in the future. Thus, while early action may carry a risk, inaction and delays also pose a risk of being "'too late"' (IPCC, 2023). Climate strategies therefore need to consider both the risks of deploying mCDR early and the risks of delaying mitigation. This does not necessarily mandate hasty deployment without sufficient scientific evidence, notably on the effectivity of the methods, but underscores the importance of engaging early in a discussion on the role mCDR should play and, ultimately, the importance of establishing a robust regulatory framework anticipating future developments. Given the difficulties faced by our societies in reducing their emissions and the increasing recognition that emission reduction alone may no longer be sufficient to achieve the Paris target, such endeavour is crucial (IPCC, 2022). Acting early is essential not only to bolster mitigation efforts but also to pre-empt time-constrained decision-making in the future.

Following a legal methodology, this chapter first provides a glance at the legal challenges faced by mCDR (Sect. 7.2) before proposing avenues for improvement (Sect. 7.3). Since the law of mCDR is still in its infancy, the arguments made here remain abstract and holistic, and only intend to provide high-level guidance for future law-making in the field. Thus, this chapter does not propose concrete regulations. Doing so still depends, in part, to the place accorded to mCDR in the political agendas of States which will ultimately shape the law. In addition, while mCDR regulation faces a plethora of legal questions, only a selection of particularly illustrative issues is covered in Sect. 7.2 to demonstrate the need for a tailored approach in Sect. 7.3.

International Law and Marine CDR

While it can be piloted domestically, mCDR will be at its maximum efficiency when coordinated on a global level. Firstly, because mCDR activities at scale may have massive transboundary effects and therefore require an unprecedented level or cooperation (Reynolds, 2018). And secondly, because in order to reach global targets, efforts must be truly global. Coordination is thus also crucial so that local efforts are spent wisely, avoiding unnecessary overlaps and contradicting results. This situation, which characterises climate action in general, is acknowledged in the UNFCCC, whose preamble prescribes: "the global nature of climate change calls for the widest possible cooperation by all countries and their participation in an effective and appropriate international response." Unfortunately, the UNFCCC and its descendants (Kyoto Protocol and Paris Agreement) which constitute the only set of dedicated international documents on combating climate change do not directly address mCDR. This absence of dedicated regime leaves us with a plethora of potentially applicable sources of law without a definite answer as to the permissibility of mCDR (a). In this context, principles of international law could serve as fillers or at least as a compass for interpreting the legality of mCDR but instead, blur the picture even more (b).

Charting a Course for Legal Clarity

It would be untrue to claim that international law does not regulate mCDR at all. In fact, multiple regimes ranging from broad customary obligations to precise procedural obligations can apply to mCDR. Yet, it is clear that these rules do not yet constitute a satisfying legal regime applicable to the field. They may either be ill-suited to deal with as the dilemma presented by mCDR or simply be leaving major regulatory gaps requiring new laws.

Naturally, the foremost question remains what potential environmental harm mCDR activities cause directly or indirectly. Indeed, while international law prohibits significant harm to the environment, science is currently unable to determine the exact level of harm mCDR activities could have at scale (Trail Smelter Case, 1941; ICJ, 1949, 1996, 2010, 2015; UN, 1972, 1992). On the other hand, a precise determination of the type and level of harm still necessitates a common understanding on what is considered 'significant.' This crucial threshold, which marks the boundary between what can be considered 'acceptable' and 'prohibited' harm, remains a societal criterion rather than a scientific one and would need clarifications to perform its role.

Yet, beyond this primary question, other regulatory gaps of critical importance exist. While these can obviously vary between methods of mCDR, with some methods benefitting from more advance regimes than others, common issues can also be identified. Before engaging in any mCDR activities, States must conduct environmental impact assessments (EIAs) to demonstrate that a proposed activity will not harm the environment beyond a certain threshold. EIAs encapsulating the complexity and transboundary effects of mCDR are thus highly important. Deployment of mCDR also often bring forth the problem of illegal dumping. As many mCDR methods rely on introducing material in the water column for various benefits, the rules of the London Convention and London Protocol regulating the matter are often raised. While the material envisaged would be used to mitigate climate change, the legality of it remains arguable due to its potential polluting effect on the marine environment. Finally, multiple gaps can be identified ensuing the deployment of mCDR. These include *inter alia,* monitoring, reporting and verification (MRV) which, in the same way as EIAs, would benefit from further regulations setting common standards; accounting, which is also important for dressing accurate inventories of emissions and removals under the UNFCCC rules; but especially, rules of accountability in order to satisfyingly deal with harm done to the environment of other States and areas beyond national jurisdiction.

To illustrate the gaps in the regulation of mCDR and the uncertainty surrounding it, this sub-section will therefore take the example of three areas in need of development: EIA, dumping and, accountability.

(a) **EIA**

In current international law, two legal rules play a crucial role before undertaking any activity with potential transboundary effects. They are the obligation to prevent transboundary damage and the obligation to conduct an EIA (ICJ, 2015). Deeply interconnected, both rules rely on scientific knowledge to be effective. Scientific knowledge is crucial to assess the damaging potential of risky activities but also to set up standards of best practice. Yet, in general, international law does not set any criterion regarding the content, depth or length of EIAs (ICJ, 2010). The scientific soundness of EIAs is therefore left up to the discretion of States and in extreme cases of bad faith, EIAs can be entirely irrelevant but still in compliance with international law (PCA, 2016). Fortunately, in certain instances, international law does set qualitative standards for EIAs thereby creating minimum thresholds to be met (e.g., the UNCLOS). In other cases, the law can even generate models of EIAs to be followed by States (e.g., Espoo convention or the Protocol on the Environmental Protection of the Antarctic). A satisfying regime for mCDR would follow this example and propose, at least, minimum standards to be met ensuring for example the use of best practices and transparency. At best, models of EIAs tailored to each method of mCDR could also be proposed to guarantee that mCDR activities around the globe are conducted in the best possible and coherent manner. The Parties to the London Convention and Protocol (LC-LP) relating to dumping at sea explored this possibility and adopted in 2010 a model of assessment framework for the purpose of ocean fertilisation activities (IMO, 2010). This document was followed, in 2013, by a more generic model of assessment framework potentially applicable to all forms of mCDR (IMO, 2013). While the 2010 model EIA for ocean fertilisation is in effect, the broader 2013 EIA is not, due to a lack of ratification of the amendment which it is a part of. These two instruments still constitute, however, useful examples for regulating EIAs in the context of marine mCDR.

(b) **Dumping**

According to the current theories, mCDR would often rely on the introduction of materials into the sea. Ocean fertilisation, for example, relies on releasing iron dust into the sea in order to create a phytoplankton bloom and enhance the local ocean productivity. Another popular option, ocean alkalinisation, is conducted by dissolving silicate and carbonate rocks in the water column to increase the alkalinity of the sea and ultimately its CO_2 absorption potential. In both methods, materials would be introduced in the sea in a process that could be qualified as dumping.

Dumping, which is the deliberate disposal of matter into the sea, is regulated by the London Convention and the more advanced London Protocol. While the Convention generally allows dumping except for a list of materials listed in an annex, the Protocol observes the opposite approach and prohibits dumping except for a list of materials also listed in an annex. The Protocol is therefore stricter. For States party to the Protocol, resorting to mCDR may consequently turn complicated as they would need to justify that their introduction of material in the sea is either an authorised form of dumping or, more simply, not dumping at all. In both cases, an essential point will be to demonstrate that the introduction of the material will not be significantly detrimental to the marine environment. Answering this question, however, remains extremely complex and relies on the ability of natural sciences to demonstrate, not only the non-polluting character of the mCDR material, but also, the absence of adverse consequences following the chemical or physical reactions triggered by the introduction of material into the sea (Article 1.2.2). Without such guarantees, the London Protocol, which also applies the precautionary approach, will continue to prohibit mCDR based on the introduction of materials in the sea whether it is dumping or not.

Illustrating this restrictive interpretation, Parties to both the Convention and the Protocol agreed in 2008 to consider ocean fertilisation activities, other than research, contrary to both treaties (IMO, 2008). This statement echoed a resolution of the Conference of Parties to the CBD (CBD COP) the same year which also took a precautionary stance and called for a moratorium on ocean fertilisation activities (CBD 2008). In the next years, the CBD Parties went further and extended this moratorium to all "climate-related geo-engineering activities" (CBD, 2010, 2012, 2016). While this development was not exactly followed by the London Parties (despite an attempt to bring marine geoengineering under the regulatory umbrella of the Protocol in 2013), the prohibition of ocean fertilisation for precautionary reasons may well have a ricochet effect on other mCDR methods also relying on the introduction of materials into the sea. Ultimately, with our current articulation of the law, controversies on the legality of mCDR through dumping are all contingent on the ability of natural science to prove these activities safe. Uncertainty, in this legal framework, is therefore akin to a showstopper.

(c) **Accountability**

Under international law, questions relating to responsibility and liability are covered by the term of accountability. Rules of State responsibility trigger when a primary obligation of international law is violated by a subject of international law (State or international organisation) while liability regimes trigger when a protected interest has been harmed despite the absence of legal violation (e.g., when a pollution occurs despite the best efforts to prevent it, or, when such harm is commonly accepted by international law and, therefore, not illegal). Since no legal violation exist in cases of liability, a State at the origin of a lawful harm is not obliged to pay reparation to the victim but only compensation falling often short of a full indemnity or restitution. The amount of such compensation varies from cases to cases as it can result from negotiation, courtesy or from a previously agreed amount (Tanzi, 2013). On the contrary, reparations, occurring when the international responsibility of an actor is engaged, must amount to a full indemnity for the damage caused, accompanied by a promise of non-repetition. Reparations are therefore more likely to be higher than compensations. That being said, nothing prevents a compensation to become automatic and/or reach a full amount of indemnity. These cases are however rare and usually constitute cases of 'strict liability' agreed by treaty law. Currently, only instances of damages caused by space objects (Convention on International Liability for Damage Caused by Space Objects, Article XII), nuclear damage (Vienna Convention on Civil Liability for Nuclear Damage, Article IV) and oil spill (International Convention on Civil Liability for Oil Pollution Damage, Article III) possess strict liability frameworks. The consequences following cases of responsibility and cases of liability are therefore really different.

In the context of mCDR, State responsibility could occur if a State violates its obligation to prevent significant transboundary damage through mCDR or by violating the regulations on dumping enshrined in the LC-LP (provided that the State in question is a Party to it). In such case, and according to the rules of State responsibility, the responsible State is bound to full reparation of the injury caused by the wrongful act on top of having to stop the wrongful activity (UN, 2001a). While such rule is laudable, the context of mCDR makes it extremely complex to operate. It is indeed difficult to quantify a damage done by mCDR and perhaps even more difficult to identify victim States due to the diluted effect these activities have on the ocean as a whole (Proelss and Steenkamp, 2023). Additionally, certain mCDR methods may have lock-in effects in which stopping the already damaging mCDR activity could lead to even worse consequences. Current rules of responsibility, based on irrefragable evidences, decisive causality and clearly identified victims, are therefore ill-suited to deal with cases of mCDR.

Unfortunately, rules of liability do not bring solace either. Based on the exercise of due diligence in preventing foreseeable damage, the rules of liability can constitute a satisfying remedy when the consequences of an activity are understood, and means to prevent accidents or adverse side-effects are available (UN, 2001b). Yet, since mCDR is far from fully understood, standards of best practice remain limited. Should an environmental catastrophe occur, the behaviour of the operating State would be assessed against very low standard likely to exonerate it from responsibility and the causal link between the harm and the operator is difficult to demonstrate (Proelss & Steenkamp, 2023). As a consequence, the damage would be devastating but not illegal and the State of origin would be liable but not responsible. The indemnity would be a compensation contingent on the good will of the perpetrating State likely to be short of a full reparation.

Considering our scientific uncertainties and the inherent danger of manipulating natural processes at scale, mCDR activities pose grave risks to our environment even when due diligence is exercised. Liability cases, wherefore operating States cause damages despite fulfilling what is prescribed by law, are thus likely to arise. In the absence of any dedicated framework, issues over e.g., the role of private actors, cooperating States and the amount of compensation are however left untouched, leaving a major gap in the governance of marine mCDR

(Proelss & Steenkamp, 2023). In view of the risks involved with mCDR, it would be reasonable to envisage a regime of strict liability, whereby full compensations are due even when due diligence has been exercised. After all, the three existing activities governed by strict liability regimes share crucial elements with mCDR, i.e., the use of inherently dangerous material or processes, and, a low risk of disaster affected by unpredictable natural factors (e.g., natural disasters). Even then, the difficulty to identify and quantify the damage as well as the victims, remains a problem in search of a solution; another uncertainty in need of legal framing for lack of satisfying scientific answer.

Can Principles Fill the Gaps?

Principles regularly make their apparition in international legal doctrine. Yet, they can easily be misunderstood. Indeed, no commonly accepted definition of what constitutes a principle exist despite them being considered a source of international law (Martin, 2018). Originally envisaged as the set of common rules shared between different domestic legal systems that could be used by international judges to fill gaps left by treaty and customary law, the roots of principles have since varied (Wolfrum, 2011). With the advance of environmental law, principles have emerged from the international sphere notably following international conference on the environment. Regardless of their origin, the function of principles in international law varies as some can operate more independently than others. For instance, the no-harm principle sets a concrete objective along with: a threshold of activation (when significant harm occurs) and a spatial delineation (transboundary and in areas beyond national jurisdiction). In contrast, the precautionary principle only sets an objective and its application will vary entirely from case to case (Wiener, 2018). These two eminent environmental principles may consequently play different roles in the regulation of mCDR.

(a) **The Precautionary Principle**

Often, the Rio Declaration of 1992 is used to define the precautionary principle. In essence, States engaging in activities raising threats of serious or irreversible damage, shall not postpone "cost-effective measures to prevent environmental degradation" even in the event of scientific uncertainty on the probability of the damage. In practice, it forces states to adopt strict measures so as to prevent adverse damage even when the likelihood of damage is still unclear. It is also evolutive as it can advocate for new measures as knowledge and technology evolve.

Even though this principle has definitely made its impact on international environmental law, its precise place in the law is still unclear. Indeed, for some it is an independent rule of law applicable in each and every case while for others it is a simply a guiding consideration when adopting regulations (Armeni & Redgwell, 2015). In any case, the precautionary principle has been adopted or recognised in most environmental instruments and is therefore applied or relevant most of the time. The International Tribunal on the Law of the Sea' Seabed Dispute Chamber has recognised the binding effect of the precautionary principle in the *Activities in the Area* case, the LP has recognised its application in Article 3 and the CBD in its many geoengineering-related resolutions as well. There is therefore no reason to believe that this principle will not be applied to mCDR. Yet, when applied to this type of activity, the precautionary principle presents a dual aspect, which inherently complicates its interpretation. On one hand, it can be seen as obliging states to take precaution against the risks of mCDR. But, on the other hand, it can also be interpreted as obliging states to take actions to mitigate climate change (Armeni & Redgwell, 2015). The divergent interpretations of the precautionary principle consequently lead to different outcomes.

To date, the precautionary principle has frequently served as a rationale for maintaining the existing state of affairs. This principle is grounded in the premise that the status quo represents the least hazardous choice, while introducing novel activities invariably entails heightened environmental risks. This interpretation has been illustrated by the CBD COP and of the London Convention and Protocol (LC-LP COP). The resolution X/33 of the CBD COP is particularly illustrative in this regard by declaring that:

> *in accordance with the precautionary approach and Article 14 of the Convention, that no climate-related geo-engineering activities that may affect biodiversity take place, until there is an adequate scientific basis on which to justify such activities* (CBD, 2010)

Or in the case of the LC-LP COP:

> *given the present state of knowledge, ocean fertilisation activities other than legitimate scientific research should not be allowed.* (IMO, 2008)

Considering the mandates of the LC-LP and the CBD, these statements should not come as a surprise as it would contravene their fundamental purposes to endorse activities that might pose a threat to the environment they are designed to safeguard. Nevertheless, the interpretation of the precautionary principle remains a subject open to evolution, contingent upon future scientific advancements that may determine the precise trade-offs between action and inaction and the ability of the principle to integrate more than one risk at a time (Wiener, 2018). The inevitability of climate-induced harm to biodiversity, with or without the implementation of mCDR measures, could trigger re-evaluations of this principle especially from stakeholders entrusted with the specific mandate to mitigate climate change. In such case, the precautionary principle could shift from hindering mCDR to encouraging it.

In that scenario, a last hurdle related to the 'right amount' of mCDR would need to be overcome. Indeed, the precautionary principle as an expression of the 'least risky option' is closely related to proportionality. Unfortunately, in the absence of certitude on future emission pathways, the 'right amount' is difficult to predict and, shall mCDR be conducted at scale, the risk of doing 'too much' is real and could lead to excessive harm to the environment and unnecessary economic loss (IPCC, 1995). To solve this issue without stripping the principle of its proportionality element, mCDR will therefore need to be kept to a minimum or in close relation with the most likely emission pathway.

The No-Harm Principle

The no-harm principle operates in a more straightforward manner. Essentially, it serves to prevent significant transboundary damage, which includes substantial harm to areas beyond national jurisdiction (ABNJ). This principle establishes two key premises: (1) it does not proscribe damage within one's own territory, and (2) it allows for a certain degree of damage falling below the threshold of 'significant' when affecting ABNJ or third States. In practical terms, this principle imposes an obligation on States to proactively assess the potential effects or risks of their activities before undertaking them. A role typically fulfilled by EIAs. Consequently, when an activity carries the potential for transboundary impact, an EIA must precede the activity (ICJ, 2015). If the EIA reveals that the activity will indeed result in significant damage, the no-harm principle then compels the State to prohibit the activity itself. Thereby, and in the same vein as the resolution X/33 of the CBD COP, the no-harm principle will remain subordinated to advances in the realm of science. In the absence of better conclusions on the likelihood of harm, the principle cannot serve its intended role of a demarcation line between what is deemed acceptable and what is not.

Beyond this issue, a clearer interpretation of the 'significant' threshold will also be needed and may be as well influenced by the dangers of inaction.

In conclusion, principles of international law exhibit inherent vulnerabilities stemming from either the absence of scientific certitude or the nuanced context of climate change which involves trade-offs and neutralises one-sided notions such as 'precaution' and 'prevention.' Consequently, it is unlikely that principles alone can effectively address the void left by the absence of a dedicated legal framework pertaining to mCDR.

Changing the Paradigm of Uncertainty

Based on the aforementioned considerations, it could appear that international law, while not excluding mCDR from the agenda, finds itself in a waiting period, reliant on scientific advancements to fulfil the requisites of the precautionary principle and no-harm principle. Such an approach carries a dual fallacy. Firstly, it overlooks the many, and unrelated to science, regulatory gaps that remain, i.e., defining acceptable harm in the context of climate mitigation activities or addressing harm caused to ABNJ and the importance of international coordination and harmonisation. Secondly, it should be acknowledged that early implementation of climate mitigation policies accrues greater benefits. In light of the recurrent alarming reports of the Intergovernmental Panel on Climate Change (IPCC), regulations should be developed promptly in order to be effective when the science of mCDR will have progressed. Contrary to the notion that law remains impotent in the face of scientific uncertainty, it holds the capacity to facilitate the development of scientific knowledge and best practices while managing risks arising from uncertainty by promoting safeguards.

Promoting Best Science to Reduce Uncertainty

First and foremost, the development of model EIAs as exemplified by the LC-LP is crucial to ensure the performance of quality EIAs. These model EIAs should serve as templates, offering comprehensive guidelines to evaluate the potential effects of mCDR activities. By standardising the assessment process, States can ensure that critical environmental factors are considered systematically, fostering greater scientific precision, reducing ambiguities in evaluating the environmental impact and therefore, improving the reviewability of the studies.

Naturally, transparency during and after the review process also needs to be obligatory so that, not only would actors involved be under public scrutiny, but it would also facilitate the transfer of experiences within the scientific community.

Another area in need of harmonisation is the MRV process. As activities of mCDR engage with dynamic ecosystems, constant reviewing is crucial to ensure the continuous effectiveness and safety of the activity. It is also through a robust MRV process that lessons can be learned and that potential adjustments can be planned.

Finally, international law would benefit significantly from establishing more explicit thresholds of acceptable harm in the context of climate change mitigation. Currently, evaluating the risks of mCDR activities outside of the climate change context likely speaks against them, since mCDR-induced harm is seen as an avoidable disturbance typically weighted against a baseline of status quo. Under this framework, only mCDR activities presenting an ecosystem improvement over the status quo, such as ecosystem restoration initiatives, would meet the threshold of acceptability. However, efficient and scalable mCDR may involve methods with higher disturbance potential. In such cases, thresholds of acceptable harm need to be adapted to the dynamism of our natural ecosystems and, in particular, include degradations that will naturally occur in scenarios without mCDR interventions.

Framing Uncertainties and Risks

In spite of the ongoing scientific uncertainty surrounding the precise effects and risks of mCDR, strategies can be implemented to mitigate its potential risks. Chief among these strategies are initiatives focused on enhancing predictability, coordination, and discouraging excessive reliance on mCDR solutions in order to keep the associated risks to a minimum.

In the first regard, long-term strategies formulated by States play a pivotal role in increasing predictability, providing a pathway to avoid unwarranted excesses in mCDR deployment. States are already under the obligation to submit the inventories of their emissions and removals to the UNFCCC secretariat while the IPCC is also tasked to establish pathways and scenarios accounting these data. Enhancing this system to model the requisite quantity of mCDR required to achieve internationally agreed targets would be beneficial to avert excessive mCDR. As this task relies on net emission predictions, the accuracy of inventories is pivotal and the capacity of developing States to draw them with precision should continue to be refined.

If accurate predictions lay the foundation for avoiding excessive levels of mCDR, international coordination becomes the next logical step to safely and effectively deploy mCDR activities. Strategic environmental assessment as well as spatial planning will be important tools to ensure implementing mCDR effectively without overburdening parts of the sea or overstressing ecosystems. The possibility to establish Area-Based Management Tool (ABMT) in the high seas under the newly adopted Convention on the conservation and sustainable use of marine biological diversity in areas beyond national jurisdiction (BBNJ), could provide a mean to reinforce mCDR coordination in this contested area. Furthermore, global coordination is essential in achieving collective targets and averting unilateral initiatives by affluent States, which possess the means to conduct mCDR. Neglecting this aspect would raise ethical concerns, as technologically advanced States could reduce their emissions through mCDR, while others are compelled to employ more conventional, and often challenging, emissions reduction methods.

Additionally, the ongoing research on mCDR and the evolutive conditions in which mCDR operate mandate for a flexible regime capable of adapting itself to new knowledge or natural variations (La Vina, 1993). This flexibility could be ensured through the design of the rules themselves but also through dynamic governance, either through regular meetings of States or through permanent institutions. Considering the complexity of the activity and the need to involve developing States, the governance of mCDR should also favour technological transfer, capacity building and technical assistance. Not only would it improve the acceptability of the governance, but it would also back compliance with international regulations (Vinuales, 2010).

Finally, the liability gap needs to be closed despite the difficulty to truly quantify the extend of ecosystem damages and the hardship to evidence causal relationships. Facilitated liability should be set established as it has been the case in the context of other hazardous activities (Vinuales, 2010).

Conclusion

As global interest and research on mCDR grows, the absence of a dedicated regime capable of regulating this complex activity will become more flagrant. While postponing the development of rules in areas in which considerable new knowledge is expected, is justified, scientific uncertainty should not refrain the international community from framing the intricacies of mCDR and filling gaps on what is already known. (La Vina, 1993). This effort entails engaging with the stakeholders affected by mCDR but also with the public at large to form a broad agreement on the role mCDR should play in the portfolio of climate mitigation solutions. Beyond this societal debate, governments should cooperate in the development of standards applicable to these activities in order to reach a satisfying level of harmonisation. Unilateral and substandard activities of mCDR should be avoided at all cost, and a potential governance of the field should ensure that mCDR only occurs after the maximum efforts in reducing emissions has been done.

While research is being conducted on the concrete efficacy and side-effects of mCDR, several legal gaps should be filled. These include, notably, clarifying thresholds of permissibility concerning the harm done to ecosystems. Such undertaking is naturally difficult considering the multiple economical interests involved around various affected ecosystem services but also considering the different values each society attach to the natural world. As unusual as it may seem, the cost of inaction should also be included in these calculations.

Determining the acceptable threshold of harm associated with mCDR will be valuable for improving the relevance of EIAs but it will not be sufficient (Glasson, 2008). The quality of EIAs is of extreme importance and would benefit from minimum standards applicable globally. Such harmonisation has already been impulse by the LC-LP but failed to concretise due to a lack of participation. As the international scientific cooperation to develop standards of mCDR continues, elements to consider when conducting these activities will be highlighted and could constitute the base for tailored models of EIAs corresponding to each method of mCDR.

A last gap of major importance concerns accountability and the necessity to develop rules and standards on assessing the harm done by mCDR activities. This will be crucial to develop a satisfying regime of indemnities and alleviate the fears pertaining to potential risks.

Finally, besides filling the regulatory gaps, international law should play its part in reducing and framing uncertainties. On the one hand, international should promote the development and practice of best science notably by adopting strict standards of EIA and MRV, and by promoting scientific transparency, cooperation and knowledge transfer. On the other hand, international law can also reduce the risks associated with mCDR by ensuring that the least amount of mCDR is conducted globally. This will require transparency and coordination in order to ensure that mCDR is conducted following stark emission reductions. Failing to cooperate entails the risk to see the multiplication of unilateral actions using mCDR as a 'techfix.'

The development of international law for common areas such as the ocean is notoriously slow, as the recently nineteen years of negotiations on the BBNJ agreement illustrate. Engaging in the development of rules at an early stage should not be misconstrued as hastiness. Rather, it is a prudent course of action, considering that by the time these deliberations bear fruit, the scientific understanding of mCDR will have considerably advanced. This inherent lag of the law-making process should prioritise action sooner rather than late.

References

Armeni, C., & Redgwell, C. (2015). International legal and regulatory issues of climate geoengineering governance. *Climate Geoengineering Governance Working Paper Series, 021*, 34.

CBD resolutions IX/16. (2008). X/33 (2010); XI/20 (2012); XIII/14 (2016).

Climate Action Tracker. (CAT). (2022). On: https://climateactiontracker.org/global/temperatures/

Gattuso, J. P., & Hansson, L. (2017). Ocean acidification. In A. Euzen, F. Gaill, D. Lacroix, P. Cury, (Eds.), *The ocean revealed CNRS Editions*, pp. 66–67. Paris.

Glasson, J. (2008). Principles and purposes of standards and thresholds in the EIA Process. In M. Schmidt, J. Glasson, L. Emmelin, H. Helbron, (Eds.), Standards and thresholds for impact assessment. *Environmental Protection in the European Union*, vol. 3, pp. 3–17. Springer.

Henocque, Y. (2017). Maritime spaces and their levels of governance. In A. Euzen, F. Gaill, D. Lacroix, & P. Cury (Eds.), *The ocean revealed* (pp. 182–183). CNRS Editions.

ICJ. (1949). Corfu Channel (United Kingdom v. Albania), Judgment, ICJ Reports 4. p. 22.

ICJ. (1996). *Legality of the threat or use of nuclear weapons, advisory opinion.* ICJ Reports, p. 226.

ICJ. (2010). *Pulp mills on the river Uruguay (Argentina v. Uruguay), Judgment.* ICJ Reports, p. 14.

ICJ. (2015). *Certain activities carried out by Nicaragua in the Border Area (Costa Rica v. Nicaragua) and construction of a road in Costa Rica along the San Juan River (Nicaragua v. Costa Rica), Judgment.* ICJ Reports, p. 665.

IMO. (2008). Resolution LC-LP.1 on the regulation of ocean fertilisation.

IMO. (2010). Resolution LC-LP.2 on the assessment framework for scientific research involving ocean fertilisation.

IMO. (2013). Resolution LP.4(8) on the amendment to the London Protocol to regulate placement of matter for ocean fertilisation and other marine geoengineering activities.

IPCC., Bolin, B., Houghton, J. T., Filho, G. M., Watson, R. T., Zinyowera, M. C., Bruce, J., Lee, H., Callander, B., Mosse, R., Haites, E., Moreno, R. A., Banuri, T., Dadi, Z., Gardner, B., Goldemberg, J., Hourcade, J., Jefferson, M., Melillo, J., Mintzer, I., Odingo, R., Parry, M., Perdomo, M., Quennet-Thielen, C., Vellinga, P., Sundararaman, N. (Eds.). (1995). Second assessment, climate change.

IPCC., Rogelj, J., Shindell, D., Jiang, K., Fifita, S., Forster, P., Ginzburg, V., Handa, C., Kheshgi, H., Kobayashi, S., Kriegler, E., Mundaca, L., Séférian, R., & Vilariño, M. V. (Eds.). (2018). Mitigation pathways compatible with 1.5°C in the context of sustainable development. In V. Masson-Delmotte, P. Zhai, H.-O. Pörtner, D. Roberts, J. Skea, P. R. Shukla, A. Pirani, W. Moufouma-Okia, C. Péan, R. Pidcock, S. Connors, J. B. R. Matthews, Y. Chen, X. Zhou, M.I. Gomis, E. Lonnoy, T. Maycock, M. Tignor, & T. Waterfield (Eds.), *Global warming of 1.5°C. An IPCC special report on the impacts of global warming of 1.5°C above pre-industrial levels and related global greenhouse gas emission pathways, in the context of strengthening the global response to the threat of climate change, sustainable development, and efforts to eradicate poverty*, pp. 93–174. Cambridge University Press.

IPCC., Shukla, P. R., Skea, J., Reisinger, A., Slade, R., Fradera, R., Pathak, M., Al Khourdajie, A., Belkacemi, M., van Diemen, R., Hasija, A., Lisboa, G., Luz, S., Malley, J., McCollum, D., Some, S., & Vyas, P. (Eds.). (2022). Summary for policymakers. In P. R. Shukla, J. Skea, R. Slade, A. Al Khourdajie, R. van Diemen, D. McCollum, M. Pathak, S. Some, P. Vyas, R. Fradera, M. Belkacemi, A. Hasija, G. Lisboa, S. Luz, & J. Malley, (Eds.), *Climate change 2022: Mitigation of climate change. Contribution of working Group III to the sixth assessment report of the intergovernmental panel on climate change.* Cambridge University Press, pp. 1–48.

IPCC. (2023). Summary for policymakers. In H. Lee, & J. Romero (Eds.), *Climate change 2023: Synthesis report. Contribution of working Groups I, II and III to the sixth assessment report of the intergovernmental panel on climate change.* Geneva, Switzerland, pp. 1–34.

La Vina, A. G. (1993). Climate change, scientific uncertainty and international law. *Philippine Law Journal, 68*(1), 17–61.

Martin, G. J. (2018). Principles and rules. In L. Krämer, & E. Orlando, (Eds.), *Elgar encyclopedia of environmental law*, vol. VI, Principles of Environmental Law. Edward Elgar Publishing, pp. 13–22. MCC Carbon Clock on: https://www.mcc-berlin.net/en/research/co2-budget.html

McGee, J., Brent, K., & Burns, W. (2017). Geoengineering the oceans: An emerging frontier in international climate change governance. *Australian Journal of Maritime & Ocean Affairs, 10*(1), 67–80.

Mercator Research Institute on Global Commons and Climate Change. (MCC). (2021). Remaining Carbon Budget. Available at: https://www.mcc-berlin.net/en/research/co2-budget.html

PCA. (2016). In the matter of the South China Sea (Philippines v. China), Award.

Proelss, A., & Steenkamp, R. C. (2023). Geoengineering: Methods, associated risks and international liability. In P. Gailhofer, D. Krebs, A. Proelss, K. Schmalenbach, & R. Verheyen (Eds.), *Corporate liability for transboundary environmental harm* (pp. 419–503). Springer.

Reynolds, J. (2018). International Law. In M. Gerrard & T. Hester (Eds.), *Climate engineering and the law: Regulation and liability for solar radiation management and carbon dioxide removal* (pp. 57–153). Cambridge University Press.

Tanzi, A. (2013). "Liability for lawful acts" Oxford: Max Planck Encyclopedias of International Law. Available at: https://opil.ouplaw.com/display/https://doi.org/10.1093/law:epil/9780199231690/law-9780199231690-e1065

UN. (1972). Declaration on the human environment, conference on the human environment, Stockolm, principle, p. 21.

UN. (1992). Rio declaration on environment and development. Report A/CONF.151/26 (vol. I), principle 2.

UN. (2001a). Draft articles on responsibility of states for internationally wrongful acts, U.N. Doc. A/56/10, ILC Yearbook, vol. 2, Part Two.

UN. (2001b). Draft articles on prevention of transboundary harm from hazardous activities, U.N. Doc. A/56/10, ILC Yearbook, vol. 2.

UNFCCC. (2022). Synthesis report UN doc FCCC/PA/CMA/2022/4.

Trail Smelter Case. (1941). *United States v. Canada* (Vol. 3). Award, R.I.A.A. 1905.

Vinuales, J. E. (2010). Legal techniques for dealing with scientific uncertainty in environmental law. *Vanderbilt Law Review, 43*, 437–503.

Webb, R., Silverman-Roati, K., & Gerrard, M. B. (2021). Removing carbon dioxide through ocean alkalinity enhancement: Legal challenges and opportunities. Sabin Center for Climate Change Law.

Wiener, J. B. (2018). Precautionary principle. Principles of Environmental LawIn L. Krämer & E. Orlando (Eds.), *Elgar encyclopedia of environmental law* (Vol. VI, pp. 174–185). Edward Elgar Publishing.

Wolfrum, R. (2011). Source of international law. Oxford: Max Planck Encyclopedias of International Law. Available at: https://opil.ouplaw.com/display/https://doi.org/10.1093/law:epil/9780199231690/law-9780199231690-e1471

Tony Cabus is a postdoctoral researcher in international law at the Walther Schücking Institute for International Law in Kiel, Germany.

Chapter 8
Smart Technologies and Approaches in Climate Change Mitigation and Adaptation: Implications for Policy-Making

Johannes Bhanye, Ruvimbo Hazel Shayamunda, and Abraham Matamanda

Abstract This chapter examines the implications of smart technologies and approaches for policy-making in climate change mitigation and adaptation. Innovative solutions are essential as the global community faces a pressing need to address climate change. Smart technologies and approaches offer promising opportunities to enhance efforts in mitigating and adapting to the impacts of climate change. These include renewable energy technologies, sustainable transportation technologies, climate-resilient agriculture and food production technologies, nature-based solutions, smart city technologies, climate modelling and monitoring and early warning systems, green finance and investment technologies, and circular economy and sustainable consumption technologies. Based on a document review-based approach, the chapter showcases the successful implementation of smart technologies and approaches in real-world contexts, both in mitigation and adaptation strategies. Using the analysis as its basis, the chapter offers valuable policy recommendations for integrating smart technologies into mitigation and adaptation strategies. It emphasised the importance of proactive policy adjustments to leverage the full potential of smart technologies and approaches, while addressing associated challenges and ethical considerations. The chapter calls for policymakers to embrace these smart technologies and approaches, and foster collaboration among scientists, technologists, and policymakers to drive innovation, accelerate progress, and effectively respond to the global climate crisis.

Keywords Smart technologies · Climate change · Mitigation · Adaptation · Policy-making · Innovation · Sustainability

J. Bhanye (✉)
African Climate and Development Initiative (ACDI), University of Cape Town, Cape Town, South Africa
e-mail: joebhanye@gmail.com

R. H. Shayamunda · A. Matamanda
Department of Urban and Regional Planning, University of the Free State, Bloemfontein, South Africa

W. Leal Filho et al. (eds.), *Climate Change Mitigation and Adaptation in Practice*, Climate Change Management, https://doi.org/10.1007/978-3-031-85217-6_8

Introduction

Climate change stands as one of the most pressing and complex challenges confronting humankind in the twenty-first century (Feulner, 2017). The undeniable scientific consensus reveals a planet in crisis, where rising global temperatures, melting polar ice caps, sea-level rise and extreme weather events are all irrefutable signs of an Earth undergoing rapid and unprecedented transformation (Paltsev, 2020; Siegert et al., 2020; Van Der Walt & Fitchett, 2022). The origins of this crisis can be traced back through centuries of industrialisation, fossil fuel consumption, deforestation, and unsustainable land-use practices (Burrell et al., 2020; Wadanambi et al., 2020). These human-induced activities have led to the accumulation of greenhouse gases in the atmosphere, resulting in the greenhouse effect and the disruption of the Earth's climate systems (Hugo & Bhanye, 2022).

It cannot be overstated that there is an urgency in addressing climate change. The consequences of inaction are far-reaching and devastating, with potential impacts on ecosystems, biodiversity, food security, human health, and global stability (Muluneh, 2021; Penuelas et al., 2020). Vulnerable communities worldwide are already experiencing the effects of climate change, which include droughts which are more frequent and severe, and more devastating floods and heatwaves (Atwoli et al., 2022). For example, in 2023, at least 300 deaths were documented while approximately 90,000 people were left homeless after their homes were swept away by flash floods induced by a cyclone in Malawi. In South Africa, the persistent heat waves highlight the severity of climate extremes in the country (Khan et al., 2023; Van Der Walt & Fitchett, 2022). The global community has recognised that the window of opportunity to mitigate these impacts and adapt to a changing climate is rapidly closing. Immediate, ambitious, and coordinated action is urgently needed if global warming is to be limited to below 2 °C above the pre-industrial levels, which was outlined in the Paris Agreement.

In the face of this daunting challenge, innovative solutions are essential (Matos et al., 2022). Scholars (Brears, 2020; Matos et al., 2022) have argued that smart technologies and approaches offer promising opportunities to revolutionise our approach to climate change mitigation and adaptation. These technologies encompass a wide range of fields and industries. Specifically, the technologies are employed in renewable energy sources and sustainable transportation, climate-resilient agriculture, nature-based solutions, smart city infrastructure, advanced climate modelling and monitoring systems, green finance mechanisms, and sustainable consumption practices (Brears, 2020; Loucks, 2023). However, the application of these technologies tends to be influenced by policies which regulate and guide what can or cannot be done in the respective industries. For example, the Sustainable Development Goals (SDG) #13 focus on climate action which provides targets for nations to pursue in the quest for climate change mitigation and adaptation. Thus, many governments have framed their climate change policies based on these targets.

Therefore, based on this background, this chapter examines the implications of smart technologies and approaches for policy-making in the context of climate change mitigation and adaptation. The chapter explores how these technologies can be harnessed to accelerate progress toward our climate goals and build resilience in the face of a changing climate. Through a comprehensive review of the literature, real-world case studies, and analysis, the chapter sheds light on how smart technologies and approaches can inform and shape climate policy. The chapter concludes by reiterating the critical role of smart technologies and approaches in climate change policy-making, and calls for collaborative efforts among scientists, technologists, and policymakers to drive innovation and effectively respond to the global climate crisis. Finally, the chapter offers policy recommendations for integrating smart technologies and approaches into climate policy, emphasising the importance of proactive policy adjustments while addressing associated challenges and ethical considerations.

Brief Review of Literature

The term 'Climate change' refers to long-term alterations in temperature, wind patterns, precipitation, and other atmospheric conditions on Earth (Dietz et al., 2020). While natural processes are the major drivers of climate change, human activities are also major players,, particularly the emission of greenhouse gases such as carbon dioxide (CO_2), methane (CH_4), and nitrous oxide (N_2O) (Caney, 2015). The increased concentration of these gases in the atmosphere causes the greenhouse effect, which traps heat and leads to global warming (Anderson et al., 2016). Climate change manifests in various ways, including rising global temperatures, changes in precipitation patterns, more frequent and intense extreme weather events (such as hurricanes, droughts, and heatwaves), and shifts in ecosystems and wildlife behaviour (Khan et al., 2023). The results of climate change have widespread implications for the environment, human societies, and economies, impacting agriculture, water resources, sea levels, biodiversity, and overall planetary health (Anderson et al., 2016; Caney, 2015).

Scientific consensus supports the understanding that climate change is a result of anthropogenic (human-induced) activities, particularly the burning of fossil fuels, deforestation, and industrial processes (Burrell et al., 2020; Penuelas et al., 2020). Mitigating and adapting to the impacts of climate change have become critical global challenges, requiring concerted efforts in policy-making, technological innovation, and international cooperation, to address the root causes and consequences of this complex phenomenon. Early efforts to address climate change primarily focused on raising awareness and understanding the causes and consequences of global warming. A significant milestone in international cooperation in the assessment of the science of climate change was marked by the establishment of the Intergovernmental Panel on Climate Change (IPCC) in 1988 (Seyboth, 2013).

As the understanding of climate change deepened, international and national policy frameworks were developed to address the issue. Key markers of success include the United Nations Framework Convention on Climate Change (UNFCCC) in 1992, which set the stage for international negotiations on climate action, and the

Kyoto Protocol in 1997, which introduced binding emissions reduction targets for developed nations (de Chazournes, 1998). These early policy efforts laid the foundation for subsequent agreements, such as the Paris Agreement 2015, which brought together nearly every nation to commit to limiting global warming to below 2 °C above pre-industrial levels (Dimitrov, 2016).

Despite these efforts, headway in reducing greenhouse gas emissions has been irregular, and many nations have struggled to meet their commitments (Levin & Rich, 2017). Scholars highlight the challenges of international cooperation, the complexities of setting equitable emission reduction targets, and the difficulties of balancing economic growth with environmental sustainability within existing policy frameworks (Roberts & Weikmans, 2017).

Smart technologies and approaches have introduced a transformative dimension to climate change mitigation and adaptation (Brears, 2020; Matos et al., 2022). Renewable energy technologies, such as solar and wind power, have experienced exponential growth, offering viable alternatives to fossil fuels (Suman, 2021). Advances in energy storage and grid management have enhanced the reliability and scalability of these technologies. Similarly, sustainable transportation technologies, including electric vehicles and public transit innovations, show possibilities to significantly reduce emissions from the transportation sector (Shah et al., 2021).

In agriculture, climate-resilient crop varieties and precision farming techniques are being developed to enhance food security in the face of changing climate conditions (Cvejić et al., 2022; Goswami et al., 2023). Nature-based solutions, such as reforestation and ecosystem restoration, are gaining recognition for sequestering carbon and protecting against natural disasters (Chausson et al., 2020). Smart city technologies are transforming urban planning and infrastructure, making cities more sustainable and resilient to climate impacts (García Fernández & Peek, 2020).

While smart technologies and approaches hold great promise, their integration into policy-making is challenging. Key issues include cost and investment, regulatory frameworks, infrastructure and grid integration, equity and access, and ethical considerations (Table 8.1).

Smart Technologies and Approaches for Climate Change Mitigation and Adaptation

This section presents a discussion on the smart technologies and approaches for climate change mitigation and adaptation, ranging from renewable energy technologies, sustainable transportation technologies, climate-resilient agriculture and food production technologies, nature-based solutions, smart city technologies, climate modelling and monitoring and early warning systems, to green finance and investment technologies, and circular economy and sustainable consumption technologies (Adedeji et al., 2020; Chausson et al., 2020; García Fernández & Peek, 2020;

Table 8.1 Key challenges in integrating smart technologies and approaches into policy-making

Key challenge	Explanation	Source
Cost and investment	Many smart technologies and approaches require significant upfront investments, which can be a barrier for governments and businesses. The literature explores the role of green finance mechanisms and investment incentives in facilitating technology adoption	Brears (2020), Castelnovo et al. (2016)
Regulatory frameworks	Policies and regulations often lag behind technological advancements, creating uncertainty for investors and innovators. The need for agile and adaptive regulatory frameworks is a recurring theme	Matos et al. (2022), Stratigea et al. (2015)
Infrastructure and grid integration	Scaling up renewable energy sources requires updates to infrastructure and grid systems to accommodate intermittent power generation. The literature discusses the technical challenges and opportunities associated with this transition	Stratigea et al. (2015), Visvizi et al. (2018)
Equity and access	Ensuring equitable access to smart technologies is critical to avoid exacerbating socio-economic disparities. Studies examine strategies for inclusive technology deployment	Matos et al. (2022), Stratigea et al. (2015)
Ethical considerations	As new technologies emerge, ethical dilemmas related to data privacy, environmental justice, and the unintended consequences of innovation arise. The literature addresses these ethical dimensions in climate policy	Visvizi et al. (2018)

Praveen & Sharma, 2019; Shah et al., 2021; Suman, 2021; Tong et al., 2022; see Fig. 8.1).

Renewable Energy Technologies

Renewable energy technologies represent a fundamental pillar of global efforts to combat climate change. These innovations leverage nature's abundant and clean energy resources, offering a sustainable alternative to traditional fossil fuel-based energy production (Suman, 2021). Solar photovoltaics (PV), wind turbines, hydropower, and geothermal systems are explored, their impact on climate change mitigation is explored, and notable case studies demonstrate their effectiveness.

Solar Photovoltaics (PV)

Solar photovoltaic (PV) technology is a standout performer in the renewable energy sector. It harnesses sunlight to generate electricity through the photovoltaic effect,

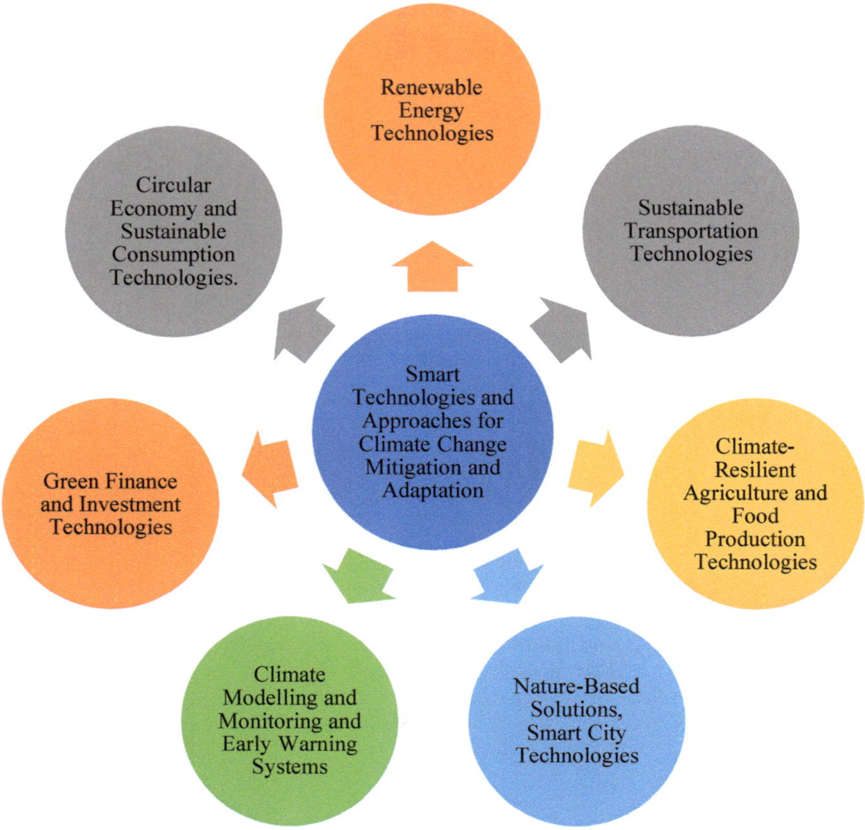

Fig. 8.1 Smart technologies and approaches for climate change mitigation and adaptation

wherein photons from the sun excite electrons in semiconductors, producing an electric current (Creutzig et al., 2017). Solar PV systems have the potential to be installed on rooftops, or larger systems can be set-up as solar farms, and even integrated into urban infrastructure. The proliferation of solar PV installations has been remarkable, with significant growth witnessed in countries like Germany, Japan, and China (Yu et al., 2016). Germany's Energiewende policy, for instance, has encouraged a vigorous uptake of solar PV systems, transforming the country into a global leader in solar energy capacity. Energiewende (energy transition) refers to Germany's policy of increasing the share of renewables and so enabling the phasing out of nuclear power. Despite the very high cost, which is acknowledged by the government, it represents a shared vision and enjoys public popularity (Von Hirschhausen, 2014). Similarly, China's ambitious solar energy targets and manufacturing capabilities have led to rapid adoption and cost reductions, making solar power more accessible and affordable.

Wind Turbines

Wind energy, harnessed through wind turbines, is another vital component of the renewable energy landscape (Kirch Kirkegaard et al., 2021). Wind turbines work by capturing kinetic energy from the wind and converting it into electricity. These structures come in various sizes, from small-scale turbines suitable for residential use to massive offshore installations. Denmark, a pioneer in wind energy, has made substantial investments in offshore wind farms. The Horns Rev and Horns Rev 2 wind farms in the North Sea are prime examples of successful large-scale offshore wind projects (Wu & Porté-Agel, 2015). Horns Rev Offshore Wind Farm was one of the first offshore wind farms in the world and is positioned in the North Sea. It consists of 80 turbines and has been expanded over the years (Wu & Porté-Agel, 2015). Horns Rev 2 is an extension of the original Horns Rev Wind Farm. Horns Rev 2 is one of the largest offshore wind farms globally and features 91 Siemens Gamesa turbines, with production exceeding 10 billion kWh at a certain point in time. These technologies have reduced carbon emissions, created jobs, and stimulated economic growth.

Hydropower

Hydropower, generated by harnessing the energy of flowing water, is a well-established renewable energy source (Saha & Idsø, 2016). Hydropower can be used to generate electricity from large dams, run-of-river systems, and even small-scale installations. Hydropower is crucial in providing consistent and reliable electricity, making it an essential part of a diversified renewable energy portfolio. With its extensive hydropower capacity, Norway is a notable example of successful utilization (Saha & Idsø, 2016). The country relies on hydropower for over 95% of its electricity needs, which has allowed it to significantly reduce greenhouse gas emissions from the energy sector (Graabak et al., 2017).

Geothermal Systems

Geothermal energy exploits the heat stored beneath the Earth's surface to generate electricity and provide heating and cooling for various applications (Ómarsdóttir, 2016). This technology offers a consistent and low-carbon energy source, making it a valuable asset in the fight against climate change (Soltani et al., 2021). Iceland's geothermal energy journey stands as a remarkable example. The country derives nearly 90% of its primary energy from geothermal and hydroelectric sources, significantly reducing its dependence on fossil fuels. Over 85% of all houses in Iceland are heated with renewable energy sources, and 66% use geothermal power. There are five major geothermal power plants active in Iceland. These five plants also produce a little over 26% of the electricity used in Iceland (Arctic Adventures, 2023). Iceland's geothermal expertise has also been shared with other nations, highlighting the potential for international collaboration in adopting this technology.

Sustainable Transportation Technologies

The transportation sector is a significant contributor to global greenhouse gas emissions, making developing and adopting sustainable transportation technologies a critical aspect of climate change mitigation and adaptation (Shah et al., 2021). This section explores various sustainable transportation innovations, and provides case studies that demonstrate their potential in reducing emissions and improving urban mobility. These innovations (electric vehicles, efficient public transit systems, and alternative fuels such as hydrogen) not only reduce the environmental impact of transportation but also enhance the quality of life in urban areas, making them critical tools for creating a sustainable and climate-resilient future.

Electric Vehicles (EVs)

Another promising solution for reducing emissions is the use of electric vehicles (EVs). These vehicles use electricity stored in batteries to power electric motors, eliminating tailpipe emissions from conventional internal combustion engines (Sanguesa et al., 2021). EVs come in various forms, from passenger cars to buses and trucks, making them versatile for urban and long-distance travel. Notable case studies highlight the success of EV integration. Oslo, the capital of Norway, stands out as a city that has made significant strides in promoting EV adoption (Schulz & Rode, 2022). A combination of incentives such as reduced tolls, free charging, and dedicated bus lanes for electric vehicles, has led to a surge in EV ownership (Schulz & Rode, 2022). This approach has reduced air pollution and positioned Oslo as a global leader in sustainable transportation.

Efficient Public Transit Systems

Efficient and accessible public transit systems are integral to sustainable urban mobility (Golbabaei et al., 2021). Well-designed public transportation reduces the number of private vehicles on the road, lower emissions and lessens traffic congestion (Shah et al., 2021). Investments in modern public transit, including buses, trams, subways, and light rail systems, are essential for sustainable urban development. Germany's public transit system (Fig. 8.2) serves as a prime example of effective urban mobility planning (Punzo et al., 2022). In Amsterdam, the city's comprehensive network of trams and buses and bike-friendly infrastructure have contributed to a modal shift away from cars. This approach has reduced carbon emissions and improved the overall quality of life for its residents.

Alternative Fuels (E.G. Hydrogen)

Alternative fuels, such as hydrogen, offer a pathway to reduce emissions from various modes of transportation, including heavy-duty trucks and buses. Hydrogen fuel cells, for instance, produce electricity by combining hydrogen and oxygen, emitting only water vapour as a byproduct (Ahmed et al., 2016). This technology holds great promise for long-haul transportation and could significantly reduce emissions from freight logistics (Xing et al., 2021). Hydrogen-powered buses in cities like London

Fig. 8.2 Tram in Germany. *Source* Authors (2022)

and Tokyo have demonstrated the viability of this technology. These buses offer zero-emission transportation, contributing to cleaner air and reduced noise pollution in urban environments.

Climate-Resilient Agriculture and Food Production Technologies

Agriculture is not only susceptible to the impacts of climate change but also a substantial contributor to the emissions of greenhouse gases (Praveen & Sharma, 2019). Climate-resilient agriculture and innovative food production technologies represent a crucial frontier in addressing these dual challenges. This section explores various technologies and practices to reduce emissions, enhance crop yields, and adapt agriculture to changing climate conditions.

Climate-Resilient Crop Varieties

Developing climate-resilient crop varieties is a cornerstone of sustainable agriculture. These varieties are bred to thrive in changing climate conditions, including increased temperatures, altered patterns of precipitation, and emerging disease and pests threats (Banga & Kang, 2014). Climate-resilient crops help ensure food security, and reduce the need for water, pesticides, and fertilisers. One remarkable example is the adoption or implementation of drought-resistant crop varieties in sub-Saharan

African countries like Zimbabwe, Zambia and South Africa (Cacho et al., 2020). In regions prone to water scarcity, crops such as drought-tolerant maize and millet have been developed and distributed to farmers. These varieties have shown significant yield improvements, even during prolonged droughts, thereby safeguarding food production for vulnerable communities.

Precision Agriculture Techniques

Precision agriculture leverages innovative technologies such as sensors, data analytics and Global Positioning System (GPS) to optimise farming practices (Shafi et al., 2019). Through precisely tailoring inputs such as water, fertilisers, and pesticides to specific areas of a field, farmers can increase crop yields while reducing resource use and environmental impact (Shafi et al., 2019). The adoption of precision agriculture in the United States' Midwest, known as the "Corn Belt," demonstrates the potential of this cutting-edge technology (Jin et al., 2019). Farmers in this region have used GPS-guided tractors and drones to monitor crop health and apply inputs precisely. This technology resulted in more efficient resource use and reduction in green house gas emissions, as fewer resources are wasted.

Sustainable Land Management Practices

Sustainable land management practices focus on soil health and ecosystem preservation. Techniques such as cover cropping, no-till farming and agroforestry, improve soil fertility, reduce erosion, and sequester carbon (Adimassu et al., 2016). These practices enhance agricultural resilience and contribute to carbon sequestration, mitigating climate change. The "Grain for Green" programme in China is an exemplary case of sustainable land management (Xian et al., 2020). This initiative aimed to combat soil erosion by converting marginal cropland into forests and grasslands. Doing so has restored ecosystems and reduced greenhouse gas emissions from soil degradation.

Nature-Based Solutions

Nature-based solutions also provide a holistic and sustainable approach to climate change mitigation and adaptation (Brears, 2020; Chausson et al., 2020). These strategies harness the power of ecosystems to combat climate change, protect against natural disasters, and promote biodiversity. Nature-based solutions include reforestation, wetland restoration, sustainable land use practices, and urban green infrastructure.

Reforestation

Reforestation involves deliberately planting trees where forests have been depleted or degraded. Forests act as carbon sinks, absorbing and storing significant amounts of carbon dioxide from the atmosphere (Moloise et al., 2023). Reforestation can help mitigate climate change by sequestering carbon, also provides habitat for wildlife and protects against soil erosion. The Great Green Wall project in Africa is an inspiring example of large-scale reforestation. Spanning 20 countries across the Sahel region, this initiative aims to combat desertification and land degradation through planting a green and lush belt of trees (Cropper, 2023). By restoring ecosystems and improving local livelihoods, the Great Green Wall showcases the potential of reforestation as a nature-based solution.

Wetland Restoration

Wetlands, encompassing marshes, swamps, and mangroves, serve as vital components in adapting to climate change. They are natural buffers against flooding, storm surges, and coastal erosion. Furthermore, wetlands are rich in biodiversity and store substantial amounts of carbon (Marambanyika & Musasa, 2023; Mack et al., 2023). The restoration of the Florida Everglades in the United States is a remarkable case study in wetland restoration (Naja et al., 2017). Efforts to reverse decades of degradation and urbanisation have revitalised this unique ecosystem and enhanced its resilience to hurricanes and rising sea levels. Communities can better adapt to climate change by conserving and restoring wetlands.

Sustainable Land Use Practices

Sustainable land use practices involve managing land for both ecological and human benefits. These practices, like regenerative agriculture and agroforestry enhance carbon sequestration promote soil health, and support biodiversity. Costa Rica's Payment for Ecosystem Services (PES) programme is a pioneering example of sustainable land use (Brownson et al., 2020). Farmers and landowners are incentivised to conserve forests and adopt sustainable practices through financial rewards. As a consequence, Costa Rica has witnessed a notable surge in forest coverage and biodiversity, all while actively contributing to mitigating climate change.

Urban Green Infrastructure

Urban green infrastructure integrates nature into urban planning and design (Van Oijstaeijen et al., 2020). It includes features like green roofs, parks, and permeable surfaces. These elements improve air quality, regulate temperatures, and enhance urban resilience to extreme weather events. Barcelona's Superblocks initiative is a notable urban green infrastructure project (Mueller et al., 2020). It transforms city blocks into pedestrian-friendly spaces with green areas, reducing traffic and pollution. This strategy promotes a more robust urban environment and helps alleviate the urban heat island effect, which is worsened by climate change.

Smart City Technologies

Smart city technologies represent a transformative approach to urban development that leverages data and connectivity to optimise urban systems, reduce resource consumption, and enhance resilience to climate impacts (García Fernández & Peek,

2020). This section explores the various facets of smart city technologies, their applications, and real-world examples that highlight their potential in addressing climate change challenges.

Energy-Efficient Buildings

Energy-efficient buildings are a fundamental component of smart city initiatives (Moazami et al., 2019). These structures are designed to minimise energy consumption through advanced insulation, efficient heating and cooling systems, and smart lighting and appliances (Hirst, 2013). They are pivotal in reducing urban carbon footprints and increasing energy resilience. Singapore's efforts in developing energy-efficient buildings stand as an exemplary case (Moazami et al., 2019). The city-state implemented strict building codes and standards requiring green building technologies. Iconic projects such as Marina Bay Sands and the Punggol Digital District showcase energy-efficient designs and sustainable building practices, contributing to Singapore's sustainability goals (Rinaldi & Tan, 2019). The Eastgate Centre in Harare, Zimbabwe, typifies the best of green architecture and ecologically-sensitive adaptation through biomimicry. For its size, through biomimicry, Eastgate Centre uses less than 10% of the energy of a conventional building (Chigwenya & Zhakata, 2020).

Smart Grids

Smart grids represent an intelligent and adaptive approach to managing electricity distribution. These grids use advanced communication technologies and sensors to optimise energy generation, distribution, and consumption (Kakran & Chanana, 2018). They enable better integration of renewable energy sources, reduce energy losses, and enhance grid resilience (Kakran & Chanana, 2018; Noussan, 2018). Chattanooga, Tennessee, a city in the United States, is a noteworthy example. Chattanooga's smart grid system, known as the "smartest city in America," has not only improved energy efficiency but also minimised power outages and increased the city's ability to recover from natural disasters (Starke et al., 2017).

Data-Driven Urban Planning

Urban planning that is data driven relies on the collection of real-time data and analysis to inform infrastructure, transportation, and land-use decisions (Engin et al., 2020). This approach helps optimise resource allocation, reduce traffic congestion, and enhance urban livability (Wu et al., 2022). The city of Barcelona's use of data-driven urban planning is a striking illustration (Bibri & Krogstie, 2020). The city employs sensors and data analytics to monitor traffic, energy consumption and air quality. This information helps to adjust traffic flow, improve public transportation, and reduce energy consumption in buildings, contributing to Barcelona's reputation as a smart and sustainable city (Bibri & Krogstie, 2020).

Intelligent Transportation Systems (ITS)

Intelligent Transportation Systems (ITS) involves the use of technology to improve transportation efficiency and reduce environmental impacts and involves systems such as real-time traffic management, connected vehicle infrastructure, and smart parking solutions (Perallos et al., 2015; Visan et al., 2022). Copenhagen, Denmark, is a leading city in the implementation of ITS. The city's intelligent traffic management system prioritises cycling and pedestrian traffic while optimising vehicle flow (Makarova et al., 2017). This approach reduces traffic congestion and promotes sustainable modes of transportation, such as cycling and walking, thereby reducing emissions and enhancing the urban quality of life (Makarova et al., 2017).

Climate Modelling and Monitoring and Early Warning Systems

Advanced climate modelling and monitoring systems and early warning systems provide essential tools for understanding climate dynamics, predicting extreme events, and supporting informed decision-making for climate adaptation and mitigation (Adedeji et al., 2020). These technologies also include disaster preparedness and resource allocation (Bhanye & Maisiri, 2023; Hoshiba & Ozaki, 2014; Kelman & Glantz, 2014; Šećerov et al., 2015; Stocker, 2011).

Climate Modelling

Climate modeling entails the utilization of mathematical and computational frameworks to simulate and forecast forthcoming climate trends (Stocker, 2011). These models consider diverse factors like atmospheric dynamics, oceanic circulation, greenhouse gas emissions, and alterations in land use. They empower scientists and policymakers to evaluate the probable ramifications of climate change and to devise strategies for both mitigation and adaptation (Flato et al., 2014). An exemplar of such models is the Community Earth System Model (CESM), crafted by the National Center for Atmospheric Research (NCAR) in the United States (Kay et al., 2015). CESM has significantly enhanced our comprehension of climate mechanisms and has been pivotal in prognosticating potential outcomes of climate change, encompassing phenomena like sea-level elevation and alterations in precipitation patterns (Kay et al., 2015).

Climate Monitoring Systems

Climate monitoring systems consist of networks of instruments and sensors that collect data on various parameters of climate, including temperature, humidity, sea level, and concentrations of greenhouse gases (Šećerov et al., 2015). These systems provide real-time information on the state of the climate, helping scientists track changes and assess the impact of human activities. The Global Climate Observing System (GCOS) is a comprehensive network that plays a vital role in climate monitoring (Zemp et al., 2021). GCOS collects data from thousands of observing stations worldwide and serves as a critical source of information for climate research, enabling scientists to identify trends and anomalies in the climate system (Zemp et al., 2021).

Early Warning Systems

Early warning systems are engineered to identify and forecast severe weather occurrences, natural calamities, and other hazards linked to climate change (Kelman & Glantz, 2014). These systems provide crucial information to communities, governments, and emergency responders, allowing them to mitigate risks. and protect lives and property proactively. The Japan Meteorological Agency (JMA) is renowned for its advanced early warning systems. JMA utilises sophisticated technology, including seismic sensors and weather radars, to detect earthquakes, tsunamis, typhoons, and other hazards (Hatsuzuka et al., 2022). Timely alerts and evacuation orders from JMA have saved countless lives in Japan and serve as a model for other regions prone to natural disasters (Hoshiba & Ozaki, 2014).

Disaster Preparedness and Resource Allocation

Climate modeling, monitoring, and the implementation of early warning systems are critical for enhancing disaster preparedness and efficiently allocating resources in regions susceptible to climate-related risks (Hoshiba & Ozaki, 2014; Šećerov et al., 2015; Stocker, 2011). By predicting extreme events and assessing their potential impact, these technologies enable governments and organisations to allocate resources effectively, develop evacuation plans, and improve disaster response efforts (Bhanye & Maisiri, 2023; Kelman & Glantz, 2014). Hurricane forecasting in the United States exemplifies this application (Guikema et al., 2014). Advanced hurricane tracking models and real-time monitoring provide accurate predictions of hurricane paths and intensities. These forecasts guide evacuation decisions, allowing coastal communities to prepare for and respond to hurricanes effectively (Guikema et al., 2014).

Green Finance and Investment Approaches

Green finance and investment approaches are pivotal in mobilising funds for climate action and fostering sustainable economic development (Tong et al., 2022). Approaches include green bonds, impact investing, carbon markets, and green banks and financial institutions.

Green Bonds

Green bonds serve as a pivotal financial tool for directing funds toward eco-conscious endeavors. These bonds are issued to support projects with evident environmental advantages, such as renewable energy initiatives, eco-friendly construction, and sustainable transport infrastructure (Fatica & Panzica, 2021). Investors in green bonds gain assurance that their investments contribute to environmentally responsible causes. The European Investment Bank (EIB) stands out as a prominent issuer of green bonds. The proceeds from these bonds have fueled diverse climate-focused projects across Europe, encompassing the installation of renewable energy sources, construction of energy-efficient public structures, and development of sustainable transportation networks (Kavvadia, 2021).

Impact Investing

Impact investing focuses on generating positive social and environmental outcomes alongside financial returns (De Angelis et al., 2023). It involves allocating capital to businesses and projects addressing climate change and other sustainability challenges (De Angelis et al., 2023). Impact investors actively seek opportunities to support climate-friendly initiatives while achieving their financial objectives. The Global Innovation Lab for Climate Finance (the Lab) is a notable initiative that promotes impact investing (Buchner, 2021). The Lab collaborates with investors, governments, and other stakeholders to develop and launch innovative financial instruments that drive investment in climate mitigation and adaptation projects (Buchner, 2021). These instruments include green credit guarantees and climate insurance products.

Carbon Markets

Carbon markets, also referred to as emissions trading systems, offer a market-driven mechanism enabling companies to trade carbon credits (Fatica & Panzica, 2021). Businesses that lower their emissions below a set threshold can sell surplus credits to those surpassing their emission allowances (Fatica & Panzica, 2021). These markets establish economic incentives for emission reduction efforts and facilitate countries in achieving their climate objectives. Among the largest and most established carbon markets globally is the European Union Emissions Trading System (EU ETS) (Flachsland et al., 2020). Encompassing various sectors such as energy, manufacturing, and aviation, the EU ETS incentivizes companies to adopt cleaner technologies and diminish their carbon footprint by assigning a value to carbon emissions (Flachsland et al., 2020).

Green Banks and Financial Institutions

Green banks and financial institutions are specialised entities that exclusively fund renewable energy projects, energy efficiency upgrades, and other climate-related initiatives (Park & Kim, 2020). They leverage public and private funds to speed up the transition to a low-carbon economy. The Green Investment Group (GIG) in the United Kingdom exemplifies the role of green banks (Stewart, 2023). GIG, originally owned by the UK government, invests in renewable energy projects worldwide. It has been pivotal in financing offshore wind farms, solar installations, and energy-efficient infrastructure, contributing significantly to the UK's renewable energy capacity (Stewart, 2023).

The Circular Economy and Sustainable Consumption Technologies

The transition to a circular economy, characterised by reduced waste and resource reuse, is integral to the mitigation of climate change and to sustainable development (Matooane et al., 2023). This section explores the circular economy concept, sustainable consumption technologies, and real-world examples that illustrate their potential to reduce emissions and promote environmental stewardship.

Circular Economy Principles

The circular economy is a holistic economic model that aims to minimise waste and maximise the use of resources throughout their lifecycle. Key principles of the circular economy include (Geissdoerfer, 2017):

- **Prioritize durability and recyclability in design**: Products are engineered for extended lifespans and straightforward disassembly for recycling.
- **Adopt a philosophy of reduce, reuse, recycle**: Focus on minimizing consumption, reusing products and materials, and maximizing recycling efforts to extract value from waste.
- **Sharing and collaborative consumption**: Sharing resources such as cars, tools, and living spaces reduces the need for individual ownership and lowers resource demand.
- **Sustainable materials**: Preference is given to materials with a lower environmental impact and which can be reused or recycled effectively.

Recycling and Upcycling Technologies

Recycling and upcycling technologies are pivotal in driving the circular economy, transforming waste materials into fresh products or resources (Matooane et al., 2023). Innovative recycling methods, including chemical and closed-loop processes, facilitate the retrieval of valuable resources while mitigating the environmental impact of waste disposal (Hugo & Bhanye, 2022). A prime illustration is the emergence of upcycled fashion. Industry leaders like Patagonia lead the charge by utilizing recycled materials to craft new clothing lines. They collect discarded garments, transform them into high-quality fibers, and utilize these fibers to create new apparel, thus reducing the need for virgin resources and curbing textile waste. Other corporations such as Dell, Adidas, Nike, Garnier, and WeWood are also embracing recycled materials in their manufacturing processes to cater to sustainability-conscious consumers.

Product Lifecycle Management (PLM) Systems

Product Lifecycle Management (PLM) systems represent digital solutions empowering enterprises to oversee and refine every stage of a product's lifecycle, spanning design, manufacturing, utilization, and end-of-life disposal (Tai, 2017). These systems aid companies in monitoring and mitigating environmental impacts, optimizing resource utilization, and crafting products conducive to circularity (Singh et al., 2020). Leading corporations such as Apple have embraced PLM systems to bolster product sustainability. For instance, Apple's "Daisy" disassembly robot efficiently disassembles and recycles iPhones, reclaiming valuable materials like aluminum and cobalt (Laser & Stowell, 2020). Such an approach resonates with the principles of the circular economy, diminishing the environmental footprint associated with electronic devices.

Sustainable Consumption Apps and Platforms

Apps and platforms for sustainable consumption empower individuals to make environmentally responsible decisions in their day-to-day routines. These digital tools offer insights into product sustainability, carbon emissions, and ethical sourcing, enabling consumers to make informed choices about their purchases and resource usage (Oliveira et al., 2020). For instance, apps such as Good On You offer information about the sustainability and ethical practices of clothing brands, assisting consumers in making more conscientious fashion selections. Similarly, platforms like Too Good To Go link consumers with surplus food from eateries and supermarkets, curbing food waste and fostering sustainable consumption habits.

Policy Recommendations

This section outlines recommendations aimed at policymakers to foster an environment conducive to innovation, advancing sustainable development, and bolstering resilience for the future. Key aspects include proactive policy adaptation, integrating technology strategies, ethical considerations, fostering international collaboration, and tackling implementation hurdles. Adopting a comprehensive approach encompassing these elements is essential for leveraging smart technologies effectively in climate change mitigation and adaptation efforts.

The Need for Proactive Policy Adjustments

Addressing climate change effectively necessitates proactive local, national, and international policy adjustments. Policymakers must acknowledge the rapidly evolving landscape of smart technologies and adapt regulatory frameworks, incentives, and strategies to harness their full potential. The role of government should be to proactively establish and update regulatory frameworks in order to promote the adoption of smart technologies in climate mitigation and adaptation. These frameworks should create incentives for innovation, provide clear guidelines for implementation, and set ambitious targets for emissions reduction and climate resilience. Policymakers must also engage in long-term climate planning, considering the dynamic nature of smart technologies. Continuous assessment and adjustment of policies are essential to ensure they remain aligned with the latest advancements and scientific findings.

Strategies for Integrating Smart Technologies into Climate Policies

To fully harness the transformative potential of smart technologies, policymakers need to craft comprehensive strategies for their seamless integration into climate policies. These strategies should entail clear targets for technology adoption, backed by financial incentives, robust support for research and development, and mechanisms facilitating public–private collaboration. Governments ought to offer financial inducements such as tax credits, grants, and subsidies to incentivize businesses and individuals to invest in smart technologies for climate initiatives. These incentives serve to offset the initial costs of technology adoption and highlight the long-term benefits. Additionally, prioritizing capacity building is imperative, necessitating investments in workforce development and training programs to ensure the availability of requisite skills for the effective deployment and maintenance of smart technologies. This encompasses training initiatives tailored for technicians, engineers, and policymakers alike. Furthermore, fostering collaboration between the public and private sectors, through Public–Private Partnerships (PPPs), is vital. Such partnerships can leverage diverse expertise, resources, and innovative approaches to accelerate the development and deployment of smart technologies, thus effectively addressing climate challenges.

Ethical Considerations and Guidelines

Adopting smart technologies brings ethical considerations to the forefront of climate policy-making. Policymakers must navigate issues related to data privacy, social equity, environmental justice, and the unintended consequences of innovation. It is also important that transparent reporting mechanisms and accountability measures are established for developing and deploying smart technologies, and furthermore ensuring that decision-making processes are inclusive so that the public can access information about technology risks and benefits, which is also very important.

International Cooperation and Collaboration

Climate change is a global challenge that demands international cooperation and collaboration. Policymakers must foster partnerships between nations, research institutions, and industry stakeholders to facilitate the exchange of knowledge, technology transfer, and joint initiatives. It is also important to support technology transfer programmes that enable developing countries to access and implement smart technologies for climate action. This can be achieved through international agreements and financial assistance.

Addressing Challenges and Barriers to Implementation

Implementing smart technologies in climate policies is not without challenges. Challenges and barriers to implementation can be addressed through research and development funding, regulatory streamlining, public awareness and education (Table 8.2).

Table 8.2 Strategies for addressing challenges and barriers to implementation

Strategy	Explanation	Sources
Research and development funding	Allocate funding for research and development to address technological gaps and overcome barriers in smart technologies. Encourage public and private sectors to invest in innovative solutions	Brears (2020), Castelnovo et al. (2016)
Regulatory streamlining	Simplify regulatory processes to reduce bureaucratic barriers to technology adoption. This includes expediting permits and approvals for clean energy projects, and providing clear guidelines for nature-based solutions	Matos et al. (2022), Stratigea et al. (2015), Visvizi et al. (2018)
Education and public awareness	Initiate public awareness campaigns aimed at educating citizens about the significance and advantages of integrating smart technologies for both mitigating and adapting to climate change. Encourage community engagement and participation in technology deployment	Matos et al. (2022), Stratigea et al. (2015), Visvizi et al. (2018)

Conclusions

In the face of the formidable challenge posed by climate change, this chapter highlighted the pivotal role of smart technologies and approaches as a ray of hope. Smart technologies and approaches offer promising opportunities to revolutionise climate change mitigation and adaptation approach. Innovation is at the forefront across various sectors, from renewable energy sources and sustainable transportation to climate-resilient agriculture, nature-based solutions, smart city infrastructure, advanced climate modelling, green finance mechanisms, and sustainable consumption practices. These technologies present the means to reduce greenhouse gas emissions and enhance our capacity to adapt to the inevitable consequences of climate change. They empower societies to reimagine their relationship with the environment, promote sustainable resource use, and create a more resilient and climate-resilient future.

In conclusion, the future of climate change policy-making is at a crossroads. The world stands on the precipice of unparalleled challenges but also unparalleled opportunities. The implications of smart technologies and approaches for policy-making are profound. Governments must incentivise innovation by providing financial support, research grants, and supportive regulatory frameworks for the development and adoption of smart technologies and approaches. Additionally, fostering cross-sectoral collaboration among technology developers, scientists, and policymakers is essential for the holistic implementation of these solutions. Capacity building and educational initiatives are crucial to equip stakeholders with the knowledge and skills necessary for the effective deployment of smart technologies and approaches. Policymakers should invest in training programmes for engineers, policymakers, and local communities to ensure these technologies are maximized to their full potential. Furthermore, achieving global impact requires policy harmonization and international cooperation. Policymakers must work collaboratively to create cohesive policies that encourage the widespread adoption of smart technologies and approaches, fostering a united front against the complex and interconnected challenges posed by climate change.

Looking ahead, the integration of smart technologies and approaches into climate change mitigation and adaptation strategies opens up numerous avenues for future research. One critical area of exploration involves advancing the capabilities of climate modelling through the incorporation of machine learning algorithms and artificial intelligence. Improved modelling techniques can enhance the understanding of complex climate systems, providing more accurate predictions and insights into the potential impacts of various mitigation and adaptation measures. Furthermore, understanding the socio-economic implications of smart technology adoption is crucial for effective policy-making. Research should delve into the distributional impacts of these technologies to ensure that benefits are equitably shared, and that vulnerable communities are not left behind. Nature-based solutions, like sustainable land management and reforestation, represent another promising area for future research. Assessing the efficacy of these methods in carbon sequestration, biodiversity enhancement, and fortifying ecosystem resilience offers valuable insights into their contribution to mitigating and adapting to climate change.

References

Adedeji, O., Olusola, A., James, G., Shaba, H. A., Orimoloye, I. R., Singh, S. K., & Adelabu, S. (2020). Early warning systems development for agricultural drought assessment in Nigeria. *Environmental Monitoring and Assessment, 192*, 1–21.

Adimassu, Z., Langan, S., & Johnston, R. (2016). Understanding determinants of farmers' investments in sustainable land management practices in Ethiopia: Review and synthesis. *Environment, Development and Sustainability., 18*(4), 1005–1023.

Ahmed. A., Al-Amin, A. Q., Ambrose, A. F., Saidur, R. (2016). Hydrogen fuel and transport system: A sustainable and environmental future. *International Journal of Hydrogen Energy, 41*(3), 1369–80.

Anderson, T. R., Hawkins, E., & Jones, P. D. (2016). CO_2, the greenhouse effect and global warming: From the pioneering work of Arrhenius and Callendar to today's earth system models. *Endeavour, 40*(3), 178–187.

Arctic Adventures. (2023). Geothermal energy in Iceland. Natural Energy Sources and Hot Springs in Iceland. Available at: https://adventures.is/blog/geothermal-energy-iceland/#:~:text=WHAT%20PERCENT%20OF%20GEOTHERMAL%20ENERGY,the%20electricity%20used%20in%20Iceland. Accessed September 6, 2023.

Atwoli, L., Erhabor, G. E., Gbakima, A. A., Haileamlak, A., Ntumba, J. M., Kigera, J., Laybourn-Langton, L., Mash, R., Muhia, J., Mulaudzi, F. M., & Ofori-Adjei, D. (2022). COP27 climate change conference: Urgent action needed for Africa and the world. *The Lancet Oncology., 23*(12), 1486–1488.

Banga, S. S., & Kang, M. S. (2014). Developing climate-resilient crops. *Journal of Crop Improvement, 28*(1), 57–87.

Bhanye, J., & Maisiri, M. (2023). Environmental communication (EC) and the new media: An African context perspective. *The Palgrave handbook of global social change* (pp. 1–23). Springer International Publishing.
Bibri, S. E., & Krogstie, J. (2020). The emerging data-driven Smart City and its innovative applied solutions for sustainability: The cases of London and Barcelona. *Energy Informatics, 3*, 1–42.
Brears, R. C. (2020). *Nature-based solutions to 21st century challenges.* Routledge.
Brownson, K., Anderson, E. P., Ferreira, S., Wenger, S., Fowler, L., & German, L. (2020). Governance of payments for ecosystem services influences social and environmental outcomes in Costa Rica. *Ecological Economics, 174*, 106659.
Buchner, B. (2021). Scaling finance through the global innovation lab for climate finance. In *World Scientific Encyclopedia of Climate Change: Case Studies of Climate Risk, Action, and Opportunity,* vol. 1, pp. 45–50.
Burrell, A. L., Evans, J. P., & De Kauwe, M. G. (2020). Anthropogenic climate change has driven over 5 million km^2 of drylands towards desertification. *Nature Communications., 11*(1), 3853.
Cacho, O. J., Moss, J., Thornton, P. K., Herrero, M., Henderson, B., Bodirsky, B. L., Humpenöder, F., Popp, A., & Lipper, L. (2020). The value of climate-resilient seeds for smallholder adaptation in sub-Saharan Africa. *Climatic Change., 162*, 1213–1229.
Caney, S. (2015). Climate change. In *The Routledge handbook of global ethics* (pp. 384–398). Routledge.
Castelnovo, W., Misuraca, G., & Savoldelli, A. (2016). Smart cities governance: The need for a holistic approach to assessing urban participatory policy making. *Social Science Computer Review., 34*(6), 724–739.
Chausson, A., Turner, B., Seddon, D., Chabaneix, N., Girardin, C. A., Kapos, V., Key, I., Roe, D., Smith, A., Woroniecki, S., & Seddon, N. (2020). Mapping the effectiveness of nature-based solutions for climate change adaptation. *Global Change Biology., 26*(11), 6134–6155.
Chigwenya, A., & Zhakata, T. I. (2020). Adopting green building technologies for sustainable development: Insights from Harare, Zimbabwe. *International Journal of Real Estate Studies., 14*(2), 8–17.
Creutzig, F., Agoston, P., Goldschmidt, J. C., Luderer, G., Nemet, G., & Pietzcker, R. C. (2017). The underestimated potential of solar energy to mitigate climate change. *Nature Energy., 2*(9), 1–9.
Cropper, J. (2023). 'Growing a World Wonder': The great green wall and the history of environmental decline in the Sahel, pp. 1450–2022. *Environment and History.*
Cvejić, S., Jocić, S., Mitrović, B., Bekavac, G., Mirosavljević, M., Jeromela, A. M., Zorić, M., Radanović, A., Kondić-Špika, A., & Miladinović, D. (2022). Innovative approaches in the breeding of climate-resilient crops. *Climate Change and Agriculture: Perspectives, Sustainability and Resilience., 18*, 111–156.
De Angelis, T., Tankov, P., & Zerbib, O. D. (2023). Climate impact investing. *Management Science., 69*(12), 7669–7692.
de Chazournes, L. B. (1998). Kyoto protocol to the United Nations framework convention on climate change. UN's Audiovisual Library of International Law (http://untreaty.un.org/cod/avl/ha/kpccc/kpccc.html).
Dietz, T., Shwom, R. L., & Whitley, C. T. (2020). Climate change and society. *Annual Review of Sociology., 46*, 135–158.
Dimitrov, R. S. (2016). The Paris agreement on climate change: Behind closed doors. *Global Environmental Politics., 16*(3), 1–1.
Engin, Z., van Dijk, J., Lan, T., Longley, P. A., Treleaven, P., Batty, M., & Penn, A. (2020). Data-driven urban management: Mapping the landscape. *Journal of Urban Management., 9*(2), 140–150.
Fatica, S., & Panzica, R. (2021). Green bonds as a tool against climate change? *Business Strategy and the Environment., 30*(5), 2688–2701.
Feulner, G. (2017). Global challenges: Climate change. *Global Challenges, 1*(1), 5.

Flachsland, C., Pahle, M., Burtraw, D., Edenhofer, O., Elkerbout, M., Fischer, C., Tietjen, O., & Zetterberg, L. (2020). How to avoid history repeating itself: The case for an EU emissions trading system (EU ETS) price floor revisited. *Climate Policy., 20*(1), 133–142.

Flato, G., Marotzke, J., Abiodun, B., Braconnot, P., Chou, S. C., Collins, W., Cox, P., Driouech, F., Emori, S., Eyring, V., & Forest, C. (2014). Evaluation of climate models. In *Climate change 2013: the physical science basis. Contribution of Working Group I to the Fifth Assessment Report of the Intergovernmental Panel on Climate Change* (pp. 741–866). Cambridge University Press.

García Fernández, C., & Peek, D. (2020). Smart and sustainable? Positioning adaptation to climate change in the European smart city. *Smart Cities., 3*(2), 511–526.

Geissdoerfer, M., Savaget, P., Bocken, N. M., & Hultink, E. J. (2017). The circular economy—A new sustainability paradigm? *Journal of Cleaner Production., 143*, 757–768.

Golbabaei, F., Yigitcanlar, T., & Bunker, J. (2021). The role of shared autonomous vehicle systems in delivering smart urban mobility: A systematic review of the literature. *International Journal of Sustainable Transportation., 15*(10), 731–748.

Goswami, M., Gupta, A. K., Kishan, R., Baidya, S., Khan, Y. I., Prakash, S., Premkumar, A., & Nautiyal, S. (2023). An evaluation of climate resilient agricultural practices in India: A narrative synthesis of literature. *Environmental Sustainability., 6*(1), 7–23.

Graabak, I., Jaehnert, S., Korpås, M., & Mo, B. (2017). Norway as a battery for the future European power system—Impacts on the hydropower system. *Energies, 10*(12), 2054.

Guikema, S. D., Nateghi, R., Quiring, S. M., Staid, A., Reilly, A. C., & Gao, M. (2014). Predicting hurricane power outages to support storm response planning. *IEEE Access., 2*, 1364–1373.

Hatsuzuka, D., Kato, R., Shimizu, S., Shimose, K. I. (2022). Verification of forecasted three-hour accumulated precipitation associated with "Senjo-Kousuitai" from very-short-range forecasting operated by the JMA. *Journal of the Meteorological Society of Japan. Ser. II. 100*(6), 995–1005.

Hirst, N. (2013). Buildings and climate change. *Design and management of sustainable built environments*, pp. 23–30.

Hoshiba, M., Ozaki, T. (2014). Earthquake early warning and tsunami warning of the japan meteorological agency, and their performance in the 2011 off the Pacific Coast of Tohoku Earthquake (9.0). In *Early warning for geological disasters: Scientific methods and current practice* (pp. 1–28). Springer Berlin Heidelberg.

Hugo, M., & Bhanye, J. (2022). The influence of classical 'green' theories on contemporary industrial planning practices: A review study. *Town and Regional Planning., 81*, 39–52.

Jin, Z., Archontoulis, S. V., & Lobell, D. B. (2019). How much will precision nitrogen management pay off? An evaluation based on simulating thousands of corn fields over the US Corn-Belt. *Field Crops Research., 240*, 12–22.

Kakran, S., & Chanana, S. (2018). Smart operations of smart grids integrated with distributed generation: A review. *Renewable and Sustainable Energy Reviews., 81*, 524–535.

Kavvadia, H. (2021). The European Investment Bank's quantum leap to become the world's first international climate bank. *Politics and Governance., 9*(2), 185–195.

Kay, J. E., Deser, C., Phillips, A., Mai, A., Hannay, C., Strand, G., Arblaster, J. M., Bates, S. C., Danabasoglu, G., Edwards, J., & Holland, M. (2015). The community earth system model (CESM) large ensemble project: A community resource for studying climate change in the presence of internal climate variability. *Bulletin of the American Meteorological Society., 96*(8), 1333–1349.

Kelman, I., Glantz, M. H. (2014). Early warning systems defined. *Reducing disaster: Early warning systems for climate change*, pp. 89–108.

Khan, A. R., Ratele, K., Dery, I., & Khandaker, S. (2023). Men and climate change: Some thoughts on South Africa and Bangladesh. *NORMA, 18*(2), 137–153.

Kirch Kirkegaard, J., Cronin, T., Nyborg, S., & Karnøe, P. (2021). Paradigm shift in Danish wind power: The (un) sustainable transformation of a sector. *Journal of Environmental Policy & Planning., 23*(1), 97–113.

Laser, S., & Stowell, A. F. (2020). Thinking like Apple's recycling robots: Towards the activation of responsibility in a post environmentalist world. *Ephemera: Theory and Politics in Organization, 20*(2), 163–94.

Levin, K., & Rich, D. (2017). *Turning points: Trends in countries' reaching peak greenhouse gas emissions over time.* World Resources Institute.

Loucks, D. P. (2023). Meeting climate change challenges: Searching for more adaptive and innovative decisions. *Water Resources Management, 37*(6–7), 2235–2245.

Mack, S. K., Lane, R. R., Deng, J., Morris, J. T., Bauer, J. J. (2023). Wetland carbon models: Applications for wetland carbon commercialization. *Ecological Modelling, 476*, 110228.

Makarova, I., Shubenkova, K., Mavrin, V., Boyko, A., & Katunin, A. (2017). Development of sustainable transport in smart cities. In *2017 IEEE 3rd International Forum on Research and Technologies for Society and Industry (RTSI)* (pp. 1–6). IEEE.

Marambanyika, T., & Musasa, T. (2023). Contestations and struggles in the use and protection of wetland ecosystems in communal areas of Zimbabwe: A case study of Intunjambili wetland in Matobo district. *Wetlands Ecology and Management, 31*(5), 611–627.

Matooane, L., Matamanda, A., & Bhanye, J. (2023). Urban food systems, and the circular economy: exploring synergies for sustainable development of African cities. In Theme: 6. Sustainable Cities Track: 6B. Future challenges in the urban transition Urban Planning.

Matos, S., Viardot, E., Sovacool, B. K., Geels, F. W., & Xiong, Y. (2022). Innovation and climate change: A review and introduction to the special issue. *Technovation, 10*, 102612.

Moazami, A., Carlucci, S., Nik, V. M., & Geving, S. (2019). Towards climate robust buildings: An innovative method for designing buildings with robust energy performance under climate change. *Energy and Buildings., 202*, 109378.

Moloise, S. D., Matamanda, A. R., & Bhanye, J. I. (2023). Traditional ecological knowledge and practices for ecosystem conservation and management: The case of savanna ecosystem services in Limpopo, South Africa. *International Journal of Sustainable Development & World Ecology., 27*, 1–4.

Mueller, N., Rojas-Rueda, D., Khreis, H., Cirach, M., Andrés, D., Ballester, J., Bartoll, X., Daher, C., Deluca, A., Echave, C., & Milà, C. (2020). Changing the urban design of cities for health: The superblock model. *Environment International., 134*, 105132.

Muluneh, M. G. (2021). Impact of climate change on biodiversity and food security: A global perspective—A review article. *Agriculture & Food Security., 10*(1), 1–25.

Naja, M., Childers, D. L., & Gaiser, E. E. (2017). Water quality implications of hydrologic restoration alternatives in the Florida Everglades, United States. *Restoration Ecology., 25*, S48-58.

Noussan, M. (2018). Performance based approach for electricity generation in smart grids. *Applied Energy., 220*, 231–241.

Oliveira, T., Tomar, S., & Tam, C. (2020). Evaluating collaborative consumption platforms from a consumer perspective. *Journal of Cleaner Production., 273*, 123018.

Ómarsdóttir, M. (2016). The role of geothermal in combating climate change. In *Proceedings of the Short Course I on Sustainability and Environmental Management of Geothermal Resource Utilisation and the Role of Geothermal in Combating Climate Change, Organized by UNU-GTP and LaFeo, Santa Tecla, El Salvador,* pp. 4–10.

Paltsev, S. (2020). Projecting energy and climate for the 21st Century. *Economics of Energy & Environmental Policy., 9*(1), 43–63.

Park, H., & Kim, J. D. (2020). Transition towards green banking: Role of financial regulators and financial institutions. *Asian Journal of Sustainability and Social Responsibility., 5*(1), 1–25.

Penuelas, J., Janssens, I. A., Ciais, P., Obersteiner, M., & Sardans, J. (2020). Anthropogenic global shifts in biospheric N and P concentrations and ratios and their impacts on biodiversity, ecosystem productivity, food security, and human health. *Global Change Biology., 26*(4), 1962–1985.

Perallos, A., Hernandez-Jayo, U., Onieva, E., & Zuazola, I. J. (Eds.). (2015). *Intelligent transport systems: Technologies and applications.* John Wiley & Sons.

Praveen, B., & Sharma, P. (2019). A review of literature on climate change and its impacts on agriculture productivity. *Journal of Public Affairs., 19*(4), e1960.

Punzo, G., Panarello, D., & Castellano, R. (2022). Sustainable urban mobility: Evidence from three developed European countries. *Quality & Quantity., 56*(5), 3135–3157.

Rinaldi, B. M., & Tan, P. Y. (eds.). (2019). Urban landscapes in high-density cities: Parks, streetscapes, ecosystems. *Birkhäuser*.

Roberts, J. T., & Weikmans, R. (2017). Postface: Fragmentation, failing trust and enduring tensions over what counts as climate finance. *International Environmental Agreements: Politics, Law and Economics., 17*, 129–137.

Saha, P., & Idsø, J. (2016). New hydropower development in Norway: Municipalities' attitude, involvement and perceived barriers. *Renewable and Sustainable Energy Reviews., 61*, 235–244.

Sanguesa, J. A., Torres-Sanz, V., Garrido, P., Martinez, F. J., & Marquez-Barja, J. M. (2021). A review on electric vehicles: Technologies and challenges. *Smart Cities. 4*(1), 372–404.

Schulz, F., & Rode, J. (2022). Public charging infrastructure and electric vehicles in Norway. *Energy Policy, 160*, 112660.

Šećerov, I., Savić, S., Milošević, D., Marković, V., & Bajšanski, I. (2015). Development of an automated urban climate monitoring system in Novi Sad (Serbia). *Geographica Pannonica., 19*(4), 174–183.

Seyboth, K. (2013). Intergovernmental panel on climate change (IPCC). *Encyclopedia of Energy, Natural Resources and Environmental Economics.*

Shafi, U., Mumtaz, R., García-Nieto, J., Hassan, S. A., Zaidi, S. A., & Iqbal, N. (2019). Precision agriculture techniques and practices: From considerations to applications. *Sensors., 19*(17), 3796.

Shah, K. J., Pan, S. Y., Lee, I., Kim, H., You, Z., Zheng, J. M., & Chiang, P. C. (2021). Green transportation for sustainability: Review of current barriers, strategies, and innovative technologies. *Journal of Cleaner Production., 326*, 129392.

Siegert, M., Alley, R. B., Rignot, E., Englander, J., & Corell, R. (2020). Twenty-first century sea-level rise could exceed IPCC projections for strong-warming futures. *One Earth., 3*(6), 691–703.

Singh, S., Misra, S. C., & Chan, F. T. (2020). Establishment of critical success factors for implementation of product lifecycle management systems. *International Journal of Production Research., 58*(4), 997–1016.

Soltani, M., Kashkooli, F. M., Souri, M., Rafiei, B., Jabarifar, M., Gharali, K., & Nathwani, J. S. (2021). Environmental, economic, and social impacts of geothermal energy systems. *Renewable and Sustainable Energy Reviews., 140*, 110750.

Starke, M., Ollis, B., Glass, J., Melin, A., Liu, G., & Sharma, I. (2017). Analysis of electric power board of Chattanooga smart grid investment. *ORNL Internal Report ORNL/TM-2017/246*.

Stewart, R. (2023). Using green investment banks to finance low-carbon pathways. *IMFG Perspectives, 34*.

Stocker, T. (2011). *Introduction to climate modelling*. Springer Science & Business Media.

Stratigea, A., Papadopoulou, C. A., & Panagiotopoulou, M. (2015). Tools and technologies for planning the development of smart cities. *Journal of Urban Technology., 22*(2), 43–62.

Suman, A. (2021). Role of renewable energy technologies in climate change adaptation and mitigation: A brief review from Nepal. *Renewable and Sustainable Energy Reviews., 151*, 111524.

Tai, Y. M. (2017). Effects of product lifecycle management systems on new product development performance. *Journal of Engineering and Technology Management., 46*, 67–83.

Tong, L., Jabbour, C. J., Najam, H., & Abbas, J. (2022). Role of environmental regulations, green finance, and investment in green technologies in green total factor productivity: Empirical evidence from Asian region. *Journal of Cleaner Production, 380*, 134930.

Van Der Walt, A. J., & Fitchett, J. M. (2022). Extreme temperature events (ETEs) in South Africa: A review. *South African Geographical Journal, 104*(1), 70–88.

Van Oijstaeijen, W., Van Passel, S., Cools, J. (2020). Urban green infrastructure: A review on valuation toolkits from an urban planning perspective. *Journal of Environmental Management. 267*, 110603.
Visan, M., Negrea, S. L., & Mone, F. (2022). Towards intelligent public transport systems in Smart Cities; Collaborative decisions to be made. *Procedia Computer Science., 199*, 1221–1228.
Visvizi, A., Lytras, M. D., Damiani, E., & Mathkour, H. (2018). Policy making for smart cities: Innovation and social inclusive economic growth for sustainability. *Journal of Science and Technology Policy Management., 9*(2), 126–133.
Von Hirschhausen, C. (2014). The German "Energiewende"—An introduction. *Economics of Energy & Environmental Policy., 3*(2), 1–2.
Wadanambi, R. T., Wandana, L. S., Chathumini, K. K., Dassanayake, N. P., Preethika, D. D., & Arachchige, U. S. (2020). The effects of industrialisation on climate change. *Journal of Research Technology and Engineering, 1*(4), 86–94.
Wu, M., Yan, B., Huang, Y., & Sarker, M. N. (2022). Big data-driven urban management: Potential for urban sustainability. *Land., 11*(5), 680.
Wu, Y. T., & Porté-Agel, F. (2015). Modeling turbine wakes and power losses within a wind farm using LES: An application to the Horns Rev offshore wind farm. *Renewable Energy, 75*, 945–955.
Xian, J., Xia, C., & Cao, S. (2020). Cost-benefit analysis for China's Grain for Green Program. *Ecological Engineering., 151*, 105850.
Xing, H., Stuart, C., Spence, S., & Chen, H. (2021). Alternative fuel options for low carbon maritime transportation: Pathways to 2050. *Journal of Cleaner Production., 297*, 126651.
Yu, H. J., Popiolek, N., & Geoffron, P. (2016). Solar photovoltaic energy policy and globalisation: A multiperspective approach with case studies of Germany, Japan, and China. *Progress in Photovoltaics: Research and Applications., 24*(4), 458–476.
Zemp, M., Eggleston, S., Míguez, B. M., Oakley, T., Rea, A., Robbez, M., & Tassone, C. (2021). *The status of the global climate observing system: The GCOS status report.*

Johannes Bhanye is a Researcher and Academic with a Ph.D. in Migration and Land Settlement, an MSc in Social Ecology and a BSc in Urban and Regional Planning. He is very passionate about urban sustainability, complexity and systems thinking, cities and social change, migration, governance, urban public health, climate change and other development-related subjects in Africa. He is currently a Post-Doctoral Research Fellow with the University of Cape Town, African Climate and Development Initiative (ACDI). He has been invited as a speaker to several International Conferences in countries like China, Belgium, Switzerland, Senegal, Zambia, Germany, South Africa, Lesotho, Morocco, and Ghana.

Ruvimbo Hazel Shayamunda is a Researcher and Doctoral Student with the University of the Free State. She holds a Master of Arts Degree in Conflict, Peace, and Governance from the University of Zimbabwe, and a Bachelor of Science (Honours) Degree in Peace and Governance from Bindura University of Science Education. Her research interests cut across conflicts and peacebuilding, climate change, leadership, media and society, gender and promotion of women's rights, civil society strengthening, and community capacity building.

Abraham Matamanda (Ph.D Urban & Regional Planning) is an Associate Professor and an NRF Y2-rated Urban and Regional Planner who also trained as a social ecologist. Abraham lectures at the University of the Free State (UFS) in the Department of Urban and Regional Planning. Abraham is currently the editor of the *Town Planning Journal* published by UFS and also serves on the academic editorial board of *Plos Water Journal*. He is a fellow of the Department of Higher Education and Training Future Professorate Programme Phase 1, third Cohort. His research focuses on urban governance and planning, climate change adaptation and resilience, informal Global South urbanism, urban food systems, and housing studies.

Chapter 9
The Future of Green FinTech Under Climate Change: The Relation of Cryptocurrencies and Sustainability

Julia M. Puaschunder

Abstract Green FinTech is based on green finance innovations. New technologies, such as robotics, artificial intelligence, big data insights, the internet of the things, and blockchain technology are contributing to the rising trend in Green FinTech. FinTech developments have the potential to enhance the efficiency of financial activities and lower the costs of financial global transactions, which helps in the accomplishment of CO_2 emissions reductions in light of climate change mitigation policies. Green FinTech in mobile banking and connected to shared transportation methods have been acknowledged to support the Sustainable Development Goals (SDGs). Green FinTech also imposes question about sustainability of cryptocurrencies. For instance, Bitcoins are generated through mining, which is an activity that requires a significant amount of energy. While Green Fintech pursues noble goals of greening the financial world, the reality is that many green innovations are not energy efficient. The relationship of sustainability and the Sustainable Development Goals is therefore yet to be established. Theoretical and empirical research of the future should define Green FinTech and evaluate the different Green FinTech solutions and their implementation's carbon footprint and greater sustainability. One of the most neglected areas of attention concerns the ethical imperatives of cryptocurrencies' use for the private exploration of space. Space exploration has become attributed as public good in recent decades. Most recent years now have seen a rise in commercialized space travel. Both trends open critical legal questions, also from the international law perspective. Space travel as a fairly novel endeavor includes safety and security as well as environmental risks but also moral dilemmas, which were solved by governmental central planners in the past. The new trend of commercialized space travel outsourcing these risks to private sector entities and insurances will likely take care of risks. But overall one may point at the enormous costs and unknown risks alongside a legal and policy vacuum on the impact of space travel on the space, earth and human ecosystem. Documented are already health impairments after space travel, which could well be only one side of a larger biosphere homeostasis imbalance that

J. M. Puaschunder (✉)
Columbia University, Graduate School of Arts and Sciences, New York, United States
e-mail: Julia.Puaschunder@columbia.edu
URL: http://blogs.cuit.columbia.edu/jmp2265; https://juliampuaschunder.com

W. Leal Filho et al. (eds.), *Climate Change Mitigation and Adaptation in Practice*, Climate Change Management, https://doi.org/10.1007/978-3-031-85217-6_9

could be triggered if space travel becomes the norm. Crowdsourcing funding for space travel and opening the gates for commercial activities in outer space therefore may hold long-term implications, such as human health risks, environmental harm and potential biological contamination of earth and space. All these trends should be examined by research and potential regulatory action be proposed.

Keywords Blockchain · Commercialized space travel · Cryptocurrencies · Environmental risks · Health risk · Space exploration

Introduction

Resilient finance includes green banking and technology. International developments in Green FinTech have prospered over the last decade. FinTech developed out of financial innovation in the areas of robotics, big data insights, Artificial Intelligence (AI), blockchain and decentralized finance as well as mobile Apps and internet banking (Kabaklarlı, 2022). The Green FinTech revolution is closely related to the advent of cryptocurrencies.

Green FinTech adds to technologically-enhanced finance solutions an environmentally-conscientious way to operate finance (Kabaklarlı, 2022). Green FinTech thereby employs cutting-edge technologies for finance in the pursuit to reduce environmental resource consumption of the finance industry and lower the carbon footprint of modern finance (Mi & Coffman, 2019).

While the working definition of Green FinTech stresses the reduction of energy consumption, the reality is that we hardly have any standardized way to measure the impact of Green FinTech on the economy and natural resources. Attention to CO_2 emissions has become essential in light of climate change. In the contemporary effort to measure natural resources as the main ingredient of economic growth and wealth of nations, our understanding of the environmental impact of Green FinTech on natural resource consumption patterns is scarce (Onder, 2023; Reamer, 2023; The White House of the United States, 2023a, 2023b). The European Union Finance Taxonomy accounts for a measurement of the carbon footprint of the finance industry, yet does not provide a clear reference to the environmental impact of cryptocurrencies yet (European Union Parliament, 2022). Climate change policies address the need to reduce carbon emissions, embracing all industries. This chapter explores the role of Green FinTech in relation to sustainable development and climate policies, with particular attention to the role of cryptocurrencies in energy consumption in light of carbon emission reductions and space exploration.

For one, Green FinTech accounts for one of the most innovative financial solutions of our lifetimes. The connection of technology—as a resource-saving innovation—with a green economy focus appears as a natural win–win solution for society. Many examples exist where technology-slimmed lean management processes achieve better performance regarding sustainability. All these features make Green FinTech a vital climate change policy support.

Yet on a closer look, the relationship between green conscientiousness and technology does not appear to always be most efficient and sustainable. In the concrete example of cryptocurrencies, different aspects have to be considered. Cryptocurrency use can be energy-consuming in the production of the currency (e.g., mining), in the individual storage of the currency (e.g., electronic wallets have to be backed by data centers that are risk prone and have energy consumption and space needs) and as a means of exchange (e.g., in the risk involved with cryptocurrency exchange), as well as in the ends funded by cryptocurrency crowdsourcing (e.g., in the use of cryptocurrencies to fund futuristic endeavors, such as space travel). In light of these features, Green FinTech's role for climate change mitigation and adaptation strategies appears questionable and worth a scientific investigation.

All these positive aspects but also the potential risks and energy consumption concerns have to be weighed against each other in order to make an informed decision on how to proceed with the regulation, use and consumer protection of Green FinTech regarding cryptocurrencies.

This chapter attempts to serve as the first step towards an evaluation of Green FinTech as an innovative and futuristic means of financialization and monetary exchange. The first part introduces the concept of Green FinTech. The second chapter outlines the advantages of Green FinTech in the modern economy with particular attention to climate change. The third chapter highlights potential risks and ethical considerations when it comes to Green FinTech cryptocurrencies' use to finance bold and future-oriented endeavors—such as the current trend of commercialized space travel. The discussion provides a summary of the advantages and potential risks of cryptocurrencies as well as an outlook for future areas of research.

Green FinTech

Green FinTech solutions are innovations enabled by new technology in the finance world that also pay attention to green economy goals and sustainable development. Green FinTech comprises, for instance, digital payment and accounting for sustainability and greening of the economy, green digital investments related to socially responsible investments and environmental, social and governance (ESG) conscientious funding, digital environmental, social and governance (ESG)-data and analytical solutions, digital crowdfunding for greening the economy, digital risk assessment in the environmental domain, digital deposit and lending for green projects, digital asset solutions that are sustainable, digital divestiture and green regulation technology solutions with an environmental touch, to name a few most recent cutting-edge green FinTech innovations (Green Finance Platform, 2022).

FinTech has had a tradition of applying technological innovations in finance ever since. The novel edge of Green FinTech now combines the worlds of cutting-edge technology solutions for the finance sector with advancements in sustainability and greening of the economy.

Already the aftermath of the 2008/09 World Financial Recession prepared the market for socially-conscientious financial market solutions. The financial crisis grew the general interest of the population, including a wide range of stakeholder in financial social responsibility (Puaschunder, 2019). The finance market for Socially Responsible Investment (SRI) developed qualitatively and quantitatively as the call for bringing finance back into the service of society became louder.

The COVID-19 pandemic was the most unexpected economic external shock in our lifetimes. The worldwide economic turmoil required drastic economic measures and financial resilience aid. Concurrent with the largest-ever outpour of governmental financial aid to soothe the economic disruption, the finance versus real economy inequality gap strengthened the understanding of governmental aid being pegged to higher societal causes, such as the alleviation of inequality within society. COVID-19 recovery funds were targeted as higher social goals that offer an unprecedented opportunity to sustainable societies (Puaschunder, 2022). As never before in the history of humankind, governmental aid targeted to alleviate social downfalls and thereby imbuing social and environmental causes at the forefront of finance. The economy is thereby meant to create a lasting value for humanity (The White House of the United States of America, 2021a, 2021b, 2021c).

The growing climate change crisis and the pressing urgency outlined during the yearly United Nations Conferences of the Parties (COP) meetings, all prepared greening of the economy pledges. What developed was the United States Green New Deal and the European Green Deal (The United States Congress, 2019; United Nations Environment Programme, 2009). The Green New Deal in the United States and the European Green Deal in the European Union peg economic recovery to societal causes and development (Puaschunder, 2023). Green FinTech is thereby at the forefront of combining financial technological advancement with the wish to produce sustainably and in harmony with the wider public (The White House of the United States of America, 2021a, 2021b, 2021c).

Green FinTech Solution Advantages

FinTech solutions use technological innovations for improving finance and banking around the world. The advancement of traditional finance via technological solutions helps to make finance more efficient and stimulate economic growth. Green FinTech now marries the idea of technological advancement for the improvement of the finance sector with higher social causes.

The 2008 World Financial Crisis, the COVID-19 external economic shock, as well as the lurking climate change crisis but also the United Nations SDGs, led to Green FinTech being at the forefront of finance innovation today. Innovative Green FinTech solutions have helped to reduce inequality, poverty and environmental destruction around the world (Kabaklarlı, 2022).

The clear advantages of FinTech developments are enhanced efficiency in financial activities and processes, which reduce monetary and environmental costs (Kabaklarlı,

2022). Concrete technologies used to improve financial processes include blockchain technology, mobile phones, open banking, big data analysis and other technological innovations that bring banking services and money transfer to the general population at low costs (Kabaklarlı, 2022). Green FinTech has thereby established itself as a viable international development tool and productive source of societal betterment.

Challenges online and with internet banking or mobile banking solutions are of security nature. Internet activities raise international liability uncertainties. For instance, mobile banking data security breaches with cloud storage in third countries raise the question of who is responsible for data leaks—the local banks, the mobile server companies or the cloud storage providers abroad.

Green FinTech innovations of the future are further calling for environmentally-conscientious finance solutions (Finextra, 2023). Green investments and renewable energy solutions in decentralized energy grids are believed to boost the already growing trend. Alone in 2022, green investments have grown 17% to reach USD 495 billion in investment capital globally according to Statista (Finextra, 2023).

Sustainable architecture—as hyped during the COVID pandemic in Agrohoods and Biophilia Trends—as well as green infrastructures are additional inspirations for eco-friendly digital platforms that are mentioned as Green FinTech innovations of the future (Finextra, 2023).

Carbon tracking and neutralization through innovations, such as filtering and binding CO_2 or commercializing it for concrete and carbonized soft drinks, are Green FinTech solutions that raise funds via crowdsourcing with the highest market innovation capitalization potential (Finextra, 2023).

Green redistribution mechanisms to avert climate inequalities within society, around the world, and over time are additional innovative Green FinTech solutions that are currently discussed to aid the climate goals and SDGs (Puaschunder, 2023).

Green FinTech as a Socially Responsible Investment (SRI) adds to a booming financial social responsibility market. The Environmental, Social, and Governance (ESG) pledges of governments around the world and oversight commissions—such as the United States Securities and Exchange Commission (SEC)—have encouraged the trend towards SRI in the last decade. Most recently, political divestiture has been discussed as a trend toward carbon emissions-heavy industries and governments.

One of the most vibrant FinTech solutions are digital payments, most innovatively in decentralized payment schemes, such as digital currencies and cryptocurrencies. Eco-friendly digital wallets promise to cut the CO_2 emissions of conventional fiat money transactions and checkout methods via credit cards, checks, etc.

Online payment revolutions are sustainability solutions in the Western world that raise hope to find secure mobile payment methods in fragile territories of the world. Contactless payments have been linked successfully to public transportation means. Tracking Apps have allowed to monitor carbon footprint and make the environmental impact of every consumption transparent to consumers in a palpable way.

Renewable online solutions are therefore Green FinTech innovations of the highest interest to push for financial solutions, also in the realm of the SDGs. With the rising prominence of greening the economy entering FinTech, the relation of environmental resources and energy consumption to Green FinTech, however, is not so clear when it comes to cryptocurrencies.

Green FinTech and Climate Policies

Given the novelty of the Green FinTech trend, the relation of Green FinTech and climate policies remains unclear. Green FinTech can enable positive and negative climate impacts.

The cost reduction of FinTech developments stands for a positive climate policy impact. Digital solutions promise to cut unnecessary red tape and carbon intensive activities. The green mandate also warrants climate conscientiousness and clear accountability of the carbon footprint of Green FinTech solutions. The innovative component of Green FinTech promises to offer constant innovation for betterment towards environmental resource constraints. The Green FinTech solution therefore will likely take over legal tenders in currencies and offer additional ways to transfer funds and invest in clean energy solutions.

At the same time, one can argue that Green FinTech is a fairly novel market solution with unknown outcomes. It could well be that Green FinTech becomes a costly fad that vanishes in importance and is at the end of the business cycle more a transaction cost than a climate policy adherent solution. In addition, Green FinTech raises new cost and liability concerns that are yet to be determined in the future to come. Further, the security, protection and data storage of Green FinTech solutions may become unexpected carbon burdens in the future if the prominence of the Green FinTech solutions rises. Lastly, Green FinTech may embrace stakeholders around the world and grant additional access to financial services to untapped clientele. It may therefore, in net terms, increase the carbon footprint of the finance industry.

Overall, Green FinTech's use and rising prominence are questionable impact factors in a world of climate change. The relation of Green FinTech and climate policies is not straightforward and demands future research.

Method: Green FinTech in Cryptocurrencies

This chapter addresses the role of Green FinTech in relation to climate policies with attention to cryptocurrencies in a theoretical analysis.

Cryptocurrencies emerged in 2009 with Bitcoin entering markets. The 2008 World Financial Crisis inspired alternative banking solutions with the help of decentralized distribution systems that are based on technological solutions employing blockchains. Blockchain enable distributed ledger enforced by a disparate network of computers (Frankenfield, Murry & Kvilhaug, 2023). Since the beginning of cryptocurrencies via Bitcoin, the market has experienced exponential growth in options—e.g., today such as Ethereum, Binance Coin, Solana, Cardano, Marscoin, and trading innovations, such as digital wallets, Ripples' XRP, meme coins etc.

To this day, legislation and regulation around cryptocurrencies vary around the globe. For instance, as of January 2023, El Salvador and the Central African Republic accept Bitcoin as a legal tender for monetary transactions; while at the same time in the U.S. and other Western countries, Bitcoin ATMs emerge (Frankenfield et al., 2023). The European Union and Japan acknowledge the legal character of Bitcoin as an alternative FinTech market solution. Other countries and territories, foremost China, have banned cryptocurrency exchange or are skeptical of mining, with India currently developing a concerted framework for the use of cryptocurrencies (Frankenfield et al., 2023).

Today, cryptocurrencies are all digital and virtual currencies that are secured by cryptography. The unique character enabled by cryptography secures from counterfeit currencies and double-spending problems. Cryptocurrencies are decentralized for not being issued by a central authority, such as a governmental central bank or world bank. The network of computers enabling cryptocurrencies is not controlled by a single entity. The independence from governmental authority but also assumed faster and individualized money transfers in streamlined remittances via decentralized systems create the advantages of cryptocurrencies over standard fiat money or monetary exchange via central banks. In this light, Green FinTech and digital payments appear as more sustainable market solutions than paper money and the central control of fiat currency. Green FinTech pledges promote digital payment as a more convenient solution that is also eco-friendly. In the concrete case of remittances, cryptocurrencies can serve as an intermediate to streamline money transfers across borders.

Disadvantages are high price volatility due to unstable regulatory frameworks regarding cryptocurrencies, energy consumption for mining activities and the attraction of criminal activities via these rather unregulated payment channels (Frankenfield et al., 2023).

While ample literature exists and is emerging on the financial, speculative and economic risks as well as societal costs of cryptocurrencies, hardly any information is provided on the sustainability of cryptocurrencies in light of the SDGs. While cryptocurrencies offer a fast and secure market transaction opportunity in territories with underdeveloped financial sector protection, the environmental costs of decentralized finance with the possibility to mine money oneself are hardly covered in contemporary literature. In the use of cryptocurrencies for innovative endeavors, such as space travel, the ethical boundaries of this innovation are not covered. This chapter therefore sets out to discuss the potential environmental downfalls and ecologic shortcomings of Green FinTech in the case of cryptocurrencies, with special attention to cryptocurrencies' novel use of crowdsourcing funding for commercialized space invasion.

Cryptocurrencies

Cryptocurrencies are an asset market solution, which started in response to the 2008 World Financial Recession. By now, the cryptocurrency market has ballooned into a 2-trillion dollar market. Cryptocurrencies are decentralized market solutions with no intermediaries needed to trade that offer different tokens. Cryptocurrencies are sometimes seen critical as for the investor exploitation risk of scams, blackmail and fake services. Pump-and-dump schemes appear to be common with certain providers hyping products only to sell them once a price rise has occurred. Cryptocurrencies' use is also seen as critical when it comes to the relatively high level of criminal activities connected to cryptocurrencies as the anonymity of the concept. Cryptocurrencies have—for a long time—been referred to as risky options, which may fail or crash. Positive aspects include the tradability across borders, the security tokenism that makes them a perfect substitute for conventional stocks as well as the utility token feature of automated obligations. Negative aspects include the relative volatility, the relative short time this market option is around and the unknown regulatory framework, by which cryptocurrencies are governed.

United States President Biden's executive order on cryptocurrencies of March 2022 mentions the energy consumption of cryptocurrencies among many other aspects, such as consumer protection, investor security, business impacts, data privacy and security, financial stability and systemic risks, crime use, national security and the ability of humans to exercise rights in overseeing markets as well as financial inclusion and equity in society.

Similarly, the European Union Markets in Crypto-Assets (MiCA) regulation was adopted in 2023 and entered into force in 2024. The regulation mentions the need of support of cryptocurrencies in the legal certainty, innovation support, consumer and investor protection as well as financial stability. Areas of concern are the market power and dominant position of some cryptocurrencies over other financial instruments as well as asymmetric information between market actors.

Given all these regulatory attempts and research conduct on the implications of a rising cryptocurrency market, the connection between Green Fintech and the Sustainable Development Goals will be covered in the following.

Cryptocurrencies' Sustainability

Green FinTech innovations are currently vibrant in cryptocurrencies, which recently triggered concern about the environmental impact and carbon footprint, all raising doubts about cryptocurrencies' sustainability (Kabaklarlı, 2022). Environmentally problematic appears that Bitcoins are generated through mining, which demands a significant amount of electrical energy (Kabaklarlı, 2022).

One of the innovations in Green FinTech of cryptocurrencies is the decentralized aspect of anyone with an internet connection being able to produce, hence mine. Producing personal currency units is independent of a central agency, such as a central bank, that produces fiat money. The mining operation, however, requires considerable amounts of energy, more than corporations or even some countries consume. Kabaklarlı (2022) estimates Bitcoin mining of cryptocurrencies has a notable impact on carbon footprinting. The mining activities let cryptocurrencies appear to be not in line with Green FinTech's environmental friendliness mandates. BTC (Bitcoin) miners' revenue and the Bitcoin electricity consumption are positively correlated. This unsustainable consumption of energy lets the use of cryptocurrencies for FinTech solutions appear not to be environmentally sustainable and not sustainability conscientious. The environmental impact of traditional cryptocurrency mining as an energy-intensive process is estimated to consume alone 198 TWh of electricity annually just for Bitcoin alone. This energy consumption level equals that of Thailand (Finextra, 2023).

As a response of the viability to the trend, the so-called mining pools help bundle energy for mining activities, which are estimated to be 98% by the beginning of 2023 (Frankenfield et al., 2023). The volatility in the cryptocurrency market, for instanced between October 2014 and April 2017, downgraded this market option (Kabaklarlı, 2022). In addition, Jabotinsky and Sarel (2023) find in 100 leading tokens over January 2020 and March 11, 2020 during the outbreak of the COVID-19 pandemic, that COVID increased the value of tokens and conclude that there is a negative correlation between cryptocurrencies and S&P500 trade but a positive correlation between cryptocurrencies and S&P500 value. This leads the authors to conclude that cryptocurrencies are market substitutes and not safe havens (Jabotinsky & Sarel, 2023). In most writings about cryptocurrencies, attention to externalities is missing (Jabotinsky & Sarel, 2023). Potential externalities mentioned are of risk nature. Investors trading in cryptocurrencies during times of crisis cause systemic risk by trading away funds from traditional financial instruments (e.g., governmental T-Bonds or conventional shares) (Jabotinsky & Sarel, 2023). This adds additional uncertainty into markets during turmoil times. The use of cryptocurrencies for crime activities is another mentionable negative externality imposed on society (Jabotinsky & Sarel, 2023). Problematic appears the regulations of cryptocurrencies as for the global decentralized market of the cryptocurrencies' trade.

Green cryptocurrency developments counter the energy-consumption problematics. The Green FinTech innovation of green cryptocurrencies uses consensus mechanisms in order to minimize energy consumption levels (Finextra, 2023). Consensus mechanisms are speeding up the process of transactions and gain on popularity as the market becomes aware of the energy consumption levels of cryptocurrencies. An example mentioned in the literature is Ethereum, which employs proof-of-stake in contrast to proof-of-work algorithms, which is believed to reduce energy consumption drastically (Finextra, 2023). This sustainability-enhancing technique is to avoid computation and use verifying transactions—for instance, let users use their own currency as collaterals for heightening accuracy and security of the blockchain (Finextra, 2023). Mining alternatives and carbon credits in blockchain technologies use are other Green FinTech solutions to green cryptocurrencies (Finextra, 2023).

While especially practitioners' literature and governmental regulation emerge to make cryptocurrencies secure, foremost from fraud and hacking, and the volatility is believed to improve as cryptocurrency options widen and mature, little is known about the innovative use of cryptocurrencies for funding space travels and the ethical implications of such endeavors. While Elon Musk warned about the environmental aspect of mining, the use of cryptocurrencies to raise funds for one of the most innovative endeavors of humankind—space travel—is currently still ongoing and flourishing. The topic of cryptocurrencies' use for commercial space invasion is hardly subject to critical reflection, especially in terms of this strategy's sustainability, potential risks and long-term externalities (Finextra, 2023).

Cryptocurrencies' Ethical Boundaries

Today society is striving to go to outer space as never before. In the most recent decade, space exploration has become a public good. In addition, commercialized space travel is evolving. Goals are—for instance—the travel to Mars and sell the experience of outer space travel to consumers. Governmental efforts push for a permanent space station and commercialized space travel. While space invasion plans date back as early as the 1960s, novel is the commercialized plan to offer space travel to consumers with plans even including hotels in space (SpaceX, 2021a, b).

As for the financialization and massive market capitalization needed for commercialized space travel and due to the high-risk nature of space travel endeavors, these innovations are often funded by innovative financial means, most recently via crowdsourcing of funding in cryptocurrencies, e.g., such as Marscoin.

From the financialization aspect, FinTech thereby becomes used as innovative crowdsourcing of financial means. Cryptocurrencies allow for tapping into novel and younger market segments than—for instance—commercialized traditional bonds or index funds. On the ethical boundaries of cryptocurrencies, critical use for innovative purposes with unknown outcomes arises. Cryptocurrencies being used for funding private space travel implies legal challenges, environmental questions and carbon footprint pressures alongside moral and ethical concerns, which are hardly discussed in the contemporary scientific literature (Pierson, 2021). The use of cryptocurrencies for space travel is accompanied by demand for legal and policy regulations that guide humankind on space travel.

Commercialized space travel could raise monetary costs and substantial risks. However, no account discusses the role of cryptocurrencies for space travel from sustainable development and SDG perspective. Possible advantages of conquering space should thereby be weighted compared to the protection of human safety and security, ecological impacts as well as moral and ethical risks.

Environmental impacts of space travel include the carbon footprint as well as debris in outer space that are left eternally to circle around in space. Space travels are conducted in the celestial ecosystem, of which we do not have enough information how it may impact human beings and the earthly biosphere (Pierson, 2021).

Producing space travel ships is associated with high carbon emissions. Space travels may also emit to outer space and information is missing how this will impact the balance of the celestial space. There is also the space debris problem, which is currently unsolved. Every space travel leaves space debris that will grow substantially if space travel becomes commercial and accessible to a wide range of population. Old satellites and space travel emissions are causing an unprecedented space waste problem, that raises the risk to collusion or further unforeseen risks in the future.

Environmental negative aspects of space exploration are weighted against the enormous benefits human innovation in the wake of space exploration offers. From the environmental aspect, the National Oceanic and Atmospheric Administration estimates that a rocket launch injects about 1,000 tons of soot into the upper atmosphere per year (Kluger, 2023). While the amount of fossil fuels burned by rocket launches is still only 1% of conventional aviation emissions, the trends are rising for space rocket launches way faster than general aviation levels change (Piesing, 2022). If commercialized space travel becomes the norm and prices drop opens commercialized space travel to the broader public, then the ratio will change as well towards more carbon impact of space rocket launches.

The debris left in outer space with satellites and other shelved flight vehicles in outer space are an additional sustainability problem in outer space travel, which appears to get worse over time and with rising travel levels. In all these considerations, cryptocurrencies' use for outer space travel therefore passively implies environmental degradation and risks to humans living on Earth as well.

There are also considerations about human health risks when considering space travel. Activities in outer space raise questions of public health and safety concerns. Cosmic radiation and microgravity cause changes to the blood and the functioning of the immune system, which is not clearly understood. First attempts to procreate animals in outer space have pointed at reproductive issues and even infertility in outer space. There are additional socio-psychological impairments, which led to issues of sabotaging space missions in outer space due to the elevated stresses in space.

So far not fully understood, yet noticeable is the fact that the immune system seems to be weakened after space travels. In astronauts, a changed gene activity in the white blood cells is believed to decrease the functionality of the immune system once astronauts get into space (Sullivan, 2023). The reactivation of dormant viruses in the body was already noticed in space missions (Sullivan, 2023). Microgravity is now also believed to cause the blood flow to change, triggering changes to white blood cells and gene expressions that suppress the overall immune system functionality (Sullivan, 2023). While the mechanism is not fully understood yet and the long-term duration of the effect is currently studied, one can already imagine the long-term implications of commercialized space travel and doubt the sustainability of space travel for human health reasons. This effect is even driven further by the current ban to procreate in space as the first attempts to breed animals in space failed and led to unsustainable results.

Direct environmental harm could occur in radiation but also contamination of space ecosystem components being exchanged with earth ecosystem components. The potential biological impact of travel from earth to space and returning to earth is a risk with unknown outcome. To this day, it remains unclear what species and microbiological systems may be prevalent in outer space. The bringing back of unknown biological components may impose ecosystem threats and contagion risks that are novel and dangerous for not being fully understood and therefore not fully controllable.

Commercial space travel is currently governed by the United Nations' "*Treaty on Principles Governing the Activities of States in the Exploration and Use of Outer Space, including the Moon and Other Celestial Bodies*" (Pierson, 2021; United Nations Office for Outer Space Affairs, 1967). International Law Convention standards and outdated maritime law norms that exist since decades are currently applied to govern territorial claims in outer space (Pierson, 2021). These legal frameworks and customary law conventions potentially will require updating to outer space territories—an endeavor that the United Nations International Law Commission should consider tackling. Being bound by the 1967 "*Treaty on Principles Governing the Activities of States in the Exploration and Use of Outer Space, including the Moon and Other Celestial Bodies*" leads to corporations being tied to the nation they represent in their outer space travels, which could lead to international tensions and global inequalities who can conquer outer space (Pierson, 2021). Corporations being pegged to the countries they represent in outer space also pits corporations against governmental regulators and could lead to unanticipated national risks and liability claims. With the internet and social online media being partially controlled by corporate entities and cryptocurrencies being decentralized financial tools to crowdsource financial resources, these new dynamics could tilt authority from governmental control of space more into the direction of private entities controlling space invasion and the financial resources for space travel as well as the social online media representation of these endeavors. This also implies that space invasion is a more commercial endeavor under the wing of governmental protection. Corporate action on public goods creation—for instance in large-scale infrastructure development—has already in the past shown benefits of efficiency but often traded off from social and ethical focus. These conflicts are yet to be addressed. The sustainability of cryptocurrencies should therefore also be studied in regards to externalities and volatility around space exploration and commercialized space travel (Puaschunder, 2023).

From an economic cost-benefits analysis perspective, space travel is still extraordinarily expensive. For instance, commercial space travel is estimated to cost hundreds of billions of U.S. dollars (Levchenko et al., 2018 in Pierson, 2021). Innovative entrepreneurial FinTech financing appears as necessary step to continue this groundbreaking basic research endeavor. From this standpoint, the viability of cryptocurrencies' use for commercial purposes rather than international development endeavors appears somewhat critical.

From the political aspect, critical ethics concerns arise out of potential risks involved in travel between earth and space. The celestial setting could impose risks of contamination and transfer of ecosystem components with disastrous outcomes to the biological harmony on earth with long-term and widespread impacts causing irreversible lock-ins (Pierson, 2021).

Overall, future-oriented cryptocurrencies' funding of space travel appears critical when it comes to sustainability and is ethically questionable. Future research is called for to discuss and debate the ethical boundaries and framework conditions to have space invasion innovation prosper under the wing of ethical protection.

Discussion

Green FinTech solutions are technology innovations in finance that promise international development advancements based on efficient and non-corrupt means to engage audiences and cater to their needs on a global basis. Green FinTech can also be a vital driver of SDG endeavors in its ICT for development features in its international outreach. In the contemporary awareness of climate change around the world, finance redistribution mandates are advocating for offsetting some of the starkest and earliest losses due to global warming (Puaschunder, 2020). Green FinTech offers an international way to alleviate climate inequality and provides a universal approach to mobilize finance for climate stabilization.

One of the newest, most vibrant emerging streams of Green FinTech are cryptocurrencies. In the most recent decade, no other financial innovation than cryptocurrencies has steered such enormous growth while—at the same time—revolutionizing the way currencies are understood and perceived by the larger society. To this day, cryptocurrencies are widely debated topics of interest for finance professionals, governmental regulators and global governance experts. The potential and overall goal of cryptocurrencies appears promising and noble—namely, to democratize and stabilize finance, and put currencies back at servicing people. For international development, cryptocurrencies hold the key to offering efficient and safe means of remittance transfers. At the same time, cryptocurrencies have become gateways of illegal activities and the large-scale mining operations to create assets have raised questions of the sustainability of this method.

This chapter addressed cryptocurrencies' role in Green FinTech. The positive aspects of cryptocurrencies in the independence of central banking authorities and the flexibility in efficient money transfers across borders were highlighted. Credit was given to cryptocurrencies' role in revolutionizing energy grids for decentralization efforts of shared energy economies. The experimental role of digital payments pegged to shared transportation methods was outlined.

At the same time, the critical role of cryptocurrencies regarding energy consumption has to be acknowledged. Positive ways out of the energy crisis surrounding cryptocurrency mining activities are shared consensus mechanisms and mining pools that appear to dominate the mining market already. Future positive advancements are innovations to create green cryptocurrencies that refrain from mining and put crypto-generated economic gains to work for the overall greening of the economy.

The chapter also paid attention to a topic that is currently on the rise, yet hardly discussed in the scientific discourse of Green FinTech, namely the relation of cryptocurrency use in ethically-challenging domains, such as space travel. In the last decade, space travel has become a reality with commercial interest. Technology entrepreneurs and globally operating space travel agencies have started to offer commercialized space travel. With the rise of rocket launches connected are cryptocurrencies that raise funds to establish space exploration missions. With this particular use of cryptocurrencies, the question arises if cryptocurrencies' use in space exploration is sustainable?

The chapter pointed at several areas of concern when it comes to cryptocurrency use for space travel of legal, economic, political, financial, environmental and human health risk nature. Overall, one can say that cryptocurrency-backed space travel is legal in the Western world but should be closely monitored for potential long-term dangers and threats to humankind. Like with bonds, a risk premium should be calculated that also integrates the ethical boundaries and raises funds for offsetting some of the potential losses and anticipated damages due to risks that are yet unknown. Legal and regulatory changes to the landscape of cryptocurrencies, financial failures of shared currency projects, political frictions over space territory disputes, private sector space commercialization being at clash with public interests are some of the anticipated externalities that one should prepare for in advance. When it comes to sustainability concretely, the environmental impact of space debris but also the potential long-term health risks due to space travel long haul aftereffects as well as the risk of contamination due to invasive living organisms being brought back from abroad are futuristic, yet potentially likely threats that are imposed by space travel. Innovation should be funded by crowdsourcing, yet vigilantly and with respect for potential harms that should be anticipated and set-asides created to curb potentially harmful developments.

Future research areas that can be envisioned from the current state-of-the-art of Green FinTech should include a wide-reaching stakeholder analysis and disparate impact considerations that could aid in shedding important light on the concrete implications and variegated risks of Green FinTech for society. Highlighting sustainability aspects in regard to Green FinTech today also opens attention to the realm of beneficiaries and victims for future world inhabitants. Today's decisions in terms of the regulation and innovation funding for Green FinTech appear to hold the key to prosperity of future generations given the lasting effects of financialization of sustainability but also when it comes to long-term unknown risks.

Cryptocurrencies' use for crowdfunding space exploration should be scrutinized for their manifold implications. For legal scholars, the international law of space occupation should become subject to international treaties with respect for an inclusive approach. Economically, the risks of cryptocurrencies should be outlined in order to create more stable crowdsourcing options as collaterals if cryptocurrencies fail. Regulatory research should focus on creating a clear framework of cryptocurrencies' use and governmental role in stabilizing cryptocurrencies, which are still considered highly volatile due to instringent or missing governmental regulatory guidelines throughout the world. The environmental community can help with transparency about space exploration's carbon footprint and space debris' long-term implications. Contemporary efforts to include natural resources in governmental national accounting as well as Finance Taxonomies could also highlight the attention to measuring the sustainability and carbon footprint of Green FinTech. Healthcare specialists and the medical profession should turn their attention to the impact of space travel on human bodies. Not only in terms of the immune system being weakened by space travel but also a gender-sensitive analysis is recommended to shed light at the fertility aspects in space. Preliminary evidence exists but further studies and open discussions are recommended to follow on the long-term impact of traveling to space on human procreation. From the biosphere aspect, the homeostasis aspects of space transfers and exchanges of fauna and floras should be discussed. When first explorations sailed across the oceans in the early ages of discovery on Earth, not all transfers of fauna and flora as well as bacteria and viruses were a positive exchange. Many extinctions and human hardships can be found in transfers from one part of the world to another. In those aspects, history will hopefully not repeat itself, but we can certainly learn from success stories to uphold ethics but also be conscientious of potential ethical downfalls in the future.

Conclusion

This chapter prepared the ground floor to discuss the ethical implications of space travel funded by cryptocurrencies in an attempt to explore space better than we used to invade the world and with contemporary respect for sustainability ethics. Space travel is the innovation of our lifetimes. Exploring space holds the potential to tab into new resources and gain most unknown insights about the orbit. At the same time, unfamiliar risks open up in the ecological, human, legal and ethical domains. Commercialized space travel therefore warrants general caution and interdisciplinary scientific oversight of the process. On a wider level, the chapter serves as a first account of calling for research attention to Green FinTech's resilience and viability in the overall framework of climate stabilization and sustainable development.

References

European Union Parliament. (2022). *Cryptocurrencies in the EU: New rules to boost benefits and curb threats*. Press Release. Accessed March 30, 2023, <https://www.europarl.europa.eu/news/en/press-room/20220309IPR25162/cryptocurrencies-in-the-eu-new-rules-to-boost-benefits-and-curb-threats>

Finextra. (2023). *FinTech operations going green: Five trends shaping the financial landscape in 2023*. Accessed July 3, 2023, <https://www.finextra.com/blogposting/24034/fintech-operations-going-green-five-trends-shaping-the-financial-landscape-in-2023>

Frankenfield, J., Murry, C., & Kvilhaug, S. (2023). Cryptocurrencies. *Cryptocurrency explained with pros and cons for investment*. Investopedia. Accessed July 3, 2023, <https://www.investopedia.com/terms/c/cryptocurrency.asp>

Green Finance Platform. (2022). *Green FinTech classification*. Accessed July 3, 2023, https://www.greenfinanceplatform.org/research/green-fintech-classification

Jabotinsky, H. Y., & Sarel, R. (2023). How crisis affects crypto: Coronavirus as a test case. *Hastings Law Journal, 74*, 433–488.

Kabaklarlı, E. (2022). Green FinTech: Sustainability of Bitcoin. *Digit Finance, 4*, 265–273. https://doi.org/10.1007/s42521-022-00053-x

Kluger, J. (2023). There's no way to make space travel good for planet earth right now. Time. Accessed July 3, 2023, https://time.com/6273065/space-travel-climate-impact/

Levchenko, I., Shuyan X., Stéphane M., Michael K., & Kateryna B. (2018). Mars colonization: Beyond getting there. *Global Challenges, 3*(1). https://doi.org/10.1002/gch2.201800062

Mi, Z., & Coffman, D. (2019). The sharing economy promotes sustainable societies. *National Communication, 10*, 1214. Accessed May 15, 2023 https://doi.org/10.1038/s41467-019-09260-4

Onder, St. (2023). The changing wealth of nations (CWON). The World Bank Group. Presentation delivered to the U.S. Committee for IIASA at the National Academy of Sciences, Washington D.C.

Pierson, N. (2021). A multi-planet species: Ethical and environmental impacts of privatized space travel. In *Proceedings of the 23rd International Research Association for Interdisciplinary Studies (RAIS) Conference on Social Sciences and Humanities.*

Piesing, M. (2022). *How to make rocket launches less polluting*. BBC. Accessed June 30, 2023, https://www.bbc.com/future/article/20220713-how-to-make-rocket-launches-less-polluting

Puaschunder, J. M. (2019). *Intergenerational equity: Corporate and financial leadership.* Edward Elgar.

Puaschunder, J. M. (2020). *Governance and climate justice: Global south and developing nations.* Palgrave Macmillan, Springer Nature.

Puaschunder, J. M. (2022). *Ethics of inclusion: The cases of health, economics, education, digitalization and the environment in the post-COVID-19 era.* Bury St. Edmunds, Ethics International.

Puaschunder, J. M. (2023). *The future of resilient finance: Finance politics in the age of sustainable development.* Palgrave Macmillan.

Reamer, A. (2023). *National strategy to develop statistics for environmental-economic decision: OSTP/Commerce/OMB.* Accessed May 9, 2023, https://www.aeaweb.org/forum/3447/national-strategy-statistics-environmental-economic-decisions

SpaceX. (2021a). *Falcon heavy: The world's most powerful rocket.* SpaceX, Accessed April 23, 2021, https://www.spacex.com/vehicles/falcon-heavy/

SpaceX. (2021b). *SpaceX—Mission.* SpaceX, Accessed April 23, 2021, https://www.spacex.com/mission/

Sullivan, W. (2023). Why astronauts have a weaker immune system in space. The Smithsonian. Accessed June 30, 2023, https://www.smithsonianmag.com/smart-news/why-astronauts-have-weaker-immune-systems-in-space-180982423/

The United States Congress, 116th Congress, 1st Session, House Resolution 109. (2019). *Recognizing the duty of the Federal Government to create a Green New Deal.* Washington, D.C. Accessed January 9, 2021, https://www.congress.gov/116/bills/hres109/BILLS-116hres109ih.pdf

The White House. (2023a). Fact sheet: Biden-Harris Administration releases national strategy to put nature on the nation's balance sheet. Accessed May 9, 2023, https://www.whitehouse.gov/ostp/news-updates/2023/01/19/fact-sheet-biden-harris-administration-releases-national-strategy-to-put-nature-on-the-nations-balance-sheet/

The White House. (2023b). National strategy to develop statistics for environmental-economic decisions: A U.S. system of natural capital accounting and associated environmental-economic statistics. Office of Science and Technology Policy. Office of Management and Budget. Department of Commerce. Accessed May 9, 2023, https://www.whitehouse.gov/wp-content/uploads/2023/01/Natural-Capital-Accounting-Strategy-final.pdf

The White House of the United States of America. (2021a). President Biden announces American Rescue Plan. Accessed May 16, 2023, https://www.whitehouse.gov/briefing-room/legislation/2021/01/20/president-biden-announces-american-rescue-plan/
The White House of the United States. (2021b). President Biden and G7 Leaders launch Build Back Better World (B3W) Partnership. Accessed April 23, 2023, https://www.whitehouse.gov/briefing-room/statements-releases/2021/06/12/fact-sheet-president-biden-and-g7-leaders-launch-build-back-better-world-b3w-partnership/
The White House of the United States. (2021c). The American Jobs Plan. Accessed April 23, 2023, https://www.whitehouse.gov/briefing-room/statements-releases/2021/03/31/fact-sheet-the-american-jobs-plan/
United Nations Environment Programme.(2009). *Global green new deal policy brief*. Accessed February 9, 2021, https://www.unep.org/resources/report/global-green-new-deal-policy-brief-march-2009#:~:text=UNEP%20outlines%20a%20GGND%20with,carbon%20dependency%20and%20environmental%20destruction
United Nations Office for Outer Space Affairs. (1967). Treaty on principles governing the activities of states in the exploration and use of outer space, including the moon and other celestial bodies. Opened for signature. United Nations Office for Outer Space Affairs. https://www.unoosa.org/oosa/en/ourwork/spacelaw/treaties/outerspacetreaty.html

Julia M. Puaschunder is a post-doctoral researcher in the Interuniversity Consortium of New York at Columbia University. As a member of the OMNES Education Group, Julia M. Puaschunder serves on the faculty of the International University of Monaco in the Department of Economics & Finance. Previously, she was a faculty associate at Harvard University. Julia Puaschunder holds PhDs in Economics and Natural Sciences from the University of Vienna, Vienna University of Economics and Business as well as the Interuniversity Consortium of New York. Julia Puaschunder has authored 13 books on various topics in Responsible Finance and Governance of Climate Justice.

Chapter 10
Climate Change and Artificial Intelligence: Leveraging Machine Learning for Informed Policy-Making and Sustainable Solutions

Gisleine Aver, Valdemiro da Rocha Júnior, Jorge Gustavo Sandoval Simão, Artur Marques, Ana Regina de Aguiar Dutra, Leandro dos Santos Coelho, and José Baltazar Salgueirinho Osório de Andrade Guerra

Abstract In the wake of the alarming 2021 Intergovernmental Panel on Climate Change (IPCC) report, the role of Artificial Intelligence (AI) and Machine Learning (ML) in addressing climate change challenges is gaining prominence. The IPCC's findings underscore the urgent need for global collaboration, policy reforms, and innovative solutions that align with the Sustainable Development Goals (SDGs). AI, originally defined as replicating tasks at which humans excel, has evolved considerably, with ML enabling computers to learn from data patterns. Recent research has emphasized the alignment of AI models with climate change mitigation strategies and sustainable development strategies, especially in forecasting, climate modeling, and environmental protection. In Brazil, the Ministry for Science, Technology and Innovation (STI) is mandated to allocate a budget for technological advancement, highlighting the potential to incorporate AI and ML in their climate change strategies. This chapter aims to address: How can governments utilize AI and ML for informed decision-making and effective public policies to counteract climate change impacts?

G. Aver (✉) · V. da Rocha Júnior · A. Marques · A. R. de Aguiar Dutra ·
J. B. S. O. de Andrade Guerra
Centre for Sustainable Development/Research Group On Energy Efficiency and Sustainability (GREENS), University of Southern Santa Catarina (UNISUL), Santa Catarina, Brazil
e-mail: gisaver@yahoo.com.br

J. B. S. O. de Andrade Guerra
e-mail: jose.baltazarguerra@animaeducacao.com.br

J. G. S. Simão
Pontifícia Universidade Católica of Paraná (PUC/PR), Curitiba, Brazil

L. dos Santos Coelho
Pontifícia Universidade Católica of Paraná (PUC/PR), Universidade Federal Do Paraná (UFPR), Curitiba, Brazil

J. B. S. O. de Andrade Guerra
Cambridge Centre for Environment, Energy and Natural Resource Governance, (CEENRG), University of Cambridge, Cambridge, UK

W. Leal Filho et al. (eds.), *Climate Change Mitigation and Adaptation in Practice*, Climate Change Management, https://doi.org/10.1007/978-3-031-85217-6_10

By integrating these technologies, governments can optimize resource use, enhance predictive modeling, and enact timely interventions. The chapter introduces scalable solutions using open-source tools, underscoring the transformative potential of AI and ML in global climate change mitigation efforts.

Keywords Climate change · Mitigation · Adaptation · Artificial intelligence · Machine learning · Policy-Making · Sustainable development goals

Introduction

In 2021, the Intergovernmental Panel on Climate Change (IPCC) published the sixth report, which highlighted the current state of climate change research worldwide. The report has heightened concerns about the impacts of these changes, especially in terms of the extension of planetary boundaries and their consequences caused by extreme weather events, loss of biodiversity and global warming (IPCC, 2022).

The Secretary-General António Guterres, in his speech on the Report above mentioned, emphasized: "The jury has reached a verdict. And it is damning. This report from the Intergovernmental Panel on Climate Change is a long list of unfulfilled climate promises" (UN, 2022).

The report's revelations have not only increased awareness but also underscored the urgency for international collaboration, policy reforms, and innovative solutions to address the multifaceted challenges posed by climate change.

The 2030 Sustainable Development Agenda, adopted by all Member States of the United Nations in 2015, recognizes we must tackle climate change, preserve our oceans and forests, while working on ending poverty, improve health and education, reduce inequality and promote economic growth. The Agenda has 17 Sustainable Development Goals (SDGs), with 169 targets which are an urgent call for action by all countries in a global partnership, considering that the increased advancements of modern society and the rise in unregulated consumption of finite resources have a global impact. Furthermore, each country has developed its own targets to reach the global Agenda at local level (UN, 2015).

Governments play a crucial role in the mitigation and adaptation of climate change, they have power to impose rules, regulate, give guidelines, and they are also the most important actors when a disaster occurs, helping people in need, and applying the necessary resources, economics and personnel, including the use of the newest technologies and tools available. With all the development in technology in the last few years, especially towards Artificial Intelligence (AI), many resources are already being used to mitigate and adapt climate change, or to diminish the carbon footprint (Leal Filho et al., 2022; Nordgren, 2022; Yakubu, 2018).

Artificial Intelligence was defined as "the study of how to make computers do things at which, at the moment, people are better" (Rich, 1983). According to her, this field does not just aim to replicate human intelligence but seeks to understand and model it in a way that can be executed by a machine. The ultimate goal is to create systems capable of undertaking activities that typically demand human intelligence, such as speech recognition, language translation, visual perception and decision-making. These systems are designed to learn and adapt to new information or stimuli from their environment, much like the human mind. As AI continues to evolve, the boundary between tasks that only humans can do and those that computers can replicate becomes increasingly blurred.

Machine Learning (ML), a branch of AI, refers to the scientific exploration of algorithms and statistical models employed by computer systems to execute a particular task autonomously, without relying on explicit instructions. Instead, they rely on patterns and inference derived from data (Mitchell, 1997). In essence, ML allows computers to learn from and make decisions based on data. This learning process can be supervised, where the algorithm is trained on a predefined set of examples, or unsupervised, where the algorithm identifies patterns and relationships in the data without any prior labels. The goal of ML is to develop models that can generalize patterns from data to make predictions or decisions about new, unseen data (Alpaydin, 2020; El Naqa e Murphy, 2015; Mitchell, 1997).

AI and ML are increasingly being recognized as pivotal tools in addressing the multifaceted challenges posed by climate change. It's important to note that its use also affects greenhouse gas (GHG) emissions, emphasizing the need to align AI technology with climate change mitigation strategies. On this subject a study presented a systematic framework that evaluates the direct and indirect impacts of AI on GHG emissions. This framework emphasizes the alignment of AI technologies with climate change mitigation, highlighting the immediate and system-level changes induced by ML applications. This holistic understanding is crucial for assessing and forecasting impacts, ensuring that AI and ML are harnessed effectively to combat climate change (Kaack et al., 2022).

Moreover, the integration of AI and ML in climate research has the potential to revolutionize our understanding and response to global warming. A research underscores the importance of leveraging these technologies to gain deeper insights into climate patterns and anomalies, which can subsequently aid in the formulation of timely interventions (Gupta et al., 2022). Furthermore, the application of automated ML in climate change AI has been explored, with a focus on high-leverage applications such as climate modeling, wind power forecasting, and catalyst discovery (Tu et al., 2022). ML can improve the accuracy and efficiency of these models by dealing with the nonlinearities and complexity that are built into climate systems. This can lead to more accurate predictions (Chantry et al., 2021).These advancements highlight the transformative potential of AI and ML in enhancing our predictive capabilities and formulating effective strategies to mitigate the adverse effects of climate change.

One study pointed to the role of digitalization and the application of AI/ML in driving automation for low-carbon transitions within the built environment. The integration of these technologies has the potential to foster fresh prosperity from a knowledge-based growth standpoint, while also acknowledging the crucial assumptions, constraints, and risks involved (Manfren, 2023). The application of AI and ML in waste fire management, a significant contributor to environmental degradation, is explored in another study, according to Jakhar et al (2023), "these technologies can significantly improve the accuracy of identifying, monitoring, and ultimately mitigating waste fires, making them indispensable tools in the fight against this complex issue". Hojageldiyev (2019) discusses the potential of AI systems in the oil and gas industry, particularly in addressing environmental protection challenges. Presently, AI technologies provide a diverse range of decision-making models for the oil and gas sector. These models support environmental preservation efforts, climate change mitigation, biodiversity conservation, water security, waste reduction, and the maintenance of clean air.

Another study employed ML classifiers to detect and map the impacts of selective logging using Unmanned Aerial Vehicle (UAV) imagery. The study showed how ML can provide accurate and efficient post-logging impact analysis by finding individual stumps and forest gaps caused by logging activities. This can help with long-term forest management (Kamarulzaman et al., 2022). ML was also used to map evidence and attributions across 100,000 climate impact studies. Such comprehensive mapping can provide policymakers and researchers with a holistic view of climate impacts, enabling more informed decision-making (Callaghan et al., 2021).

In conclusion, the convergence of AI, ML, and other digital technologies—especially the ones that are open source—presents a promising avenue for devising and implementing effective strategies to mitigate the adverse effects of climate change by any government. As these technologies continue to evolve, their role in climate change mitigation is expected to become even more significant (Leal Filho et al., 2022).

In Brazil we have a Ministry for Science, Technology and Innovation (STI) that is responsible for the National policies of STI, coordinating with the governments of states and municipalities, with society and with federal government agencies. Over the years, several laws have been enacted to ensure the promotion and advancement of STI by the government. These laws mandate the allocation of a minimum percentage of the annual budget specifically for this sector, underscoring its significance in the government's strategic priorities. Such a dedicated focus on Science, Technology, and Innovation not only fosters a culture of research and development but also positions the nation at the forefront of global technological advancements. Within this domain, there lies a significant opportunity to harness the rapid advancements in AI and ML (Leite et al., 2021; Tenorio et al., 2017).

These technologies, with their transformative potential, can be crucial in addressing one of the most pressing challenges of our time: combating and mitigating the effects of climate change. By integrating AI and ML into climate change strategies, governments can enhance predictive modeling, optimize resource allocation, and develop more effective interventions, thereby demonstrating a proactive approach to a global concern (Reichstein et al., 2019; Rolnick et al., 2019).

In this chapter are presented some proposals addressing the central inquiry: How can governments leverage the capabilities of Artificial Intelligence, specifically through Machine Learning, to facilitate informed decision-making and formulate public policies tailored to adapt to and mitigate the impacts of climate change? The objective is not to exhaust the field, but to illustrate the potential applications of technologies for the mitigation and adaptation to climate change in a straightforward, efficient, and cost-effective manner. We will introduce ideas that can be further developed using open-source tools.

Methodology

The question that motivates the writing of this chapter seeks to update what is being done regarding how governments can leverage the capabilities of Artificial Intelligence, specifically through Machine Learning, to facilitate informed decision-making and formulate public policies tailored to deal with and mitigate the impacts of climate change. A convenience literature review was conducted, followed by a presentation of proposals based on the found documents.

The review was carried out by searching for research articles published in peer-reviewed academic journals, without excluding other literature sources, such as review articles, books, book chapters, programs, and projects. For the search, the filters applied were: English language and the time frame (2019–2023). The databases used were ScienceDirect, SpringerLink, and Web Of Science. The selected search terms that covered the highest number of publications related to the topic were: "Artificial Intelligence" AND "Climate Change" AND (Mitigation OR Adaptation) AND ("Policy Making" OR "Government").

It is worth noting that this work, inherently about the applications of AI and ML, has reciprocally employed these technologies in its conceptualization, in the execution of complementary semantic searches and the linguistic translation from Portuguese to English of this chapter.

Semantic searches were carried out using ChatGPT-4 (Generative Pre-trained Transformer 4) integrated with the PaperPile plugin, using the above-mentioned key-words set, and from the results amplifying the search on specific topics. This iteration of the Chat interface allows searches in consolidated databases. The search in Paperpile is designed to be comprehensive, pulling from a wide range of sources to provide the most relevant results. The outcomes of this method were juxtaposed with those obtained from traditional database searches, with the consolidated findings subsequently curated for inclusion in the portfolio, determined by convenience.

The Fig. 10.1 shows a schematic of the aforementioned methodological design.

Based on findings from the database exploration and documentary analysis, it is evident that while initiatives leveraging technology and innovation are underway, the potential of AI and ML as instruments for climate change mitigation remains underexplored. Therefore, this chapter suggests six distinct proposals that use the capabilities of AI, specifically through ML for both mitigation and adaptation to climate change, with the aim of expanding the field of application of these technologies in public policies.

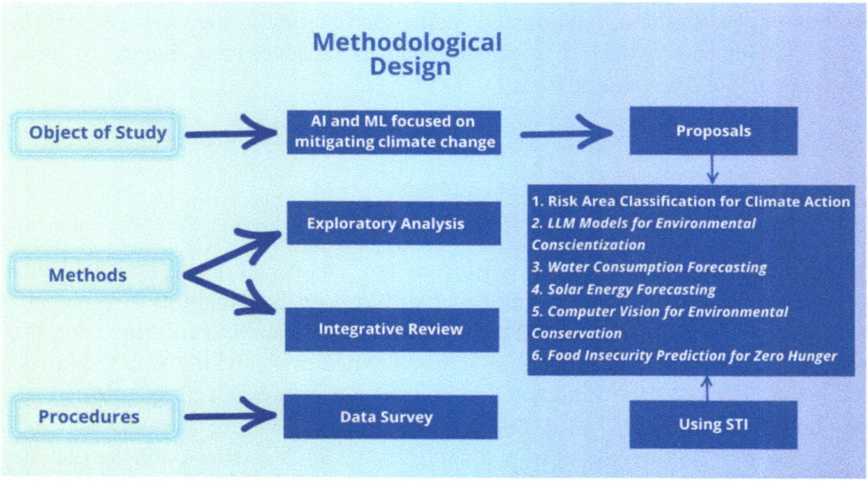

Fig. 10.1 Methodological design. *Source* Prepared by the authors (2023)

The Proposals

We developed proposals with the understanding that they could be seamlessly integrated into governmental public policies for mitigating and/or adapting to climate change. These proposals are particularly cost-effective, as we rely on open-source solutions that can be freely replicated.

The main focus was based on the Sustainable Development Goal 13 (SDG 13) that is Climate Action. However, our approach also intersects with other SDGs, given their interconnected nature. In the following section, we will introduce our ideas, provide concise descriptions, outline pathways for their realization, and suggest policy guidelines. We will bring references and similar studies, when possible.

The six proposals are Risk Area Classification for Climate Action, Large Language Models for Environmental Conscientization, Water Consumption Forecasting, Solar Energy Forecasting, Computer Vision for Environmental Conservation and Food Insecurity Prediction for Zero Hunger.

Risk Area Classification for Climate Action

Use machine learning algorithms to classify regions based on their vulnerability to climate change impacts.

Utilizing machine learning (ML) algorithms can significantly enhance our ability to classify regions based on their vulnerability to the impacts of climate change. By gathering comprehensive data on weather patterns, soil quality, vegetation, and human activities, we can train these algorithms to predict areas that are most susceptible to climate change risks. For example, machine learning has been used to figure out how vulnerable ecosystems are in the western Himalayas and how climate change is changing the way water flows (Ahlawat & Roy, 2023; Ivanov et al., 2022).

To ensure the effectiveness and accuracy of these predictive models, it's crucial to regularly update them with new data. Implementing preventive measures in identified high-risk areas can mitigate potential damages. Additionally, cooperation with local communities is crucial for the successful implementation of these measures. Such collaborations can leverage local knowledge and ensure that interventions are culturally and contextually appropriate.

From a policy perspective, these machine learning-driven insights can be invaluable. By identifying areas with a higher risk of climate change impacts, policymakers can prioritize interventions and allocate resources more effectively. This proactive approach not only safeguards vulnerable regions but also ensures that resources are used efficiently, maximizing the benefits of preventive measures and minimizing potential damages.

LLM Models for Environmental Conscientization

Use large language models (LLMs) to raise awareness about environmental issues.

LLMs have demonstrated significant potential in various domains, from generating code to assisting in biomedical research. Leveraging this potential, there's an opportunity to use LLMs to raise awareness about environmental issues, a pressing concern in today's world. By developing chatbots or virtual assistants powered by LLMs, users can have their environmental queries addressed in real-time. Such platforms can serve as interactive mediums, educating users about sustainable practices and the importance of environmental conservation (Kumar et al., 2023a, 2023b).

To further enhance the reach and impact of these LLM-driven tools, collaboration with educational institutions is paramount. Integrating these tools into curricula can provide students with a hands-on experience, allowing them to interact, question, and learn about environmental issues in an engaging manner. This not only fosters a sense of responsibility but also equips the younger generation with knowledge that's crucial for the planet's future (MacNeil et al., 2022).

However, the dynamic nature of environmental research necessitates that these LLMs be regularly updated with the latest findings and data. This ensures that the information provided is current and accurate. Moreover, as LLMs like ChatGSE have been used in biomedical applications, a similar approach can be adopted for environmental awareness, ensuring that the AI-readiness of environmental education is enhanced, making LLMs a valuable asset in promoting sustainability and environmental consciousness (Lobentanzer & Sáez-Rodríguez, 2023).

Water Consumption Forecasting

Predict future water consumption patterns using machine learning to ensure sustainable water management.

Predicting future water consumption patterns is crucial for sustainable water management. By leveraging machine learning techniques, we can forecast these patterns with higher accuracy and precision. The first step in this process involves collecting comprehensive data on historical water consumption, weather patterns, and population growth. Such data-driven approaches have been emphasized in recent studies, highlighting the importance of real-time data analysis platforms for water consumption forecasting (Boudhaouia & Wira, 2021).

Some machine learning models, like CEEMD-ARIMA and NARX neural networks, have been found to be good at making accurate predictions (Zhu & Zhang, 2021). For instance, the use of models like LSTM and BPNN has shown promising results in predicting water consumption without requiring prior knowledge about the data (Boudhaouia & Wira, 2021). Furthermore, federated learning has been introduced as a novel approach for water consumption prediction in smart cities, ensuring privacy and reducing data transmission overhead (El Hanjri et al., 2023).

Once high consumption periods are predicted, it becomes imperative to implement water-saving measures. This not only ensures efficient water usage but also aids in managing resources effectively. Additionally, educating the public based on forecasted data can play a crucial role in promoting water conservation. By integrating these strategies, we can ensure a more sustainable approach to water management, addressing both immediate and long-term challenges.

Solar Energy Forecasting

Predict solar energy production using deep learning techniques to optimize energy consumption and reduce reliance on non-renewable sources.

Harnessing solar energy efficiently requires accurate predictions of energy production, and deep learning techniques offer a promising approach to achieve this. To optimize energy consumption and reduce reliance on non-renewable sources, it's essential to predict solar energy production accurately. By gathering data on solar irradiance, weather conditions, and solar panel efficiency, we can create a robust dataset to train deep learning models. New developments in the field have shown that models like Long Short-Term Memory (LSTM) and hybrid deep learning methods, like those that combine Convolutional Neural Network (CNN), LSTM and Transformer, are good at accurately predicting energy needs (Al-Ali et al., 2023; Hari & Jisha, 2022; Ziyabari et al., 2022).

Adjusting energy consumption based on forecasted solar energy production is a strategic move. By doing so, we can ensure that energy is consumed efficiently, especially during periods when solar energy production is predicted to be high. Furthermore, these forecasts, when combined with prediction models of future scenarios consumption and trends in power generation using, for instance, the software developed by the Stockholm Environment Institute of Long-range Energy Alternatives Planning System (LEAP) can guide decisions about where to install new solar panels. Areas with high solar energy potential can be identified, and installations in these regions can be prioritized.

Promoting the use of solar energy also involves public awareness and education. By providing insights derived from deep learning models, we can educate the public about the benefits of solar energy and its potential to meet energy demands sustainably. Integrating with an analytical model like LEAP addresses key debates concerning transformative shifts within energy infrastructures, and enhances the understanding of how these interventions quantify impacts on national policies. This integration facilitates the creation of various scenarios based on defined socioeconomic contexts and foundational energy configurations (Andrade Guerra et al., 2015).

As more solar energy is added to power grids, accurate forecasting methods like the CNN-LSTM-Transformer model will be very important for making sure a smooth transition to a more sustainable energy future (Al-Ali et al., 2023).

Computer Vision for Environmental Conservation

Computer vision can be used to monitor and protect ecosystems, detect illegal activities like deforestation, and track wildlife.

The application of computer vision in environmental conservation has the potential to revolutionize the way we monitor and protect ecosystems. By deploying cameras and drones in vulnerable areas, we can capture real-time data that can be analyzed to detect anomalies or illegal activities, such as deforestation or poaching. For instance, studies have demonstrated the use of computer vision techniques for automated bird detection from drone footage, achieving high accuracy in identifying and tracking wild bird species (Mpouziotas et al., 2023).

Training computer vision models to recognize these anomalies is pivotal. By collecting data on various activities and wildlife movements, models can be trained to differentiate between normal and suspicious activities. For example, Kumar et al., (2023a, 2023b) suggested a system that could use deep learning models to identify and re-identify elephants. This system could be used for real-time monitoring to keep animals from going into certain areas. Another study uses computer vision for monitoring adequate areas for bush harvesting in Namibia (Marggraff & Venter, 2020).

Collaboration with local authorities is essential to ensuring a rapid response to detected threats. Real-time alerts can be sent to enforcement agencies, allowing for immediate action. Furthermore, the data collected can be invaluable in shaping conservation strategies and policies. For instance, the Smart Wildlife Sentinel system was designed to prevent wildlife-vehicle collisions and monitor road ecology using Internet of Things (IoT) and ML, emphasizing the potential of technology in conservation efforts (Alan Ma, 2022). Computer vision is also being used to identify sea turtles, as they are an endangered species, and contribute to protection projects (Attal & Direkoğlu, 2019).

Computer vision, combined with other technologies, can play a transformative role in ecosystem protection. By integrating these tools into conservation strategies, we can ensure more effective monitoring and protection of our planet's precious ecosystems.

Food Insecurity Prediction for Zero Hunger

Utilize machine learning algorithms to measure and predict food insecurity, targeting interventions to ensure zero hunger.

Utilizing ML algorithms to measure and predict food insecurity can play a pivotal role in achieving the global goal of zero hunger. By analyzing data on food production, consumption, and socio-economic indicators, it is possible to gain insights into regions that are most vulnerable to food shortages. For instance, a study by Dharmawan et al. (2022) utilized machine learning algorithms on a food insecurity dataset, highlighting the importance of such data-driven approaches in achieving zero hunger. The study emphasized the effectiveness of algorithms like Random Forest in predicting food insecurity based on various household characteristics.

Once regions at risk of food insecurity are identified using machine learning models, targeted interventions can be implemented. These interventions, informed by predictive analytics, can be more effective in addressing the root causes of food insecurity in specific regions. Collaborating with local authorities ensures that these interventions are tailored to the unique needs of each community and can be implemented rapidly in response to detected threats.

In this proposal ML can also be used to improve agriculture, a sector particularly vulnerable to climate change. A study demonstrated the use of ML in classifying high-resolution nanosatellite images to monitor agricultural practices in the Al-Qassim region of Saudi Arabia. Such monitoring can provide insights into the interplay between climatic parameters, water requirements, and vegetation, leading to more sustainable agricultural practices (Haq, 2022).

A continuous feedback loop is essential to ensuring the effectiveness of these interventions. By updating the ML model based on intervention outcomes and changing data, we can refine our predictions and strategies over time. This iterative approach, combined with the power of ML, can pave the way for more informed and effective policies aimed at ensuring food security for all.

Proposals resume and link to the SDGs:

Proposal	Risk area classification	Environmental education	Water forecasting	Solar energy forecasting	Computer vision conservation	Food insecurity prediction
How to achieve	Gather data on weather patterns, soil quality, vegetation, and human activities; Train ML models to predict areas most at risk	Develop chatbots or virtual assistants to answer environmental queries; Create interactive platforms for users to learn about sustainable practices	Collect data on historical water consumption, weather patterns, and population growth; Use models for accurate forecasting; Implement water-saving measures during predicted high-consumption periods	Gather data on solar irradiance, weather conditions, and solar panel efficiency; Use models for accurate energy forecasting; Adjust energy consumption	Deploy cameras and drones in vulnerable areas; Train computer vision models to detect anomalies or illegal activities; Collaborate with local authorities for rapid response to detected threats	Analyze data on food production, consumption, and socio-economic indicators; Use machine learning algorithms to predict regions at risk of food insecurity, and to identify the best agriculture areas
	Regularly update the model with new data for improved accuracy					
	Collaborate with local communities for effective implementation					
Policies	Implement preventive measures in high-risk areas	Collaborate with educational institutions to integrate these tools into curricula	Educate the public about water conservation based on forecasted data	Encourage the installation of solar panels in areas with high solar energy	Use the data to inform conservation strategies and policies	Implement policies and interventions to combat food insecurity
SDGs	13 Climate Action; 17 Partnerships for the Goals	13 Climate Action; 17 Partnerships for the Goals	6 Clean Water and Sanitation; 13 Climate Action; 17 Partnerships for the Goals	7 Affordable and Clean Energy; 13 Climate Action; 17 Partnerships for the Goals	13 Climate Action; 14 Life Below Water; 15 Life on Land; 17 Partnerships for the Goals	2 Zero Hunger; 13 Climate Action; 17 Partnerships for the Goals

Conclusion

The escalating challenges posed by climate change, as underscored by the IPCC's sixth report, necessitate innovative and proactive approaches to both mitigation and adaptation. The integration of AI and ML into climate change strategies offers a promising avenue to address these challenges. As demonstrated in this chapter, the application of AI and ML spans a wide range of areas, from predicting water consumption patterns and solar energy production to leveraging computer vision for environmental conservation and predicting food insecurity. These technologies, when harnessed effectively, can significantly enhance our predictive capabilities, optimize resource allocation, and facilitate the formulation of timely and effective interventions.

Furthermore, the role of governments in leveraging these technologies cannot be understated. As primary actors in disaster response and policy formulation, governments have the unique capability to integrate AI and ML into broader climate change strategies, ensuring a holistic and coordinated approach.

In essence, the convergence of AI, ML, and digital technologies presents a transformative potential in our collective endeavor to combat the adverse effects of climate change. As these technologies continue to evolve, it is imperative for governments, researchers, and policymakers to collaborate, ensuring that the benefits of AI and ML are maximized while aligning with global sustainability goals. The proposals presented in this paper, grounded in the latest research and technological advancements, serve as a testament to the potential of AI and ML in shaping a sustainable and resilient future. As we move forward, it is crucial to continuously refine these strategies, ensuring that they remain adaptive and responsive to the ever-evolving challenges of climate change.

Comparative studies across various national implementations of AI could shed light on best practices and policy impacts. Concurrently, the environmental and ethical implications of AI itself demand scrutiny, especially concerning the carbon footprint of AI operations and the potential biases within AI systems when applied to climate data. The integration of AI into climate policies, both at the national and international levels, offers another rich vein for exploration, examining how AI can bolster the enforcement of environmental regulations and support the achievement of the SDGs.

In the realm of public engagement, the role of AI in disaster response and resource management is pivotal, offering a potential field for studies focusing on AI's contribution to improving disaster response times and optimizing the management of finite resources like water and energy. Additionally, research into AI's role in renewable energy, particularly in energy forecasting and storage, alongside its use in precision agriculture, could provide insights into building resilience against climate change impacts. These research avenues not only promise to expand the capabilities of AI in addressing climate challenges but also offer a pathway to harnessing AI for sustainable development and a more resilient future.

Acknowledgements This study was conducted by the Center for Sustainable Development (Greens), the Graduate Program in Administration (PPGA) from the University of Southern Santa Catarina (Unisul) and Ânima Institute, with support of the Coordination of Superior Level Staff Improvement (CAPES) and the National Council for Scientific and Technological Development (CNPq).

References

Ahlawat, A., & Roy, A. (2023). Assessing the vulnerability of the western Himalayan ecosystem to climate change using machine learning algorithms. *Proceedings of SPIE 12727, Remote Sensing for Agriculture, Ecosystems, and Hydrology, XXV*, 127270E. https://doi.org/10.1117/12.2680118

Al-Ali, E. M., Hajji, Y., Said, Y., Hleili, M., Alanzi, A. M., Laatar, A. H., & Atri, M. (2023). Solar energy production forecasting based on a hybrid CNN-LSTM-transformer model. *Mathematics, 11*, 676. https://doi.org/10.3390/math11030676

Alan Ma. (2022). Smart Wildlife Sentinel (SWS): Preventing wildlife-vehicle collisions and monitoring road ecology with embedded IoT systems and machine learning. In IEEE MIT Undergraduate Research Technology Conference (URTC), Cambridge, MA, USA, pp. 1–4, https://doi.org/10.1109/URTC56832.2022.10002174

Alpaydin, E. (2020). *Introduction to machine learning* (4th ed.). MIT Press.

Andrade Guerra, J. B. S. O. D., Dutra, L., Schwinden, N. B. C., Andrade, S. F. D. (2015). Future scenarios and trends in energy generation in Brazil: Supply and demand and mitigation forecasts. *Journal of Cleaner Production, 103*, 197–210. ISSN 0959–6526. https://doi.org/10.1016/j.jclepro.2014.09.082

Attal, Z., & Direkoğlu, C. (2019). Sea Turtle species classification for environmental research and conservation. In R. Aliev, J. Kacprzyk, W. Pedrycz, M. Jamshidi, M. Babanli, F. Sadikoglu, (Eds.), 10th International Conference on Theory and Application of Soft Computing, Computing with Words and Perceptions—ICSCCW-2019. *Advances in Intelligent Systems and Computing* (vol. 1095). Springer, Cham. https://doi.org/10.1007/978-3-030-35249-3_74

Boudhaouia, A., & Wira, P. (2021). A real-time data analysis platform for short-term water consumption forecasting with machine learning. *Forecasting, 3*(4), 682–694. https://doi.org/10.3390/forecast3040042

Callaghan, M., Schleussner, C., Nath, S., Lejeune, Q., Knutson, T., Reichstein, M., Hansen, G., Theokritoff, E., Andrijevic, M., Brecha, R., Hegarty, M., Jones, C., Lee, K., Lucas, A., van Maanen, N., Menke, I., Pfleiderer, P., Yesil, B., & Minx, J. (2021). Machine-learning-based evidence and attribution mapping of 100,000 climate impact studies. *Nature Clinical Practice Endocrinology & Metabolism, 11*, 966–972. https://doi.org/10.1038/s41558-021-01168-6

Chantry, M., Christensen, H., Dueben, P., & Palmer, T. (2021). Opportunities and challenges for machine learning in weather and climate modelling: Hard, medium and soft AI. *Philosophical Transactions of the Royal Society A: Mathematical, Physical and Engineering Sciences, 379*(2194). https://doi.org/10.1098/rsta.2020.0083

Dharmawan, H., Sartono, B., Kurnia, A., Hadi, A. F., & Ramadhani, E. (2022). A study of machine learning algorithms to measure the feature importance in class-imbalance data of food insecurity cases in Indonesia. *Communication in Mathametical Biology and Neuroscience* https://doi.org/10.28919/cmbn/7636

El Hanjri, M., Kabbaj, H., Kobbane, A., & Abouaomar, A. (2023). Federated learning for water consumption forecasting in smart cities. Arxiv 2301.13036. https://doi.org/10.48550/arXiv.2301.13036

El Naqa, I., & Murphy, M. J. (2015). *What is machine learning?* (pp. 3–11). Springer International Publishing.

Gupta, S., Kaila, A., & Gupta, P. (2022). Understanding climate change using AI/ML. In Proceedings of the International Conference on Innovative Computing & Communication (ICICC), Available at SSRN: https://ssrn.com/abstract=4029356 or https://doi.org/10.2139/ssrn.4029356

Haq, M. A. (2022). Planetscope nanosatellites image classification using machine learning. *Computer Systems Science and Engineering, 42*(3), 1031–1046. https://doi.org/10.32604/csse.2022.023221

Hari, N. G., & Jisha, G. (2022). Solar irradiance prediction using deep learning-based approaches. In 2022 IEEE Asia-Pacific Conference on Computer Science and Data Engineering (CSDE) (pp. 1–5). IEEE. https://doi.org/10.1109/CSDE56538.2022.10089282

Hojageldiyev, D. (2019). Artificial intelligence opportunities for environmental protection. Paper presented at the SPE Gas & Oil Technology Showcase and Conference, Dubai, UAE, October 2019. https://doi.org/10.2118/198616-ms

IPCC. (2022). Climate change 2022: Impacts, adaptation, and vulnerability. contribution of working Group II to the sixth assessment report of the intergovernmental Panel on climate change. In H.-O. Pörtner, D.C. Roberts, M. Tignor, E.S. Poloczanska, K. Mintenbeck, A. Alegría, M. Craig, S. Langsdorf, S. Löschke, V. Möller, A. Okem, B. Rama (Eds.), Cambridge University Press, p. 3056. https://doi.org/10.1017/9781009325844

Ivanov, A. M., Gorbarenko, A. V., Kireeva, M., & Povalishnikova, E. (2022). Identifying climate change impacts on hydrological behavior on large-scale with machine learning algorithms. *Geography, Environment, Sustainability, 15*(3), 80–87. https://doi.org/10.24057/2071-9388-2022-087

Jakhar, R., Samek, L., & Styszko, K. (2023). A Comprehensive study of the impact of waste fires on the environment and health. *Sustainability, 15*(19), 14241. https://doi.org/10.3390/su151914241

Kaack, L., Donti, P., Strubell, E., Kamiya, G., Creutzig, F., Rolnick, D., Neiswanger, W. (2022). Aligning artificial intelligence with climate change mitigation. *Nature Climate Change, 12*, 518–527. https://doi.org/10.1038/s41558-022-01377-7

Kamarulzaman, A. M. M., Wan Mohd Jaafar, W. S., Abdul Maulud, K. A., Saad, S. N. M., Omar, H., & Mohan, M. (2022). Integrated segmentation approach with machine learning classifier in detecting and mapping post selective logging impacts using UAV imagery. *Forests, 13*(1), 48. https://doi.org/10.3390/f13010048

Kumar, H., Wang, Y., Shi, J., Musabirov, I., Farb, N., & Williams, J. J. (2023). Exploring the use of large language models for improving the awareness of mindfulness. In CHI EA '23: Extended Abstracts of the 2023 CHI Conference on Human Factors in Computing Systems. April 2023 Article No 129, pp. 1–7. https://doi.org/10.1145/3544549.3585614

Kumar, L., Renuka, D., & Kumar, S. S. (2023). Computer vision based knowledge distillation model for animal classification and re-identification using Siamese Neural Network. *Journal of Intelligent & Fuzzy Systems, 44*(4), 5731–5743. https://doi.org/10.3233/JIFS-222672

Leal Filho, W., Wall, T., Mucova, S. A. R., et al. (2022). Deploying artificial intelligence for climate change adaptation. *Technological Forecasting and Social Change, 180*. https://doi.org/10.1016/j.techfore.2022.121662

Leite, A. C., Paiva, D. M. S., & Souza, J. S. (2021). Innovation public policies: Brief study on conceptual frameworks, challenges and their perspectives in Brazil. *Journal of Entrepreneurship and Management of Micro and Small Enterprises, 6*(2), 41–61.

Lobentanzer, S., & Sáez-Rodríguez, J. (2023). A Platform for the biomedical application of large language models. [q-bio.QM]. https://doi.org/10.48550/arXiv.2305.06488

MacNeil, S., Kim, J., Leinonen, J., Denny, P., Bernstein, S., Becker, B. A., Wermelinger, M., Hellas, A., Tran, A., Sarsa, S., Prather, J., & Kumar, V. (2022). The implications of large language models for CS teachers and students. In SIGCSE 2023: Proceedings of the 54th ACM Technical Symposium on Computer Science Education. V. 2. https://doi.org/10.1145/3545947.3573358

Manfren, M. (2023). Innovation and knowledge-based growth for low carbon transitions in the built environment. Challenges and open research questions. *TEMA, Technologies Engineering Materials Architecture, 9*(1). https://doi.org/10.30682/tema0901c

Marggraff, P., & Venter, M. (2020). Monitoring of Namibian encroacher bush using computer vision. *AgriEngineering, 2*(2), 213–225. https://doi.org/10.3390/agriengineering2020013

Mitchell, T. M. (1997). *Machine learning*. McGraw-Hill.

Mpouziotas, D., Karvelis, P., Tsoulos, I., & Stylios, C. (2023). Automated wildlife bird detection from drone footage using computer vision techniques. *Applied Sciences, 13*(13), 7787. https://doi.org/10.3390/app13137787

Nordgren, A. (2022). Artificial intelligence and climate change: Ethical issues. *Journal of Information, Communication and Ethics in Society*. https://doi.org/10.1108/jices-11-2021-0106

Reichstein, M., Camps-Valls, G., Stevens, B., Jung, M., Denzler, J., Carvalhais, N., & Prabhat. (2019). Deep learning and process understanding for data-driven Earth system science. *Nature, 566*(7743), 195–204. https://www.nature.com/articles/s41586-019-0912-1

Rich, E. (1983). *Artificial intelligence*. McGraw-Hill.

Rolnick, D., Donti, P. L., Kaack, L. H., Kochanski, K., Lacoste, A., Sankaran, K., & Bengio, Y. (2019). Tackling climate change with machine learning. ArXiv: 1906.05433. https://arxiv.org/abs/1906.05433

Tenorio, M., Mello, G. A., & Viana, A. L. (2017). Policies for fostering health science, technology and innovation in Brazil and the role of clinical research. *Ciência & Saúde Coletiva, 22*(5), 1441–1454. https://doi.org/10.1590/1413-81232017225.33342016

Tu, R., Roberts, N., Prasad, V., Nayak, S., Jain, P., Sala, F., & White, C. (2022). AutoML for climate change: a call to action. https://doi.org/10.48550/arXiv.2210.03324

UN. (2015). Transforming our world: The 2030 agenda for sustainable development. *Resolution Adopted by the General Assembly on 25 September 2015, 42809*, 1–13.

UN. (2022). UN Secretary-General António Guterres' video message on the launch of the third Intergovernmental Panel on Climate Change (IPCC) report, in New York. Available at: https://press.un.org/en/2022/sgsm21228.doc.htm

Yakubu, R. N. (2018). The role of local government in the adaptation and mitigation of climate change in the agricultural sector in the Northern Region of Ghana. *Journal of Agricultural Extension and Rural Development, 7*(2). https://doi.org/10.15640/jaes.v7n2a11

Zhu, Y., & Zhang, Z. (2021). A CEEMD-NARX Model of Sino-U.S. Bilateral Trade Forecast and the Empirical Analysis Based on Public Data. In ICAIIS 2021: 2nd International Conference on Artificial Intelligence and Information Systems. May 2021. Article No.: 50, pp. 1–5. https://doi.org/10.1145/3469213.3470250

Ziyabari, S., Du, L., & Biswas, S. K. (2022). Short-term solar irradiance forecasting based on self-attentive transformers. In 2022 IEEE Power & Energy Society General Meeting (PESGM) (pp. 1–5). IEEE.

Gisleine Aver is a PhD Student at Unisul, and researcher at Greens.

Valdemiro da Rocha Júnior is a PhD Student at Unisul, and researcher at Greens.

Jorge Gustavo Sandoval Simão is a PhD Student at PUC-PR.

Artur Marques is a Master Student at Unisul, and researcher at Greens.

Ana Regina de Aguiar Dutra is Professor at Universidade do Sul de Santa Catarina (UNISUL) and researcher at Greens. She has experience in the field of Production Engineering, with an emphasis on Ergonomics, working mainly on the following topics: ergonomics, sustainable and healthy products and production processes, and education. Professor Ana is part of the faculty of the Graduate Program in Environmental Sciences (PPGCA/UNISUL) and coordinates the project "Change the Climate: Assuring the Quality of Environmental Strategies in Latin-American Higher Education" of the Erasmus + Programme of the European Union. She is also part of the Graduate

Program in Administration/PPGA (master's and doctorate) with the theme Sustainable Organizations.

Leandro dos Santos Coelho is Professor at Universidade Federal do Paraná (UFPR) in Electrical Engineering and at Pontifícia Universidade Católica do Paraná (PUC-PR) in Mechatronics Engineering/ Production. He serves as a permanent professor of PPGEPS/PUCPR (Graduate Program in Production Engineering and Systems—master's and doctorate) and is also a collaborating professor of PGEE/UFPR (Graduate Program in Electrical Engineering—master's, Research Group in Operation and Planning of Electric Power Systems) and of UnB (Graduate Program in Mechatronic Systems of the University of Brasília—doctorate). He has also acted as a consultant and speaker at training events in artificial intelligence, machine learning, deep learning, forecasting, identification, optimization, and advanced control algorithms. Leandro is a member of the editorial board of the journals Engineering Applications of Artificial Intelligence (Elsevier), Expert Systems with Applications (Elsevier), International Journal of Innovative Computing and Applications, Journal of Engineering, The Scientific World Journal, Telecom (MDPI), International Transactions on Electrical Energy Systems (Wiley), Journal of Advances in Applied Computational Mathematics, and The Chinese Journal of Artificial Intelligence. Leandro is a member of the Technical Committee (TC) on Soft Computing of the IEEE Systems, Man and Cybernetics Society and Research Fellow at CNPQ/Brazil.

José Baltazar Salgueirinho Osório de Andrade Guerra is the Dean of the Graduate Program in Administration and holds the Chairs of Sustainable Development and Global Environmental Change at the University of Southern Santa Catarina, Unisul—Ânima Institute, Brazil. He is the Director and Founder of the Centre for Sustainable Development/Research Group in Energy Efficiency and Sustainability (Greens, Unisul). Professor Baltazar is a Research Fellow at the Cambridge Centre for Environment, Energy, and Natural Resource Governance, University of Cambridge, has over 25 years of experience in sustainable development research and boasts more than 200 publications to his credit.

Chapter 11
Towards Adopting Action Research as an Underpinning Methodology for Addressing Climate Change

Helena Kettleborough

Abstract This paper arises from the perspective of an educator, community activist, manager and policy maker and argues for a radical change in the methodologies used to teach and share ideas for climate change mitigation and adaption to 'bend down the curve' on greenhouse gas emissions. Humanity is destroying the safe operating space for itself in planet Earth's boundaries framework. Alongside science and technological advancements to mitigate further damage, learning in society needs curiosity, inquiry and an openness to community, planet and Cosmos. This chapter explores how action research can contribute to the development of such learning. The chapter calls for action research to be adopted as a transformational learning methodology, an orientation towards inquiry and as an underpinning methodology across science, technology and policy making.

Keywords Climate · Hope · Paradigms · Action · Research

Introduction

The "World Symposium on Climate Change Policies: Science and Technology in Support of Policymaking in Climate Change Mitigation and Adaptation", held online in Manchester, UK on 19th September 2023, brought together an international group of academics to explore how science and technology could help mitigation and adaption to the climate emergency. The Symposium sought to influence the debates on mitigation and adaption and contribute to the effectiveness of the process. This work remains of outstanding importance. Despite initiatives by nations, the current context in terms of the state of planet Earth is highly challenging and policy often remains just that and does not translate into practical action.

H. Kettleborough (✉)
Department of Strategy, Enterprise and Sustainability, Manchester Metropolitan University Business School, Manchester, UK
e-mail: h.kettleborough@mmu.ac.uk

W. Leal Filho et al. (eds.), *Climate Change Mitigation and Adaptation in Practice*, Climate Change Management, https://doi.org/10.1007/978-3-031-85217-6_11

Carbon dioxide in the atmosphere is at unprecedented levels. The monthly average emissions at the Mauna Loa monitoring station for September 2023 were 418.51 ppm, up from 415.91 ppm a year ago.[1] The UN General Secretary, Antonio Gutérres, concluded that "it is no longer global warming it is global boiling" or more graphically still "we have opened the gates of hell" (Gutérres, 2023a, b). Veteran climate advocate James Hansen writing with seventeen other scientists argues in an open access paper:

> Thus, under the present geopolitical approach to GHG [Greenhouse gas] emissions, global warming will exceed 1.5C in the 2020s and 2C before 2050. (Hansen et al., 2023: 1)[2]

Such temperature increases will impact massively on humans and on nature. Humans with resources may be able to adapt initially. Much of nature cannot survive or mitigate to the higher temperatures. Thus baby swifts were dying in Seville because their nest were too hot (Kassam, 2023) and bats dying in Australia in the extreme heat (Mao, 2019). Water scarcity increases in the Andalusia region of Spain (Cabot, 2023) while the continued melting of the West Antarctic ice sheet is likely to be inevitable, driving up sea levels even when emissions stop (Carrington, 2023). The Earth's natural support systems, on which humanity depends, are in dire trouble, with six of the nine planetary boundaries now exceeded, meaning that the 'Earth is well outside the safe operating space for humanity' (Richardson et al., 2023).

The need for wise and urgent action now

Cosmologist Carl Sagan observed that from the vantage point of the planet Saturn, planet Earth is 'a pale blue dot' in the overwhelming vastness of space and that everyone we know and love lives on this planet (Sagan, 1995). Sagan recommended that humanity learn from this viewpoint, seeing the futility of war and fighting over a corner of a pixel of light and learn to treat each other and the planet in a more kindly manner. Learning is one method which can help humanity to take care of the world and recognize how special it is.

Within the university field, Principles for Responsible Management Education's (PRME's) mission is to transform management education and develop the responsible decision-makers of tomorrow to advance sustainable development through the framework of six principles and the United Nations Sustainable Development Goals. The Federation of Eco Schols (FEE) brings environmental education to schools in over 80 nations. In establishing the Green Belt Movement in Kenya, Wangari Maathai fought for the principles of empowerment and action within communities (Maathai, 2008). Yet the Director of the Worldwatch Education Institute argues that "the gap between earth education and where schools are today is about as wide as the gap between current human civilization, current climate change policies and what science requires us to do to get us to a sustainable future." (Assadourian, 2017: 19). The Independent Review of the UK's Research, Development and Innovation Organizational Landscape found that:

---

[1] As of figures last updated on October 5th 2023: https://gml.noaa.gov/obop/mlo/

[2] They include in their analysis the cooling effect of aerosols and conclude that the 1970–2010 global warming rate of 0.18C per decade has increased to a post-2010 rate of at least 0.27 C per decade.

> The natural sciences are developing new technologies towards goals such as net zero, and these require insights from the social sciences to assess potential economic, psychological and social impacts on society, and the arts and humanities to ensure wide usage and acceptance (Nurse, 2023: 122).

This chapter argues that despite the extensive work by scientists and academics in support of policy making, not enough is being done within the learning side of the field. The chapter argues for the urgent development of an 'orientation towards inquiry' across and between all fields. Such an orientation towards inquiry needs to take place across individual research, across how researchers work together and across wider society. This can be translated into the "I, We and the Bigger We"[3] or first, second and third person action research (Reason & Bradbury, 2001 and 2008). The chapter identifies the use of the intellectual tradition of action research, not to replace other methodologies which researchers or teachers are using, but to be a dynamic part of the research world, an underpinning methodology alongside other methodologies being used in the search for urgent and wise action. The chapter is not arguing for an "ism" but for the creation of deep orientations towards inquiry within science, technology and teaching in the quest for a sustainable world.

The pivotal effects of using action research

In examining remarkable efforts to adopt renewable energy and to mitigate climate change in South Australia State, the New Scientist reporter pointed to the importance of political courage amongst politicians, in order to make progress (Klein, 2023). James Hansen, in examining the role of scientists in the climate emergency, argues of the dangers of scientific reticence and gradualism in respect to the challenges facing the planet (Hansen et al., 2023: 24). The author of this chapter argues that in order to make the gains that are needed to mitigate these effects—every tenth of a degree of global warming matters immensely—transformational learning across all of society is needed. Such transformational learning as offered by action research, enabling students and researchers to think differently (O'Mahoney, 2023). The author herself uses the transformational learning methods of action research in a variety of settings, in both community and work, in academic and NGOs, with small and large groups, in formal educational settings and in among the trees in a woodland and by a river. It is not possible in the space of this chapter, to examine all aspects of action research nor all the settings but the reader is referred to her book (Kettleborough, 2023) for an in-depth review. This chapter seeks to argue a case and to provide as much evidence as possible to the effectiveness of the method.

Growing from the work of Reason and Bradbury (2001 and 2008), Bradbury (2015) and Marshall et al. (2011), this chapter posits nine characteristics of action research which can help scientists, policy makers and educators take wise and effective action. Given space constraints, the chapter explores four of these characteristics and offers signposts to the others:

[3] This term was coined by Manchester Community Activist Jane Morris in a co-operative inquiry facilitated by the author in 2017.

1. **Extended ways of knowing** and how researchers can experience and know the world more holistically in order to take wise action.
2. **First person action research**, where the researcher learns to examine their own practice and values. Examples within the community and academia explore the first person practice of bearing witness to what is happening to the planet to deepen understanding and empathy.
3. **Second person action research** or learning together as equals is introduced and explored in more depth in one example. Co-operative inquiry enables all participants to work together as equals and three examples are briefly examined from community and teaching.
4. Finally, action research argues for a recognition of the **paradigms within which researchers work seeking to extend and broaden**. Heron and Reason (1997) argue for an emerging participatory paradigm. The author argues for developing wider holistic paradigms which can enable researchers to understand and appreciate that they are part of community, the planet and Cosmos. Stories from daily life are used to illustrate being part of the wider universe and a dynamic model offered for researchers to examine.

The chapter also offers a critique of the methodology and argues for further action and research to take the next steps to broaden knowledge of action research in the context of climate change.

In exploring each of these in turn a start can be made in the transformational learning needed to bring our planet Earth back within the safe operating limits for humanity.

The main body of the chapter therefore explores four of the characteristics of action research in turn, to illustrate how action research can be seen as an underpinning methodology to address the climate emergency. The four characteristics explored are: extended ways of knowing; first person action research; second person action research and developing wider holistic paradigms for action research. These characteristics are explored through the literature and stories from the author's own practice in developing action research in different communities. This is followed by some critiques and consideration of ethics and values. Finally, the chapter concludes with implications for next steps in research and collaboration to explore the value of action research and how it can be developed more extensively in a wide variety of settings.

Exploring Action Research as an Underpinning Methodology

1. Extended ways of knowing

This phrase was used by researcher John Heron in the 1990s to describe a different way of using and seeing knowledge. Traditionally, the most favoured form of knowledge is referred to by Heron as Propositional knowing, the scientific, logical, academic knowledge found in research experiments, policy and journal articles. Heron however, adds another three (1996:14), placing them in a hierarchy, but they can also be seen as a continuous circle, which can be played forward or backward, but always going in a continuous loop.

Experiential knowing is the lived knowing of daily life, our experience of the world in each day of our lives. In relating the story of first being introduced to action research through a master's course, the author wrote:

> The second week of the course took place…at a residential college…We were encouraged to spend time out of doors, sitting by the river, tuning into what the water was telling us, walking in the dark and listening to the sounds of nature (Kettleborough, 2023: 123).

Presentational knowing is the knowing of creativity, of art, journalling, drawing, music, story, colour, painting, cooking, gardening, singing, pottery, craft activities and making. Chris Seeley, for example, writes of how she used drawing to help her understand her research and developed it as a form of action research practice (Seeley & Reason, 2008; Seeley, 2011) and Satish Kumar bakes bread! (Kumar, 2019).

Propositional knowing is the knowing of research, policy and academia and is the most widely recognised and taught.

Practical knowing is the summation of all three above and then going out into the world and acting. Such knowledge is the knowledge of the activists, sometimes the spiritual activists as McIntosh and Carmichael (2016) describe, sometimes the tree planter or the neighbourhood community activist. Social entrepreneur and leader Jane Riddiford describes practical action with her inner city students as part of teaching about action research and the wider planet and Cosmos, using experiential, presentational and propositional knowledge (Riddiford, 2021).

In Heron's view, practical knowing is the most important as it takes all the other ways of knowing with it. For Heron and later for his co-researcher Peter Reason, the most important use of knowledge is not for its own sake but, in action research, its use in creating a flourishing world for humans and more-than-humans alike. Such extended ways of knowing can be added to the forms of research needed to undertake first and second person action research. In terms of the climate emergency, extended ways of knowing offers policy makers and scientists knowing of their own daily lives. An example of using propositional knowledge in practice comes from Manchester Metropolitan University, which has been number 1 to 3 in the People and Planet league for ten years now. In March and November 2022 and again in November 2023, in total over eighty community members, students and academic staff worked together to plant some two hundred trees, integrating their experiential, presentational and propositional knowledge into practical knowing.[4]

2. First person action research

Action research suggests that there are three types of research: first, second and third person. A definition of first person action research:

[4] See https://creativerusholme.c4cp.net/project/47-trees-for-sir-gerald-kaufmann/

> First-person action research involves the researcher thinking reflectively and deeply about his or her own practice, assumptions, values and methods, before, during and after taking action and conducting research. (Kettleborough, 2023: 131)

First person action research is therefore about examining deeply one's own values as a researcher, lecturer or policy maker. Judi Marshall, in thinking about first person action research, developed the idea of 'living life as inquiry' which is where all aspects of life come in for examination (Marshall, 1999: 155, 2016), whether at work, in the home, in relationships, recreation or civic engagement.

First person action research methods are many and varied. Marshall suggested a number which remain important (Gearty & Marshall, 2021; Marshall, 2001). The first are cycles of action and reflection. These might be the weekly cycles of teaching or researching within a fixed timescale. A further technique is 'inner and outer arcs of attention', where the student is aware of both their interior thoughts and the external stimuli. Students in a class for example might be paying attention to their teacher at the same time thinking 'I am really looking forward to my coffee break or scrolling on their phones.

Writing is a method of first person action research which the author has used over many decades. Different forms of writing as inquiry, include taking field or diary notes recording research undertaken (Richardson, 2000) or journalling. Over many years, the author has regularly journalled and subsequently examined them to find significant patterns and material to learn from herself or to create stories to share with others (Kettleborough, 2023: 157)—and is explored later in the chapter.

First Person Practice of Bearing Witness

Bearing witness as a practice started out with the author's children twenty-five years ago. In collecting cuttings for a Millennium scrap book on the animals, the children noticed that something was wrong with the animals, they either did not have a home, or were being poached or did not have enough to eat. The children stopped collecting once the Millennium arrived but the author continued, coming to call the practice 'Gaia's Graveyards'—a sustained practice of bearing witness. As the author wrote and reflected on the practice, it became part of her PhD and now a keystone within her book *Journey to Hopeful Futures*.

> …the practice of bearing witness instills in us empathy both for humans and for nature. (Kettleborough, 2023: 287)

There are of course many different ways to bear witness and each individual needs to find their own route. For the author, the practice developed into sharing cuttings on the floor in workshops, creating 'Pilgrimage for Gaia' workshops and creating art installations of different sorts.[5] "The Ark", a collection of cuttings from a 15 year period contained in a small chest, is used as a center piece for workshops to both reflect on what is happening and also to take action.[6] Different cuttings connect to different participants, from the plight of elephant families to the loss of curlews

[5] See https://c4cp.net/blog/project/pilgrimage-for-gaia/ for a fuller explanation.

[6] The Installation was created by artist Phil Barton: https://philbartonartist.c4cp.net/

Fig. 11.1 Gaia's Graveyards: The Ark, Barborough Kettleton, (2015). Exhibition View at the Whitaker Museum workshop, November 2023 (Photo: Stephen Johns)

and their song or from disappearing bees and insects to disappearing coral reef. At a recent workshop at the Day of Action to Remember Nature, in the Whitaker Museum, Rawtenstall, Lancashire, UK, both children and adults wrote postcards to their MPs, asking for them to act, to remember nature and to address climate change. The use of art installations is important as it allows participants to feel sad or upset about what they see in the cuttings (Fig. 11.1).

In an educational setting, a version of bearing witness are ice breaker exercises at the start of Tutorials. 'In the News', where the teacher asks students to scroll through the news feeds and identify any story which calls to them. Not to be overwhelmed by the news but to find out the social, economic, and environmental roots of the story and relate them back to the United Nations Sustainable Development Goals. A further adaptation is to ask students to search both positive and negative stories affecting the planet. One student recently found an inspiring story from the State of South Australia, which in 2003 only had 1% renewable energy and today has 73%, leading to a rich discussion about how this was achieved. Post graduate students can be asked to bear witness through finding a story a day for fourteen days in a row and to write about how they are feeling at the same time. Students are also invited to consider both social and ecological justice in the stories they follow.

First person action research also allows researchers, practitioners and students to know themselves at a more profound level:

> As we bear witness and pay deep attention, the struggles of species and nature become our struggles, the melting of the glaciers is the melting of us, the attacks and murders of rangers defending wildlife are attacks on all of us. We start to care deeply at a number of different levels… (Kettleborough, 2023: 289).

First person action research is also effective when it is used alongside second person action research.

3. Second person action research

> Second person action research involves "speaking and listening with others" (Torbert, 2001: 201) and researching together in mutually supportive ways, with an emphasis on participation. (Kettleborough, 2023: 31).

There are many examples of second person action research in group settings, and many different methodologies for understanding the research. It can be seen as a family of methods including action inquiry, action science, action learning, appreciative inquiry, participatory action research, co-operative inquiry, feminist action research and antiracist participatory action research (Kettleborough, 2023; 125). To the author, it is a question of exploring and identifying which method best suits both the researcher and the participants. Examples of using all these methods are given in the Handbooks of Action Research by Reason and Bradbury (2001 and 2008) and Bradbury (2015). Three examples are illustrated in this chapter to give an indication of the power of second person action research.

Co-operative inquiry emerged as a method to co-research together as equals, using cycles of action and reflection to delve deeper into the research question and learn together. Whilst there is often a lead facilitator to organise the group, within the group all are equals. Co-operative inquiry is fruitful and outlined in a number of exploratory articles, for example Riley and Reason (2015). Examples can be found in a range of academic, teaching and learning circles. In terms of teaching—the author used co-operative inquiry to examine creativity in teaching and learning at undergraduate and support service levels at a university. One memorable occasion involved teaching and learning in an ancient woodland on the then university campus in Crewe, UK (Wozniak et al., 2019). Whilst neither students nor staff were confident about what the trees leaves and moss could teach them, all found patterns and inspiration amongst the trees.

Another example of co-operative inquiry in university research involved staff, undergraduate students and community learning together. Eight staff and students co-wrote a book chapter exploring how transformational learning—using co-operative inquiry and nonhierarchical ways of researching together—could also translate into working in the community (Wadham et al., 2023). They explored together the ideas of an 'exploding university' and how the ideas learnt in the academic sustainability teaching units and Staff Student Sustainability Group could broaden out into the community. Three case studies were explored, involving an animal rescue project, tree planting in the local park and a Sustainability Festival (ibid: 2023). Subsequent co-operative research focused on creating new models of community/student/academic working together. Started during Covid 19, one teaching session was delivered during the pandemic. Owing to government regulations it was not possible to sit inside in the local community venue. Students, academic staff and local community members stood and sat together outside in nature, with the rain dripping off their umbrellas, debating how to grow carbon literate communities and encourage very different individuals to take action together. The result of this learning and other initiatives bringing the community into the University for workshops led to the creation of a model way of working: 'Transformational Learning, Leadership and Community'. The chapter outlining this model was co-written by ten staff and twelve students (Randles et al., 2023).

In terms of the planet Gaia, cooperative inquiry can help policy makers and scientists learn by listening, observing and touching nature to better understand how nature is living. The author participated in co-operative inquiry at Schumacher College exploring deep ecology with twenty four fellow students (Maughan & Reason, 2001). The same college, but two decades later, saw the launch of a global initiative to use co-operative inquiry to enable participants to learn to listen to rivers. This time in an online setting, students worked together in small learning groups to reflect and then spend individual time outside of the group sitting with their living water. Such is the effect of this learning that other initiatives have developed and the course itself continues to run (Kurio & Reason, 2022). The very great need for humans to learn to see themselves as an integral part of nature, rather than separate from or superior to nature is well suited to a learning methodology where all participants are equal.

Appreciative inquiry is another method of second person action research. This involves discerning the questions to be researched, discovering the good in what researchers are doing, dreaming better futures based on the good and then designing and destiny in making steps to achieve the dream. Appreciative inquiry works well in a variety of settings, including human resource management to help staff teams improve; community settings to build better networks and to help communities change direction and in research settings to identify changes needed or gaps in approach. For scientists and policy makers working around the climate emergency, it allows individuals to build from strength rather than from weakness or despair and to feel a sense of agency in their actions (Kettleborough, 2023; Stavos et al., 2018; Watkins et al., 2011). Appreciative inquiry can be added to the visionary notions of dreaming for sustainable futures, such as the Dream of the Earth (Berry, 1988).

Other second person action research methods can help in the creation of a better world. Wishing to understand further the needs of the labourers and rural workers in Colombia, Latin America and what would make their lives better, Orlando Fals-Border developed participatory action research as a means to bridge the divide between workers and academia and to help actively engage in their struggles for justice through knowledge and learning alongside participation and organisation (Fals-Borda, 1987, 2001).

4. Developing wider holistic paradigms from action research

One of the characteristics of action research identified by Peter Reason and Hilary Bradbury was exploring worldviews of the researcher, arguing for an emerging participatory paradigm (Reason & Bradbury, 2001 and 2008). In developing her own practice, the present author argues that such worldviews and paradigms can create radial new perspectives in research and policy making and lead to different forms of action in aid of Earth and the challenges facing the planet.

In particular, the author argues for a family of paradigms to help create hopeful futures. One such paradigm the author identifies as 'The Journey of the Universe'.[7] (Primack & Abrams, 2006, Swimme & Tucker, 2011). To illustrate this emerging worldview, the author shares in workshops the 1968 photograph taken when the Apolo 8 astronauts were emerging from far side of the moon which was grey and mottled beneath them. Suddenly through their window, they saw a bright blue and white shining orb and realised that they were seeing the Earth rising. It was the first time that humans had seen planet Earth from space. This incredible photograph deserves some reflective time. The viewer sees the Earth as a whole, as one, in the depth of space. As cosmologist Carl Sagan observed it is our home and our only home. In the context of first person action research, the photograph takes on a new dimension, in that it now includes feelings, values and priorities.

It can be argued that, today, humanity is locked into a number of dominant paradigms. The preparatory work for policy makers at COP 28 involves a great many different and competing paradigms. The scientific paradigm, which is explored as part of this book has led to many valuable developments for humanity. Others which are present in the world today, for example, consumer capitalism and unlimited growth on a finite planet, are destructive of the natural world, Earth's climate and our natural life support systems. Collectively, these paradigms are leading us, increasingly rapidly, to disaster.

The author argues that action research offers a way for individuals to examine how we see the world, worldviews, mindsets and paradigms. By using action research over time and through reflection and first and second person research practice, it is possible to broaden or widen the individual paradigms and make connections across and between them to include the living planet Gaia (Shiva, 2020) and an appreciation of the awesome Universe. This is a journey the author has been on leading her to create stories which illuminated how she could live into the planet, Cosmos, self and community within daily life. Figure 11.2 offers one imagining of the journey undertaken.[8]

How Can Action Research Help Researchers and Policy Makers Live into Different Worldviews?

For the author, this relates to extended ways of knowing. Research and policy makers can use their experiential and presentational knowing to live into another worldview. Using the first person practice of writing as inquiry the author offers two stories which indicate paradigm shifts in thinking, leading to the ability to see humanity as part of something bigger.

[7] The others included in the hopeful family of paradigms are: Gaia: Our Living Planet; Sacred Earth and Cosmos; Earth as Community; and Emerging Participatory Paradigm. They are explored with signposts to resources and creative learning exercises in Kettleborough (2023) Chapter 2.

[8] Other images of how the journey might look at given in chapter 6, Kettleborough (2023).

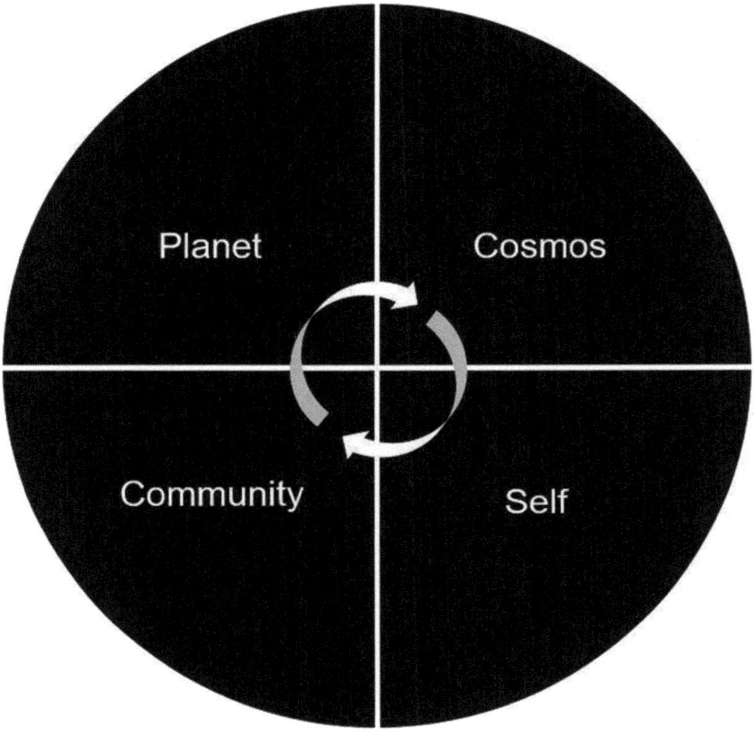

Fig. 11.2 Imagining our way into living in the Cosmos. Joining together Self, Community, Planet and Cosmos a first attempt (Kettleborough, 2023: 182)

Living in Deep Time: It involved a summer holiday day trip with the family in the beautiful scenery of the Wild Atlantic Way Ireland, up an ancient Mass path to the summit and there on the top of the mountain to eat lunch, taking in the beautiful view all down the Ferta valley to the sea. It was whilst idly watching the wildflowers and insects of this peaty, watery world at the top of the mountain, that the author found a piece of rock. Amazingly, the rock had etched on it in sandstone the same ripples found on the beach below. A visit to the local museum elucidated that this tiny piece of rock was likely to be approximately 320 million years old. The author began to sense the eons of time within which humanity is placed.

The Pale Blue Dot: At a fresher's week at a science orientated university where the author's daughter was registering for a physics degree she was given a photo of the Pale Blue Dot—a lovely replica of the famous Voyager 1 photo taken from Saturn looking back towards Earth, just visible as a single pixel almost hidden in Saturn's rings. On reflection, when turning away from the stall, her daughter turned back and asked for a copy as a gift for her mother. In studying the present, the author meditated over many years on our planet as one holistic, interconnected and precious whole.

There are also techniques to achieve this shift in educational and work settings. One example is the Deep Time Walk, where participants are enabled to walk the entire 13.8 billion year journey of the universe from the big bang to the present with stops for science and reflation on the way. Such walks are created for second person inquiry—the author's partner created one in the local park. The University of Winchester created a time walk for their students. There is also a Deep Time app to help walk the journey of the universe.[9] The NGO GreenSpirit has an excellent publication for walking the Time Walk to scale.[10] Such experiential and practical knowing can help researchers to see their work on the climate emergency as part of the history of the universe and as acquiring meaning and depth.

But there is a final part of this journey into wider paradigms which I cannot cover here, but which is essential. This is the wider paradigm of seeing our universe and Earth as sacred—in however this word is interpretated—and therefore worthy of the utmost respect and reverence. Such an attitude is beginning to be manifest in seeing that our rivers have legal rights alongside humans or that nature's creatures are equal citizens of our neighbourhoods. In broadening the parameters of the scientific research paradigm in this way, it becomes a reflective journey of loving and spiritual qualities in an awesome and wondrous Cosmos, new qualities suggest themselves for research exploration and questions, which can be explored using first and second person action research methods. Seeing the Earth as sacred belongs to indigenous and other cultures in the world but is often lost to the Western mindset. For example, Luetz and Leo (2021) explore approaches in spiritual contexts, whilst Luetz and Nunn (2023) explore approaches informed by Indigenous/Traditional worldview orientations. Satish Kumar, in his work at Schumacher College has worked tirelessly to bring the perspective of love back into the educational and research world (Rattanani et al., 2023).

In glimpsing and living into our living planet and an awesome cosmos, researchers and policy makers in the field of the climate emergency can find the spiritual and motivational energy to keep going year after year, to face disagreement, to learn together and to keep going. It will potentially give us the energy to fight against each tenth of a degree of global warming.

Critiques, Ethics and Values

Critiques of action research are explored with the three *Sage Handbooks of Action Research* by Reason and Bradbury (2001 and 2008) and Bradbury (2015) and within the *Action Research Journal*. Here the author wishes to concentrate on three.

The first relates to diversity and inclusion and that the action research literature is too much from the global north. This has been addressed in part by some of the

[9] See https://www.deeptimewalk.org/

[10] https://www.greenspirit.org.uk/blog3/2020/08/27/from-our-feet-to-our-heart-the-cosmic-walk-adventure-by-erna-colebrook/

contributors to the field over the years, for example, Fals Borda from Columbia (2006) and by articles in the *Action Research Journal,* such as the work of Taliep et al. (2023) using community asset mapping in their work to prevent interpersonal violence in South Africa. In order to further contribute to rectifying this, the action research community at the Association of Sustainability Practitioners (ASP) is working with others to produce resources which will be available online.[11]

A second critique is that the tools of action research are only for the privileged and not for the disadvantaged and marginalised. This critique in the UK relates to the fact that when much government funding was cut from 2011 for both local government and national government initiatives related to empowerment and disadvantage, most of the infrastructure supporting such communities disappeared. However, the ASP is seeking to create free resources for action research and *Action Research Journal* now has a range of articles which are available on open access. The Action Research Group Ireland offers reasonably priced attendance at an annual Symposium. And, in arguing for the adoption of action research in universities and research institutes the ideas can spread further.

The third is a specific critique of first person action research—that it is in some way self-indulgent or not for the wider good of the planet. Such a criticism implies that it is not rigorous or objective enough. However, first person action research requires and challenges the researcher to be aware of their values and ethics, of power relationships and the need to avoid hidden agendas on a continuous basis. It also holds researchers to account: are researchers acting for a flourishing world and being courageous enough, rather than thinking of the organisational benefits of their research? Are they continually reviewing and honest about their values and conduct? In a moving account, Sparkes recounts the initial hostility of his supervisor to his first person action research and how over a period of time, this changed (2002). In *Journey to Hopeful Futures: A Handbook* (2023), the author outlines how first person action research was a crucial element in the author's growing understanding that she was part of the wider planet and Cosmos as well has her human community and herself.

An exploration of the ethical issues connected with action research can be found in Mary Brandon-Miller (2008, 2015). It behoves anyone using first person and second person action research to be continuously aware of the ethical issues involved. It is particularly important that participation is not a tick box but is a genuine attempt to involve all participants as co-researchers (Casey et al., 2023).

Implications for research/practice/design and collaboration

There is much evidence as to the impact of action research on participants, groups and organisations. See, for example, thirty case studies of individuals in Marshall et al. (2011), accounts by disadvantaged young people in Riddiford (2021) and university students, academics and local communities in Randles et al. (2023) and Wadham et al. (2023).

[11] See https://sustainabilitypractitioners.org/action-research/

An October 2023 'In the News' story shared with students from the journal *Nature* gave humanity only until 2034 to cut carbon emissions to net zero if we are to hold global temperature to 1.5 degrees of warming above preindustrial levels. How can this be achieved?

Inspired by the work of Wangari Maathai, whose Green Belt movement has planted 52 million trees through grassroots activism, and from her experience in communities, organisations and academia over the past thirty years, the author suggests that society needs to plant seeds of participatory learning in every neighbourhood and organisation. The following steps for research and action are recommended for academics, public sector staff, businesspeople, faith communities and activists:

(a) Use appreciative practice methods to establish the strengths of individuals, organisations and community groups and to grow from such strengths.
(b) Plant trees of participatory learning in every neighbourhood, celebrating the achievements of local institutions, communities and individuals in the fight for our planet's survival.
(c) Explore the PRME and Civic University initiatives to create opportunities to introduce action research within all areas of research, learning and policy development.
(d) Identify local businesses, including social enterprises, SMEs and larger bodies with CSR and socially responsible business to join in partnerships to grow action research methods.
(e) Use research funding and centres for enterprise to establish collaborative alliances between University Business Schools, NGOs and voluntary organisations, public sector agencies such as the NHS and businesses practicing sustainability.
(f) Explore, share, debate the characteristics of action research as essential tools and part of transformational learning. Include within Reflective Learning modules and as part of assessment.
(g) Add the practice of bearing witness to what is happening to the planet and social justice in individuals and organisations to teaching and learning in as many different forms as possible such as journal entries, In the News, and time out sitting with nature
(h) Explore how action research can help create steps to the flourishing of human and more-than-human communities, working to bring social and ecological justice together with the UN SDGs
(i) Freely share resources through the creation of climate and book/online hubs in community centres, museums, libraries, art galleries and businesses
(j) Create joint alliances with organisations, businesses and NGOs to teach action research in accessible forms.

These research and action points are a win–win for communities, academics, scientists, students, NGOs, businesses public sector bodies—and planet Earth.

Conclusion

Using action research methodology can help policy makers, scientists and teachers understand the need for wise and urgent action within transformational learning methods and a wider hopeful family of paradigms. Action research methods can be used across disciplinary and organizational boundaries to underpin and extend current work to support policy making in climate change areas. The methodology can be used to help participants create broader paradigms arising out of faithful use of first and second person action research methods and the characteristics of action research which can help the researcher locate themselves in more hopeful mindsets. The consistent and faithful application of action research methods can help scientists and teachers to understand that our planet Gaia is living; that as humanity, human society is part of the long journey of the universe; that social and ecological justice are woven together and that the worldviews needed are interconnected, interdependent, holistic and joined up. Above all, first person action research can help us understand that the present situation facing the planet is an emergency and that time is of the essence. Second person action research can help scientists, teachers and individuals find ways to rapidly work together to save every tenth of a degree of global warming.

The implications for research practice and design and for collaboration with a wide range of stakeholders raised above have real potential to develop action research as an interdisciplinary underpinning methodology. This would enable the development of new and emerging worldviews and holistic solutions to 'global boiling'. Universities are still well-resourced powerhouses for change. Can universities work individually and together to develop forests of participatory learning for all inhabitants of Planet Earth to live together on our pale blue dot, our one and only home, far into the future (Sagan, 1995).

References

Assadourian, E. (2017). EarthEd: Rethinking Education on a Changing Planet. In *The Worldwatch Institute State of the World Earthwatch Series*, EarthEd Rethinging Education on a Changing Planet, (pp. 5–20). Washington Press.

Berry, T. (1988). *The dream of the earth*. Republished 2015. Counterpoint Press.

Bradbury, H. (Ed.). (2015). *The Sage handbook of action research*. Sage.

Bryon-Miller, M. (2008). Ethics and action research: Deepening our commitment to principles of social justice and redefining systems of democratic practice. In P. Reason & H. Bradbury (Eds.), *The Sage handbook of action research, participative inquiry and practice* (2nd ed., pp. 199–210). Sage.

Bryon-Miller, M., Aranda, A., & Stevens, D. (2015). Widening the circle: Ethical reflections in action research and the practice of structured ethical reflection. In H. Bradbury (Ed.), *Handbook of action research* (3rd ed., pp. 596–607). Sage.

Cabot, C. (2023). 'The country is becoming a desert' 13th May, France 24 https://www.france24.com/en/europe/20230513-the-country-is-becoming-a-desert-drought-struck-spain-is-running-out-of-water

Carrington, D. (2023). Rapid Ice Melt in west Antarctica now inevitable, research shows. *The Guardian*, 23rd October, https://www.theguardian.com/environment/2023/oct/23/rapid-ice-melt-in-west-antarctica-now-inevitable-research-shows?ref=lite.improvethenews.org
Casey, M., Coughlan, D., O'Carroll, A., & Stokes, D. (2023). Towards a check list for improving action research quality in healthcare contexts. *Systemic Practice and Action Research., 36*(6), 1–12.
Fals-Borda, O. (1987). The application of participatory action research in Latin America. *International Sociology., 2*(4), 329–347.
Fals-Borda, O. (2006). The North-South convergence: A 30-year first person assessment of PAR. *Action Research Journal, 4*(3), 351–358.
Fals-Borda, O. (2001). Participatory (action) research in social theory: Origins and challenges. In P. Reason, & H. Bradbury, (Eds.), *Handbook of action research* (Concise Paperback Edition, pp. 27–37). Sage.
Gearty, M., & Marshall, J. (2021). Living life as inquiry: A systemic practice for change agents. *Systemic Practice and Action Research., 4*, 441–462. https://doi.org/10.1007/s11213-020-09539-4
Guterres, A. (2023a). Secrertary-General's opening remarks at press conference on climate (United Nations, 27th July, 2023). https://www.un.org/sg/en/content/sg/speeches/2023-07-27/secrertary-generals-opening-remarks-press-conference-climate.
Guterres, A. (2023b). 'Humanity has opened the gates to hell' (United Nations, 20th September 2023). https://news.un.org/en/story/2023/09/1141082.
Hansen, J., Sat, M., Simons, L., et al. (2023). Global warming in the Pipeline. *Oxford Open Climate Change, 3*(1). https://doi.org/10.1093/oxfclm/kgad008
Heron, J., & Reason, P. (1997). A participatory inquiry paradigm. *Qualitative Inquiry, 3*(3), 274–294.
Kassam, A. (2023). 'They're being cooked': Baby swifts die leaving nests as heatwave hits Spain. *The Guardian*, 16th June: https://www.theguardian.com/world/2022/jun/16/spain-heatwave-baby-swifts-die-leaving-nest
Kettleborough, H. (2023). *Journey to hopeful futures: A Handbook*. Centre for Connected Practice.
Klein, A. (2023). How countries can go fossil free with wind and solar superpowers. *New Scientist*, 24th October https://www.newscientist.com/article/mg26034621-600-how-countries-can-go-fossil-fuel-free-with-wind-and-solar-superpowers/
Kumar, S. (2019). *Elegant simplicity, the art of living well*. New Society Publishers.
Kurio, J., & Reason, P. (2022). Voicing rivers through ontooetics: A co-operative Inquiry. *River Research and Applications, Special Issue, Voicing Rivers, 38*, 374–386.
Luetz, J., & Leo, R. (2021). Christianity, creation and the climate crisis: Ecological paradigms and perspectives. In J. Luetz, & P. Nunn (Eds.) *Beyond belief: Opportunities for faith engaged approaches to climate change in the Pacific Islands*. Springer. https://doi.org/10.1007/978-3-030-67602-5
Luetz, J., & Nunn, P. (2023). Spirituality and sustainable development: An entangled and neglected relationship. *Sustainable Science, 18*, 2035–2042. https://doi.org/10.1007/s11625-023-01347-8
Maathai, W. (2008). *Unbowed: One woman's story*. Arrow Books.
Mao, F. (2019). 'How one heatwave killed 'a third' of a bat species in Australia, BBC News, 15th January. https://www.bbc.co.uk/news/world-australia-46859000
Marshall, J. (1999). Living Life as Inquiry. *Systemic Practice and Action Research, 12*(2), 155–171.
Marshall, J. (2001). Self-reflective inquiry practices. In P. Reason & H. Bradbury (Eds.), *Handbook of action research* (pp. 335–342). Sage.
Marshall, J. (2016). *First person action research: Living life as inquiry*. Sage.
Marshall, J., Coleman, G., & Reason, P. (2011). *Leadership for sustainability: An action research approach*. Routledge (Paperback edition, 2023).
Maughan, E., & Reason, P. (2001). A co-operative Inquiry into deep ecology. *ReVision, 23*(4), 18–24.
McIntosh, A., & Carmichael, M. (2016). *Spiritual activism: Leadership as service*. Green Books.

Nurse, P. (2023). *The independent review of the UK's research, development and innovation organisational landscape, final report and recommendations.* Crown Copyright. Open Government Licence v. 3.0. https://www.gov.uk/government/publications/research-development-and-innovation-organisational-landscape-an-independent-review

O'Mahoney, C. (2023). The efficiency of interventions in reducing belief in conspiracy theories. A systematic review. *PlosOne, 5*(4),18. https://doi.org/10.1371/journal.pone.0280902

Primack, J., & Abrams, N. (2006). *The view from the centre of the universe: Discovering our extraordinary place in the cosmos.* Penguin.

Randles, S., Wadham, H., Skritsovali, K., Hart, C., Hoque, S., Kettleborough, H., Klapper, R, Marron, R., Taylor, D., Walley, L. With Clovis, A., Derbyshire, M., Devaney, M., Doig, G., Jones, D., Leigh, S., Louis, R., Margetson, M., Protopsaltis, T., Tucker, M., Twitchett, L., & Whiteley, C. (2023). Leveraging hope and experience: Towards an integrated model of transformative learning, community and leadership for sustainability action and change. In W. Purcell, & J. Haddock-Fraser (Eds.), *The bloomsbury handbook of sustainability in higher education.* Bloomsbury.

Rattanani, J., Iyengar, S., and Pearson, L. (2023). *Satish Kumar: Abundant Love.* Triarchy Press.

Reason, P., & Bradbury, H. (Eds.). (2001 and 2008). *Sage handbook of action research.* Sage

Richardson, L. (2000). Writing: A method of inquiry. In N. Denzin & Y. Lincoln (Eds.), *Handbook of qualitative research* (2nd ed., pp. 923–948). Sage.

Richardson, K, Steffen, W., Lucht, W., Bendtsen, S., et al. (2023). Earth beyond six of nine planetary boundaries. *Science Advances, 9.* Eadh2458.

Riddiford, J. (2021). *Learning to lead together: An ecological and community approach.* Routledge.

Riley, S., & Reason, P. (2015). Cooperative inquiry: An action research practice. In J. Smith (Ed.), *Qualitative psychology: A practical guide to research methods* (3rd ed., pp. 168–198). Sage.

Sagan, C. (1995). *Pale blue dot: A vision of the human future in space.* Random House.

Seeley, C. (2011). Uncharted territory: Imagining a stronger relationship between the arts and action research. *Action Research, 9*(1), 83–99.

Seeley, C., & Reason, R. (2008). Expressions of energy: An epistemology of presentational knowing in action research. In P. Liamputtong & J. Rumbold (Eds.), *Knowing differently: Arts based and collaborative research* (pp. 25–46). Nova Science Publishers.

Shiva, V. (2020). *Reclaiming the commons. Biodiversity, indigenous knowledge and the rights of mother earth.* Synergetic Press.

Sparkes, A. (2002). Autoethnography: Self-indulgence or something more? In A. Bochner & C. Ellis (Eds.), *Ethnographically speaking: Autoethnography, literature and aesthetics* (pp. 209–233). Altima Press.

Stavos, J., Tores, J., & Copperider, D. (2018). *Conversations worth having: Using appreciative inquiry to fuel productive and meaningful engagement.* Berrett-Kohler.

Swimme, B., & Tucker, M. (2011). *Journey of the Universe.* Yale University Press.

Taliep, N., Lazarus, S., Cochrane, J., Olivier, J., et al. (2023). Community asset mapping as an action research strategy for developing an interpersonal violence prevention programme in South Africa. *Action Research Journal., 21*(2), 175–197.

Tam, H. (Ed.). (2021). *Tomorrow's communities: Lessons from community transformation in the age of Global Crisis.* Bristol University Press.

Torbert, W. (2001). The practice of action inquiry. In P. Reason & H. Bradbury (Eds.), *Handbook of Action Research* (pp. 207–208). London: Sage.

Wadham, H., Hart, C., Hashmi, A., Kettleborough, H., Marron, R., Randles, S., Skritsovali, K., & Tucker, M. (2023). Sustainability education beyond the classroom: how the "exploding university" nurtures collective intelligence across local and global communities. In: *Handbook of research on implications of sustainable development in higher education* (pp. 202–229). IGI Global.

Watkins, J., Mohr, B., & Jelly, R. (2011). *Appreciative inquiry: Change at the speed of imagination.* Wiley.

Wozniak, M., Cobb, S. Kettleborough, H., & Leathlean, D. (2019). Working together collaboratively and in partnership to explore 'I (heart) learning' creative approaches to motivate and inspire students. *MMU Learning and Teaching in Action, 13*(1). https://ltiammu.files.wordpress.com/2021/05/12_wozniak_etal_working_together_i_love_learning.pdf

Helena Kettleborough is a Lecturer in Sustainability, the UN SDGs and Responsible Enterprise at Manchester Metropolitan University. With colleagues, she received three Awards this last academic year for work with the Staff Student Sustainability Group. Helena is a community activist in her local community in inner city Manchester, Rusholme. Helena is a Director of the Association of Sustainability Practitioners. Her first career was in local government, as a policy maker, manager and Head of Service for community development and neighbourhood regeneration in disadvantaged areas and Director of North West Together We Can and Neighbourhoods NW. In 2023 her book *Journeys to Hopeful Futures: a Handbook* was published.

Chapter 12
Towards Freezing Global Warming

Neil Phillips, Sarah M. Ayling, Norman M. Ratcliffe, Ben de Lacy Costello, William Foord, Thomas C. Draper, Anthony Plumridge, and Jona Razzaque

Abstract Numerous policies have been proposed to mitigate the problem of climate change as reducing emissions alone will not be sufficient. Carbon Capture and Storage (CCS), which captures carbon at the source, and Negative Emissions Technologies (NETs), which remove carbon from the atmosphere, are two of the most important strategies currently used to combat global warming. Existing technologies, however, are widely believed to be too expensive to implement on a global scale. The estimated total cost of climate change varies greatly, for example, \$178 trillion over the next 50 years. Economists have frequently emphasised the importance of minimising expenditure while meeting climate goals. According to new research, NETs may be more cost-effective and less disruptive than various forms of CCS. We anticipate that NETs based on plant growth and biomass freezing in seawater will be the most cost-effective option for capturing and storing 10 gigatonnes of carbon per year, with an operating cost in the order of \$50/t CO_2. To reverse global warming, policies at the national, transnational, and international levels will need to be overhauled.

Keywords Carbon capture · Sequestration · Composite ice · Antarctica · Terraforming

N. Phillips (✉)
College of Arts, Technology and Environment, University of the West of England, Bristol, UK
e-mail: neil.phillips@uwe.ac.uk

S. M. Ayling · W. Foord
School of Architecture and Environment, University of the West of England, Bristol, UK

N. M. Ratcliffe · B. de Lacy Costello · T. C. Draper
Institute of Bio-Sensing Technology, University of the West of England, Bristol, UK

N. M. Ratcliffe · B. de Lacy Costello
Centre for Research in Biosciences, University of the West of England, Bristol, UK

A. Plumridge
School of Accounting and Finance, College of Business & Law, University of the West of England, Bristol, UK

J. Razzaque
Bristol Law School, College of Business and Law, University of the West of England, Bristol, UK

W. Leal Filho et al. (eds.), *Climate Change Mitigation and Adaptation in Practice*, Climate Change Management, https://doi.org/10.1007/978-3-031-85217-6_12

Introduction

Our paper outlines the challenges of tackling global warming and proposes a novel solution that overcomes the financial barrier which prevents other proposals from being implemented. Following that, it provides a road map for reversing global warming as well as an update on the current state of project development. Finally, it focuses on the policy changes that will be necessary for full implementation.

Although the Earth's temperature fluctuates there is a clear trend of increasing temperature above pre-industrial levels over time (Allen et al., 2019; Guilyardi et al., 2018; IPCC, 2021). Scientific evidence indicates that human activities have warmed the Earth's surface, which in turn has impacted the Earth's climate. The potential for human activities, such as burning fossil fuels, to raise atmospheric carbon dioxide levels and thus global temperatures was recognised as long ago as 1896 by Arrhenius (Arrhenius, 1896) although he did not appreciate that there could be negative as well as positive effects of these changes.

The Intergovernmental Panel on Climate Change (IPCC), has stated that "Since systematic scientific assessments began in the 1970s, the influence of human activity on the warming of the climate system has evolved from theory to established fact." (Arias et al., 2021; IPCC, 2021). In July 2023 Antonio Guterres (UN General Secretary) stated "The era of global warming has ended; the era of global boiling has arrived. The air is unbreathable. The heat is unbearable, and the level of fossil-fuel profits and climate inaction is unacceptable. Leaders must lead. No more hesitancy. No more excuses. No more waiting for others to move first. There is simply no more time for that." (United Nations, 2023a).

This statement was prompted by a series of record climate events, including global average temperatures of 17.2 °C and Antarctic sea-ice extending 1.3 million km^2 below the previous lowest value (Met. Office, 2023a, 2023b). The impacts of climate change include both rising sea levels (Edwards et al., 2019) and more extreme weather events (Met. Office 2023b). As emphasised by UK Met. Office (2023c), even if all emissions of greenhouses gases stopped there will still be changes in the climate, but the sooner emissions are reduced the best chance humanity has of minimising the rate and extent of changes. The challenge is how to reduce emissions and slow or halt Global Warming without further disadvantaging those peoples who already lack access to the facilities considered vital for sustainable development (United Nations, 2023b).

192 nations agreed, under the Kyoto Protocol of 1997, to limit emissions of greenhouse gases and established legally binding targets for 37 industrialised countries and the EU (United Nations no date a and b). The Paris Agreement of 2015, which came into force in 2016, replaced the Kyoto Protocol and adopted a decentralised approach in which nations establish their own emission reduction targets (United Nations, no date c). Under the Paris Agreement, targets were set to slow and limit global warming to 1.5 °C above pre-industrial era temperatures, but progress is not fast enough and without dramatic reductions in levels of atmospheric greenhouse gases, climate scientists think there is a fifty percent chance that by 2040 global

warming will exceed 1.5 °C and reach 2.7 °C at the end of the century (UN Environment Programme (UNEP) 2020). The Paris Agreement is signed by 194 nations from the developed and developing world and establishes a financial mechanism aimed at delivering technical and capacity-building support to countries that exhibit heightened vulnerability to the consequences of climate change (United Nations no date c). In 2022, delegates at the COP27 talks agreed to create and put into action a loss and damage fund (United Nations, no date d). However, the idea of direct reparations is politically unpopular in some countries, especially following the economic impact of the Covid pandemic and Russian military operations in Ukraine. For example, in July 2023 the US special envoy on climate change, John Kerry, stated that the United States will not pay reparations to developing countries affected by climate-linked disasters (Reuters, 2023).

Numerous ideas to combat global warming have been proposed (Biello, 2007; Wildenborg et al., 2005). The foremost technologies can be divided into two groups as follows:

Carbon Capture and Storage (CCS) is the process of capturing or collecting carbon emissions from a significant source and then permanently storing them. For example, CCS from coal power stations and cement plants (Benson & Orr, 2008; Hunt et al., 2010). Although the capacity of CCS is increasing (e.g., from 73 million tonnes in 2020 to 111 million tonnes in 2021) (G.C. Institute 2022), it is still too small to have a major impact on global warming. Carbon capture, utilisation, and storage (CCUS) is the process by which captured CO_2 is used, rather than stored, in other industrial processes or even in the manufacture of consumer goods. The primary processes employed for the capture of CO_2 include post-combustion, pre-combustion, and oxy-fuel combustion:

- **Post-combustion technology** involves the utilisation of a chemical solvent, among other methods, to separate CO_2 from the flue gas after the combustion of fuel (Chao et al., 2021).
- **Pre-combustion techniques** encompass the conversion of the fuel into a gaseous blend comprising primarily hydrogen and carbon dioxide. The high concentration of CO_2 is easier to separate away, yielding a gas which is rich in hydrogen and can be utilised as a fuel with no further emissions (Olabi et al., 2022).
- **Oxy-fuel technology** entails the combustion of a fuel using a high concentration of oxygen, resulting in the production of CO_2 and steam. The emitted CO_2 is subsequently captured for further utilisation or storage (Yadav & Mondal, 2022).

Negative Emissions Technologies (NETs). Removal and sequestration of carbon dioxide from the atmosphere or enhancements of natural carbon sinks (Le Quéré et al., 2009; National Academies of Sciences, Engineering, and Medicine, 2019). The primary NETs can be broadly classified into two categories: synthetic technologies (◆) and natural (∞).

◆ **Sorbent materials** are utilised to capture CO_2 from the ambient air, which is drawn over them by fans (Azarabadi & Lackner, 2019; McQueen et al., 2021).

Recently, a Lewis acid–base interaction–derived hybrid sorbent with polyamine-Cu(II) complex that captures over 220 g of CO_2/kg (two to three times more than most direct air capture sorbents) has been reported (Chen et al., 2023).

◆ **Chemical adsorption** is a process that captures CO_2 molecules by passing air through a container containing an adsorbent material such as activated carbon or zeolite (Bezerra et al., 2011; Lu et al., 2008; Wang et al., 2021).

◆ **Liquids that react with CO_2 to form a solid compound**, such as monoethanolamine or potassium carbonate (Borhani et al., 2015; Grant et al., 2014; Oexmann et al., 2008).

◆ **Electrochemical reduction**, in which CO_2 is converted to carbon monoxide or other valuable chemical compounds (Moura de Salles Pupo & Kortlever, 2019; Renfrew et al., 2020; Sullivan et al., 2021).

◆ **Mineral carbonation**, a chemical reaction between CO_2 and metal oxides that results in stable carbonates (Neeraj & Yadav, 2020; Snæbjörnsdóttir et al., 2020; Thonemann et al., 2022).

◆ **Regenerable solid-supported materials** for CO_2 capture from air, such as amine tethered sorbents, resin materials that undergo humidity swing, and solid sorbents with CO_2 capturing organic groups grafted on their surface (Gelles et al., 2020; Shi et al., 2020; Varghese & Karanikolos, 2020).

∞ **Micro algae** sequestration, capture rate of 1.8 kg of CO_2 per kg of dry algal biomass formed (Baohua et al., 2020).

∞ **Bio-oil** sequestration involves converting waste biomass to bio-oil using processes such as fast pyrolysis, and then injecting the bio-oil into deep geological formations for permanent storage (Xiu & Shahbazi, 2012).

∞ **Biochar**, produced by pyrolysis of biomass, has been estimated to have the potential to reduce CO_2 carbon equivalent emissions by up to 1.8 billion tonnes per year, via a combination of increasing carbon stored in soil and replacing use of fossil fuels (Pant et al., 2023; Woolf et al., 2010)

∞ **Wood burial** entails burying large amounts of organic material, such as wood biomass, in specific locations where they can slowly decompose over time while emitting CO_2 at a reduced rate (Zeng, 2008).

Known technologies for carbon capture and storage are considered financially too expensive to be usable at a global scale (Baylin-Stern & Berghout, 2021). The dilute atmospheric CO_2 concentration (~400 ppm) makes low cost, high-capacity, direct air CO_2 capture challenging (Sabatino et al., 2021). It is predicted that the cost of current and anticipated direct air capture technologies will be $94 to $232 per tonne within 10 years (Keith et al., 2018); this is still too expensive for global implementation. Estimates vary for the global cost, for example $178 trillion over the next 50 years ("The turning point | Deloitte" 2022). Economists have consistently emphasised the importance of achieving climate goals at the lowest possible cost (Delbeke & Vis, 2019). Recent analyses found that NETs may be less expensive and less disruptive than reducing some emissions at source (National Academies of Sciences, Engineering, and Medicine, 2019). Photosynthesis by plants is the basis of many NETs; photosynthesis uses free, 'unlimited' energy from the sun to convert carbon dioxide in the atmosphere into stored carbohydrates. The UK government, amongst others, is encouraging landowners to plant trees and then to auction the 'credit' for the stored carbon on the carbon credits market (Forestry Commission, 2023). Although the growing trees will capture carbon there does not appear to be a clear end use for the timber, and once the trees reach the end of their life they will begin to decay and release carbon. Other plant-based carbon capture schemes, such as bio-oil (Xiu & Shahbazi, 2012) require energy-intensive steps to convert plant biomass into a more concentrated form of carbon, which can then be injected into suitable geological strata. Apart from the energy costs, it is unclear whether this will have any negative effects on aquifers. Growing trees, cutting them down, and burying them (returning the carbon extracted to the earth) is the most basic form of plant-based carbon capture (Zeng, 2008). The Social Value of an Offset (SVO) for a finite time can have considerable value. For example, it has been reported that timber storage for 500 years has 95% of the value of permanent storage (Groom & Venmans, 2023). The mass of carbon that plants can extract from the atmosphere and store depends on several factors including species, age, local environmental conditions and land-use type. Managed forests can store an average of ~335t per ha., traditional agroforests an average of ~145t per ha. and pastures an average of ~46t per ha (Kirby & Potvin, 2007).

We hypothesised that preventing the carbon stored in timber from re-entering the atmosphere by storing it in frozen seawater is the most affordable way of 'breaking' the carbon cycle, see Fig. 12.1.

Our novel solution combines three core principles and activities:

1. Cultivate as much biomass (specifically trees) as feasible, as photosynthesis represents the most cost-effective method for sequestering carbon from the atmosphere.

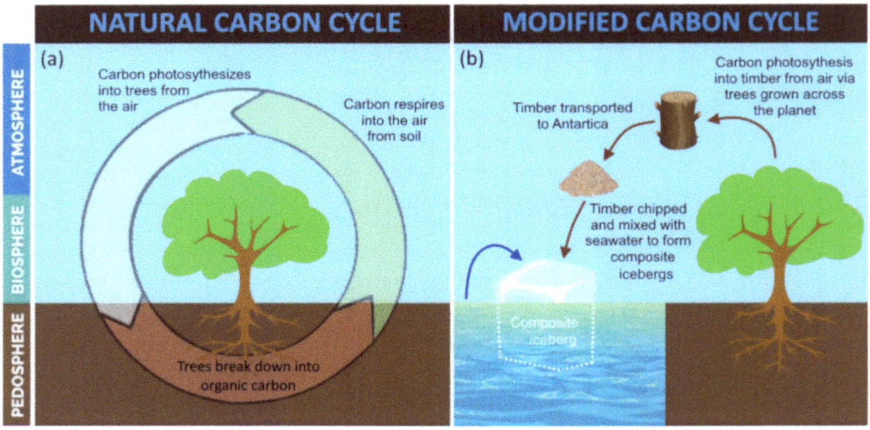

Fig. 12.1 Graphical representation of carbon cycles **a** natural carbon cycle with net zero change **b** modified carbon cycle to reduce carbon release into atmosphere. *Source*: authors

2. Ship the biomass (predominately timber) to suitable locations within the Antarctic Circle, where it is combined with seawater to create composite icebergs.
3. Transport composite icebergs to the Antarctic coast, where low temperatures prevent organic decay over geological timescales.

While there has been speculation regarding the storage of timber in Antarctica, such as the works of Holmes (2018) and Zeng (2008), to our knowledge no one has examined in detail how to implement the entire process, encompassing timber production and long-term storage on geological time scales. Additionally, we have successfully obtained Proof of Principle through an experimental testing programme that demonstrates a massive decrease of at least three-orders-of-magnitude in the rate of organic decay (Phillips et al. in preparation). The initial concept of amalgamating freshwater and wood chippings was formulated by Geoffrey Pyke in 1942 (Wikipedia, 2022). Today, the eponymously named Pykrete continues to be employed for constructing structures in areas with cold climates (Hani & Evirgen, 2022). Our research suggests that utilising seawater instead of freshwater for the formation of composite icebergs is a viable proposition. This is primarily due to the greater availability of seawater compared to freshwater, and our findings indicate that seawater exhibits enhanced efficacy in retarding organic decay. The utilisation of additive manufacturing enables the construction of 'functional' icebergs up to 400 m thick (utilising available ship building capability). This approach offers several advantages, including enhanced packing efficiency, improved durability of icebergs, and an ability to reflect sunlight back into space with snow-covered icebergs.

Obtaining Biomass

Biomass can be obtained in many ways; each with different economic, social, and environmental impacts. It is essential that production of plant material to capture carbon does not damage sensitive ecosystems (arboreal or other) or compete with food production.

Establishment of New Managed Forests

Managed forests have the highest potential to extract carbon from the atmosphere, averaging −335t per ha. (Kirby & Potvin, 2007). Only 7% of the 4 Bha. of forests worldwide are planted woodlands and plantations (World Resources Institute, 2023). Creating new managed forests is a simple way to generate environmentally and economically viable timber. In countries, such as the UK, that promote afforestation to combat climate change and have a carbon credit trading system (Forestry Commission, 2023) these policies could be expanded. Many existing tree planting schemes lack a clear end of life plan, particularly for the large volume of timber that will not be used for building construction or furniture manufacturing. Integration into composite icebergs would prevent this wood from decaying and thereby re-releasing the stored carbon. Tree planting at scale requires land use changes. In the UK, achieving net zero emissions by 2050 needs woodland coverage to increase from 13 to 17%. This expansion would require planting 1 million hectares of trees (Gambles, 2019). Planting trees on degraded agricultural land can improve soil quality, water regulation, and ecosystem services (The World Bank, 1991). Planting trees can protect established forests, link habitats, and increase biodiversity, but not as much as native forests (Wang et al., 2022). Strategic site selection will promote the positive environmental and social impacts of new woodland, and reduce the negative impacts associated with harvesting, processing and transporting timber. To deter exploitation of native and old-growth forests an effective and universally agreed monitoring system for timber production and provenance, with enforceable fines, is needed. At present timber supplied to sawmills exceeds reported timber production by 20%, and globally 60 million m^3 more timber is imported than exported (Fenning & Gershenzon, 2002). Landowners will need a legally binding agreement for timber provision from new managed forests with protective measures to protect farmers in the event of project failure.

Fruit Tree Plantations

There are 37 billion fruit trees in 2.2 million orchards worldwide (Catlin et al., 2018). If all of these are replaced at the end of their lives, 333 billion trees of various species will need to be planted within 100 years. Although many fruit plantations are monocultures, there are financial pressures on major growers to promote biodiversity, reduce agrochemical use, reduce GHG emissions, and increase sustainability (EIT-Climate no date). Many orchards are already accessible by road or rail, so collecting timber would not necessitate the construction of additional roads.

Waste Streams from Existing Timber Production

The manufacture of wood products generates a significant amount of wood waste. Up to 80% of timber in Nigeria is lost during logging operations (Dionco-Adetayo, 2001). Some timber waste is used for biomass power, which contributes to the use of renewable energy (Adhikari & Ozarska, 2018). Because there is already infrastructure in place, such as sawmills, plantations, and roads, the cost (both environmental and financial) of collecting this material is lower than it would be from new plantations. However, creating a large market for waste timber may discourage efforts to improve the efficiency of timber harvesting and use, as well as putting additional pressure on natural forests.

Utilisation of Reclaimed Timber

Germany generated around 11.9 million tonnes of wood waste in 2015 (Adhikari & Ozarska, 2018) which included wood packaging (21%), demolition and construction (26.7%), wood processing industry (14%), and municipal waste (20.7%). Some timber waste will have gone through extensive manufacturing and will require costly disassembly to extract the wood. Because the waste sources are dispersed, it is unlikely that they can be collected centrally. It would be preferable to reuse and repurpose these products locally, thereby reducing the emissions associated with their disposal. In the UK, 58.5% of waste wood was used as fuel in biomass burners in 2022, and 32% reused in other wood products (Wood Recyclers Association, 2023).

Timber from Private and Public Properties

Many people and businesses have trees on their properties that could contribute to the supply of timber. The financial cost and greenhouse gas emissions of harvesting individual or small groups of trees is a significant impediment to using this wood source. Government incentives could be used to encourage the harvesting of timber from parks and gardens, with the added benefit of encouraging urban tree planting. Increasing the number of trees in cities and suburbs has many benefits including reduced noise and visual pollution (Dwyer et al., 1991), reductions in the urban heat island, improved rainwater interception and biodiversity, and improved human mental health benefits (Woodland Trust, 2023).

Implementation

Selecting Land for New Managed Woodlands

Land on which new managed woodlands are created should be selected to cause the minimum negative impact and maximum positive impact. Areas of degraded farmland, areas of degraded woodland (particularly if adjacent to existing woodlands), and brownfield sites close to urban centres will all be considered. The knowledge held by local people can prevent the demolition of culturally or historically significant sites. Land surveys and resource studies can help identify crucial economic areas to avoid during planning. For the project to be successful, plantations must have access to seaports from which timber can be transported to the Antarctic. Any new roads should be positioned to create fire breaks within the woodland, thereby reducing the area lost from a potential wildfire. The location, scale and implementation of a tree plantation or managed woodland can affect socioeconomic well-being

(Bayle, 2019). Because they employ few people, tree plantations often have little direct socioeconomic impact. In sparsely populated areas managed woodlands can boost the local economy (Environment Agency, 2002). Active involvement of local people in the planning process can maximise benefits while minimising negative economic and social issues.

Ecological Impacts

Tree plantations, depending on factors such as type, management regime or location, have lower biodiversity than primary forests (Wang et al., 2022). Mixed plantations of native species have higher levels of biodiversity and provide more ecological benefits than monocultures of exotic species or intensive agricultural areas (Liu et al., 2018; Wang et al., 2022). While planted woodlands will never have the ecological diversity of natural forests, they can be sited to maximise biodiversity. Placed near natural forest edges and boundaries, planted woodland can buffer native forests and reduce the edge effect (Brockerhoff et al., 2008). Plantings can create wildlife corridors connecting primary forests and native ecosystems (Brockerhoff et al., 2008). Improved habitat links could help native species move and merge, improving gene flow and resilience (Correa Ayram et al., 2016).

Soil

Forestry can have a wide range of impacts on soil. Vehicles and heavy machinery used during site preparation and logging can cause soil erosion and compaction. Plantations, on the other hand, have the best soil stability of any type of agriculture (except natural forests) (The World Bank, 1991; Smith et al., 2016). Natural forests have higher levels of soil nutrients such as nitrogen, phosphorus, potassium, calcium, and magnesium than tree plantations, and similar soil pH (Zarafshar et al., 2020). When compared to natural forests and tree plantations, agricultural soils generally have the lowest levels of soil nutrients, and establishing tree plantations in agricultural soil will likely increase nutritional content (Zarafshar et al., 2020). Natural forests have higher soil carbon levels (up to 36% higher) than tree plantations, which have 60% more soil carbon than agricultural land (Zarafshar et al., 2020). Planting trees on degraded agricultural land will increase soil carbon storage. The amount of carbon stored in soil is affected by the type of trees planted as well as the cultivation method (Danise et al., 2021). Different tree species have different nutrient requirements, but careful species selection and the use of mixed plantings can help to avoid nutrient depletion issues. For example, leguminous and nitrogen-demanding species complement each other well (Sharma et al., 2022). Planting in sympathy with the site and with appropriate mixed species will help to mitigate erosional issues. As a result, if the timber used in this project comes from managed woodlands, particularly if those woodlands are planted on degraded agricultural land, not only will carbon be stored in the timber, but soil carbon levels (as well as soil stability and fertility) will increase.

Method of Tree Cultivation

All agricultural rotation cycles and harvesting methods have significant impacts. Fast-growing trees with short rotation cycles can deplete soil nutrients (World Bank, 1991). When harvesting equipment is used in short rotations, it can compact the soil. Preference for specific nutrients can result in soil nutrient imbalances (World Bank, 1991). Rotation lengths will be decided based on each species' annual carbon sequestration rate and published research (Kimberley et al., 2014; Zhou et al., 2017) to maximise carbon storage. The project will use a variety of trees and different rotation lengths to reduce any negative impact on soil nutrients and structure. Teak, for example, has a rotation length of 30–40 years compared to Neem (7–8 years); teak helps maintain soil structure while neem is harvested (Nanang et al., 1997). Coppicing can also help tree plantations survive and biodiversity thrive (Vanbeveren & Ceulemans, 2019).

Felling Trees

Clearing and log harvesting make plantations susceptible to soil erosion and instability. Removing vegetation can cause nutrient loss and increase leaching and soil disruption. No forest canopy will raise temperatures, killing soil organisms and drying the soil, making regeneration difficult. To reduce soil exposure, tree cover should be restored as soon as possible after clearing. The World Bank (1991) suggests mulching or planting faster-growing intermediate tree crops to protect exposed soils. Removal of forest cover reduces soil interception, infiltration, and water storage. Thus, rain-induced flash flooding and low flow can increase during dry months (Okullo & Muhoro, 2021; The World Bank, 1991). Suspended solids from increased water runoff and weakened soil structure can further disrupt waterways (Shah et al., 2022). Debris from felling can block nearby watercourses, increasing flood risk, and causing eutrophication (Baillie, 1996). Maintaining vegetation buffer zones around all waterways reduces runoff and sediment deposition (FAO no date; The World Bank, 1991). Planning site preparation and harvesting for drier months will reduce flash flooding and soil instability; additionally, long downhill open extraction routes should be avoided, to reduce quick runoff (Shah et al., 2022). Debris can be checked regularly in waterways during harvesting to prevent blockages. Stream sediment traps can settle sediment in runoff-prone areas.

Transportation of Timber

During the timber production life cycle, transportation is the primary source of greenhouse gas (GHG) emissions. Adhikari and Ozarska (2018) found that placing woodlands near waterways or the coast can reduce GHG emissions and other transportation impacts. River transport is one of the most GHG-efficient modes of transport available (EEA, 2023) and should be used whenever possible. Globally the shipping industry is committed to reducing GHG emissions (CE Delft 2023) with major companies, such as Maersk aiming for net zero emissions by 2040 (Maersk, 2022). Advances in the production of a new generation of wind-powered cargo ships (BBC, 2023) will further reduce GHG emissions and improve transport efficiency. Whilst the primary concern regarding maritime transportation is the environmental impact, collisions with marine mammals (as well as noise and chemical pollution) are possible. Furthermore, in a variety of climates, ballast water has been identified as a potential source of invasive species. This danger can be reduced by cleaning and filtering ballast water at ports before returning it to the ocean. Ship arrival planning and distribution of timber across multiple ports can both help to reduce the risk of harbour ecosystem degradation.

Chipping Timber

Sawmills have high energy and material requirements. GHG emissions can be reduced by using renewable energy sources such as hydroelectricity, wind power, and solar energy to power sawmills (Adhikari & Ozarska, 2018). Alternatively, waste wood can be used in carbon–neutral biomass boilers. Ash and point-source CO_2 from biomass fuel can also be combined to produce construction products, further reducing CO_2 emissions (Tripathi et al., 2019).

Composite Iceberg Formation

When the ambient temperature is below the freezing point of seawater (-1.8 °C), composite icebergs can be formed by combining chipped timber and seawater (in liquid state). Within the Antarctic Circle is the most advantageous geographic location for this endeavour. The region surrounding the Antarctic Peninsula is sufficiently cold to freeze seawater for most of the year without the more extreme lower temperatures of Antarctic mainland. Sound and light pollution are potential issues due to the constant production of composite icebergs. Light pollution can disorient or blind bird species, causing collisions with ships (Cabrera-Criz et al., 2018), while sound pollution has been found to affect Antarctic wildlife (Sordello et al., 2020). This, combined with the constant cycle of ships dropping off timber, will make it one of the most active marine areas in the world, posing a risk to wildlife. As a result, detailed screening and planning are required to determine the best locations for iceberg production. This must be done in areas where the environmental impact is minimal. Composite icebergs can be made more wildlife-friendly by including a gradual slope that aids penguins to reach the top from sea level.

Positioning of Composite Icebergs

The majority of Antarctic wildlife can be found along its coastlines and relies heavily on these areas for food and migration. It is difficult to predetermine how the positioning of composite icebergs off the coast of Antarctica will affect penguin populations, but some predictions can be made based on current sea ice trends. The use of composite icebergs may initially have a positive impact on penguin populations. As a result of climate change and declining sea ice cover, many penguin species are losing their breeding habitat on thick sea ice (which they require for 9 months of the year) (ASOC, 2022). Many emperor penguin colonies have been abandoned due to inconsistency in sea ice cover, so large icebergs of thick ice could be beneficial in mitigating population declines (ASOC, 2022). However, these positive effects may change as the icebergs accumulate, as penguin populations can suffer if there is too much sea ice (ASOC, 2022; Younger et al., 2015). Over time, monitoring (Fretwell et al., 2012) can be conducted to determine how the implementation of composite icebergs may affect penguin behaviour, and if negative impacts are found to be significant, specific areas important to penguin breeding should kept clear from composite icebergs.

Timeline

Figure 12.2 presents a project roadmap that demonstrates a positive outlook, encompassing essential activities, significant milestones, and anticipated deliverables. Proof of Principle has been achieved through an extensive period of experimental testing, the results of which will be shared in a forthcoming scientific publication (Phillips et al. in preparation). The scale demonstration is currently a Work-in-Progress (WIP), with discussions ongoing.

Financial Cost of Carbon Capture

The preliminary assessment of operational expenses does not fully account for the project's entire life cycle, as certain costs, such as seaport depreciation, require

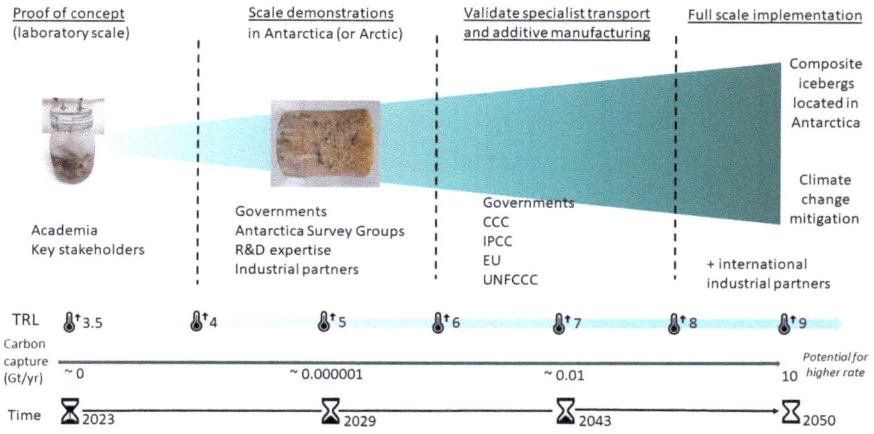

Fig. 12.2 Graphical representation of project roadmap to assist stakeholders in recognising interdependencies and critical path to full implementation. The project is currently at Technology Readiness Level (TRL) 3.5 with Proof of Concept demonstrated in the laboratory. *Source* authors

additional investigation to accurately determine. The three major activities are as follows:

(1) **Carbon removal from the atmosphere** via tree plantations has been considered in the range \$5 to \$100 per tonne of CO_2 (Austin et al., 2020). Which can be converted to \$18 to \$367 per tonne of carbon. Considering 10 gigatons (Gt) of carbon captured per year, an optimistic estimate of \$100 per tonne of carbon is suggested due to the large scale. Consequently, the total annual cost is projected to be around \$1000 billion.
(2) **Composite berg formation** will require approx. 40 manufacturing cells (with freezing rate of ca. 1 m/hr). Each additive manufacturing cell contains four ships equipped with onboard bulk handling gear, reinforced hulls of approx. 500 m length. A combined operating cost of \$40 billion per year.
(3) **Transportation of biomass** can be divided into three legs:
    (a) The transportation of biomass from plantations to chipping plants located on the Falkland Islands and/or southernmost part of South America. Approx. 46,000 cargo ships, each with a capacity of approx. 30 kt, incur an annual operating cost of approximately \$5 million per ship. Consequently, the total annual expenditure for operating these ships amounts to approx. \$650 billion.
    (b) Transportation of chipped biomass to manufacturing cells located in the vicinity of the Antarctic Peninsula. Around 5,000 shuttle cargo vessels, with reinforced hulls and 25 kt capacity. Annual operating cost ca. \$15 million (Solakivi et al., 2018) resulting in an estimated annual expenditure of approx. \$100 billion.

(c) Relocating composite icebergs to the periphery of Antarctica. Approx. 250 icebreakers/tugboats, with an estimated annual operating cost of around $100 million each, amount to a total expenditure of approx. $50 billion per year.

The total operating cost is ~$1840 billion per year, which can be expressed as ~ *$50 per tonne of carbon dioxide.*

The cost of carbon capture will vary depending on the efficiency of the shipping propulsion systems. Using current engines each cargo ship emits 5 g/t of CO_2 per km loaded and 1 g/t per km empty, assuming an average journey distance of ~8000 km from the plantation to the Antarctic Circle. Each cargo ship emits ~1440 tonnes of CO_2 on each round trip which requires an additional journey every 16 months to offset maritime transportation emission. Each round trip will take ~20 d assuming an average cargo ship speed of 15 kn when loaded and 22 kn when unloaded. Allowing for port time for loading, unloading, and vessel service/maintenance, this equates to approximately 15 return journeys per year. Overall, the project's efficiency is anticipated to further increase with the adoption of decarbonising technologies by the marine transport sector (BBC, 2023).

Communication and Dissemination

The primary goal is to effectively disseminate the project and its corresponding research outcomes to foster recognition and understanding among all stakeholders (including governments, the third sector, industry, academia, and people in general). The scope of the work needs to include the creation and implementation of visual identity and communication materials, as well as the coordination and execution of various events such as conferences, workshops, and community sessions. Additional resources (including publicly available computer-based simulations designed to model a variety of scenarios) should be employed to assist in public support of the project.

Legal Policies

To achieve full-scale implementation (which includes biomass production and transportation, as well as the fabrication and placement of composite icebergs), changes to legal policies at three distinct levels are required: national, transnational, and international.

National Level

Most nations have demonstrated their dedication to engaging in afforestation and reforestation endeavours aimed at promoting timber production. The commitments outlined in this context are informed by the United Nations Framework Convention on Climate Change (1992), the Paris Agreement (2015), and the Convention on Biological Diversity (1992). Moreover, there exist non-binding objectives that provide guidance for afforestation and reforestation endeavours on a national scale, including the Bonn Challenge (2011), the New York Declaration on Forests (2014), the Sustainable Development Goals (2015), and the United Nations Strategic Plans for Forests 2030 (2019).

Transnational Level

When biomass (predominately timber) is transported across national borders for the purpose of carbon capture, there are various transnational laws in place that govern the legality and sustainability of this activity. An example of a regulation that prohibits the introduction of unlawfully obtained timber and timber products into the European Union market is the EU Timber Regulation (2013). Australia, Japan, and the United States exhibit comparable instances of logging legislation.

Antarctica Treaty System

Antarctica encompasses designated regions that have been specifically safeguarded and regulated, resulting in their inaccessibility or imposition of particular restrictions. To gain access to specific regions, it may be necessary to obtain certain permits or specific authorization. An exemplification of this phenomenon is observable in the Antarctic Specially Protected Areas (ASPAs), which are characterised by the presence of officially designated management plans. Therefore, participation in activities within an ASPA requires the acquisition of additional permits and strict adherence to rigorous guidelines (Protocol on Environmental Protection to the Antarctic Treaty, 1991).

Research and Planning

All proposed activities must undergo thorough environmental impact assessments to assess their potential environmental effects and ensure their alignment with conservation goals and guidelines.

Exploratory Evaluation

Compliance with regulations, adherence to best practises, and the prompt implementation of corrective actions are critical. All activities must adhere to the Protocol on Environmental Protection, specifically Annex VI. Transportation, lodging, waste management, safety protocols, emergency response, and health requirements must all be implemented appropriately.

Further Considerations

Global Warming poses an existential threat to the biosphere. Halting and reversing climate change will require a decrease in overall net emissions, and implementation of methods to remove and capture carbon dioxide (and other greenhouse gases) from the atmosphere. It is imperative for humanity to store a greater amount of carbon than it releases annually. The Freezing Global Warming project presents a straightforward solution through the utilisation of natural processes, photosynthesis by plants and temperature variations across the planet.

There are no requirements for the development of groundbreaking technological advancements; rather, the implementation of a straightforward concept on a global level. All nations possessing a climate conducive to biomass cultivation have the potential to contribute to a collective global resolution. This enables developed nations, which are responsible for the highest carbon emissions, to contribute a 'fair' share without resorting to politically contentious measures such as direct reparations. Additionally, the project addresses several key issues highlighted in the Intergovernmental Panel on Climate Change (IPCC) report released in March 2023 (IPCC, 2023) to restore ecosystems, improve sustainable forest management, and reduce the cost of construction materials. The Freezing Global Warming proposal has the potential to provide a respite, allowing for the development and implementation of low-carbon energy sources such as commercial fusion (Mathew, 2022), and carbon-free systems (Phillips et al., 2022), which are crucial for ensuring the long-term sustainability of our planet.

The expansive nature of this undertaking will result in a plethora of diverse environmental ramifications, each possessing varying degrees of significance. However, by employing thorough strategic planning and enacting suitable policies, risks can be minimised. This assertion is particularly true during the initial stage of establishing new plantations, as the implementation of suitable methods and protocols can result in significant ecological benefits in terms of soil composition and the conservation of diverse biological life, if they are carried out with careful attention. There should also be a focus on the later stages of the project and the environmental consequences related to Antarctica, as a considerable portion of these impacts could be irreversible due to the continent's pristine condition.

Conclusions

To avert the forthcoming climate crisis, large-scale global action is needed to reduce net emissions. The Freezing Global Warming project presents the most efficient NET or CCS solution reported-to-date. By utilising nature's efficient carbon-capture process (tree growing) and locking the captured carbon away by perpetually freezing it, a surprisingly tractable and cost-effective system can be implemented. It is worthy of note that this solution does not rely on any future invention, as all required technologies already exist and are in widespread use today.

References

Adhikari, S., & Ozarska, B. (2018). Minimizing environmental impacts of timber products through the production process "From Sawmill to Final Products." *Environmental Systems Research, 7*(1), 1–15.

Allen, M., Antwi-Agyei, P., Aragon-Durand, F., Babiker, M., Bertoldi, P., Bind, M., Brown, S., Buckeridge, M., Camilloni, I., Cartwright, A., & Cramer, W. (2019). Technical Summary: Global warming of 1.5 C. An IPCC Special Report on the impacts of global warming of 1.5 C above pre-industrial levels and related global greenhouse gas emission pathways, in the context of strengthening the global response to the threat of climate change, sustainable development, and efforts to eradicate poverty. Intergovernmental Panel on Climate Change.

Antarctic and Southern Ocean Coalition (ASOC). (2022). Changing life: Sea ice. Available from: https://www.asoc.org/learn/changing-life-sea-ice/. Accessed 12 August 2023.

Arias, P., Bellouin, N., Coppola, E., Jones, R., Krinner, G., Marotzke, J., Naik, V., Palmer, M., Plattner, G. K., Rogelj, J., & Rojas, M. (2021). Climate change 2021: The physical science basis. In *Contribution of working group I to the sixth assessment report of the intergovernmental panel on climate change; technical summary.*

Arrhenius, S. (1896). On the influence of carbonic acid in the air upon the temperature of the ground. *The London, Edinburgh, and Dublin Philosophical Magazine and Journal of Science, 41*(251), 237–276. https://doi.org/10.1080/14786449608620846

Austin, K., Baker, J. S., Sohngen, B. L., Wade, C. M, Daigneault, A., Ohrel, S. B., Ragnauth, S., & Bean, A. (2020). The economic costs of planting, preserving, and managing the world's forests to mitigate climate change. *Nature Communications, 11*(1). https://doi.org/10.1038/s41467-020-19578-z

Azarabadi, H., & Lackner, K. S. (2019). A sorbent-focused techno-economic analysis of direct air capture. *Applied Energy, 250*, 959–975.

Baillie, B. (1996). Measuring the impacts of harvesting on waterways. *Report Logging Industry Research Organisation New Zealand, 21*(17). Available from: https://fgr.nz/documents/download/4428. Accessed 30 July 2023.

Baohua, Z., Shen, H., Li, Y., Liu, Q., Jin, G., Han, J., Zhao, Y., & Pan, K. (2020). Large-scale cultivation of spirulina for biological CO_2 mitigation in open raceway ponds using purified CO_2 from a coal chemical flue gas. *Frontiers in Bioengineering and Biotechnology, 7*, 441. https://doi.org/10.3389/fbioe.2019.00441

Bayle, G. K. (2019). Ecological and social impacts of eucalyptus tree plantation on the environment. *Journal of Biodiversity Conservation and Bioresource Management, 5*(1), 93–104.

Baylin-Stern, A., & Berghout, N. (2021). Is carbon capture too expensive? Available from: https://www.iea.org/commentaries/is-carbon-capture-too-expensive. Accessed 31 December 2022.

BBC. (2023). Pioneering wind-powered cargo ship sets sail. Technology 21 August. Available from: https://www.bbc.co.uk/news/technology-66543643. Accessed 21 August 2023.

Benson, S. M., & Orr, F. M. (2008). Carbon dioxide capture and storage. *MRS Bulletin, 33*(4), 303–305.

Bezerra, D. P., Oliveira, R. S., Vieira, R. S., Cavalcante, C. L., & Azevedo, D. C. (2011). Adsorption of CO_2 on nitrogen-enriched activated carbon and zeolite 13x. *Adsorption, 17*, 235–246.

Biello, D. (2007). 10 solutions for climate change. Available from: https://www.scientificamerican.com/article/10-solutions-for-climate-change/. Accessed 21 July 2023.

Borhani, T. N. G., Azarpour, A., Akbari, V., Alwi, S. R. W., & Manan, Z. A. (2015). CO_2 capture with potassium carbonate solutions: A state-of-the-art review. *International Journal of Greenhouse Gas Control, 41*, 142–162.

Brockerhoff, E. G., Jactel, H., Parrotta, J. A., Quine, C. P., & Sayer, J. (2008). Plantation forests and biodiversity: Oxymoron or opportunity? *Biodiversity and Conservation, 17*, 925–951.

Cabrera-Cruz, S. A., Smolinsky, J. A. & Buler, J. J. (2018). Light pollution is greatest within migration passage areas for nocturnally-migrating birds around the world. *Science Reports, 8*, 3261. https://doi.org/10.1038/s41598-018-21577-6

Caitlin, T., Shannon, L., & SageStar. (2018). Fruit Orchards Market Size—Global. Available from: https://askwonder.com/research/fruit-orchards-market-size-global-amhbwzlet. Accessed 20 August 2023.

Chao, C., Deng, Y., Dewil, R., Baeyens, J., & Fan, X. (2021). Post-combustion carbon capture. *Renewable & Sustainable Energy Reviews, 138*, 110490. https://doi.org/10.1016/j.rser.2020.110490

Chen, H., Dong, H. Shi, Z., & SenGupta, A. (2023). Direct air capture (DAC) and sequestration of CO_2: dramatic effect of coordinated Cu(II) onto a chelating weak base ion exchanger. *Science Advances, 9*(10). https://doi.org/10.1126/sciadv.adg1956

Correa Ayram, C. A., Mendoza, M. E., Etter, A., & Salicrup, D. R. P. (2016). Habitat connectivity in biodiversity conservation: A review of recent studies and applications. *Progress in Physical Geography: Earth and Environment, 40*(1), 7–37. https://doi.org/10.1177/0309133315598713

Danise, T., Andriuzzi, W. S., Battipaglia, G., Certini, G., Guggenberger, G., Innangi, M., Mastrolonardo, G., Niccoli, F., Pelleri, F., & Fioretto, A. (2021). Mixed-species plantation effects on soil biological and chemical quality and tree growth of a former agricultural land. *Forests, 12*(7), 842. https://www.mdpi.com/1999-4907/12/7/842

de Salles, M., Pupo, M., & Kortlever, R. (2019). Electrolyte effects on the electrochemical reduction of CO_2. *ChemPhysChem, 20*(22), 2926–2935.

Delbeke, J., & Vis, P. (Eds.). (2019). *Towards a climate-neutral Europe: curbing the trend* (p. 248). Routledge. https://doi.org/10.4324/9789276082569

Delft, CE-Delft. (2023). Shipping GHG emissions 2030: Analysis of the maximum technical abatement potential [prepared by Faber, J., van Seters, D., Scholten, P. Publication code: 23.230208.097]. Available from: https://www.transportenvironment.org/wp-content/uploads/2023/06/CE_Delft_230208_Shipping_GHG_emissions_2030_Def.pdf. Accessed 20 August 2023.

Dionco-Adetayo, E. A. (2001). Utilization of wood wastes in Nigeria: A feasibility overview. *Technovation, 21*(1), 55–60.

Dwyer, J. F., Schroeder, H. W., & Gobster, P. H. (1991). The significance of urban trees and forests: Toward a deeper understanding of values. *Journal of Arboriculture, 17*(10), 276–284.

Edwards, T. L., Brandon, M. A., Durand, G., Edwards, N. R., Golledge, N. R., Holden, P. B., Nias, I. J., Payne, A. J., Ritz, C., & Wernecke, A. (2019). Revisiting Antarctic ice loss due to marine ice-cliff instability. *Nature, 566*(7742), 58–64. https://doi.org/10.1038/s41586-019-0901-4

EIT-Climate-KIC. (no date). *Friendly fruit: Growing more sustainable fruit supply chains*. Available from: https://www.climate-kic.org/success-stories/friendly-fruit-sustainable-supply/. Accessed 21 August 2023.

Environment Agency. (2002). Scoping the environmental impacts of afforestation and deforestation projects. In *Handbook for scoping projects: Environmental impact assessment*. 4507 EAI B1 AW (publishing.service.gov.uk). Accessed 24 July 2023.

Environmental Protocol | Antarctic Treaty. (1991). Available at: https://www.ats.aq/e/protocol.html#:~:text=The%20Protocol%20on%20Environmental%20Protection%20to%20the%20Antarctic%20Treaty%20was,2. Accessed 29 August 2023.

European Environment Agency (EEA). (2023). Rail and waterborne - best for low-carbon motorised transport. Available from: https://www.eea.europa.eu/publications/rail-and-waterborne-transport. Accessed 24 July 2023.

Fenning, T., & Gershenzon, J. (2002). Where will the wood come from? Plantation forests and the role of biotechnology. *Trends in biotechnology, 20*, 291–296. https://doi.org/10.1016/S0167-7799(02)01983-2

Food and Agriculture Organization (FAO). (no date). 5. *Social issues in environmental assessment, training manual for environmental assessment in forestry*. Available from: https://www.fao.org/3/ac625e/ac625e00.htm. Accessed 29 July2023.

Forestry Commission. (2023). *Woodland grants and incentives overview table*. Available from: https://www.gov.uk/government/publications/woodland-grants-and-incentives-overview-table/woodland-grants-and-incentives-overview-table. Accessed 20 August 2023.

Fretwell, P. T., LaRue, M. A., Morin, P., Kooyman, G. L., Wienecke, B., Ratcliffe, N., Fox, A. J., Fleming, A. H., Porter, C., & Trathan, P. N. (2012). An emperor penguin population estimate: The first global, synoptic survey of a species from space. *PLOS ONE, 7*(4), e33751. https://doi.org/10.1371/journal.pone.0033751

G.C. Institute. (no date). CO_2 database. Available from: https://co2re.co/. Accessed 31 December 2022.

Gambles, I. (2019). *Trees for Zero? Forestry commission blog 29 October*. Available from: https://forestrycommission.blog.gov.uk/2019/10/29/trees-for-zero/. Accessed 20 August 2023.

Gelles, T., Lawson, S., Rownaghi, A. A., & Rezaei, F. (2020). Recent advances in development of amine functionalized adsorbents for CO_2 capture. *Adsorption, 26*, 5–50.

Grant, T., Anderson, C., & Hooper, B. (2014). Comparative life cycle assessment of potassium carbonate and monoethanolamine solvents for CO_2 capture from post combustion flue gases. *International Journal of Greenhouse Gas Control, 28*, 35–44.

Groom, B. & Venmans, F. (2023). The social value of offsets. *Nature, 619*(7971), 768–773. https://doi.org/10.1038/s41586-023-06153-x

Guilyardi, E., Lescarmontier, L., Matthews, R., Point, S. P., Rumjaun, A. B., Schlüpmann, J., & Wilgenbus, D. (2018). IPCC special report "global warming of 1.5 °C: Summary for Teachers.

Hani, M., & Evirgen, B. (2022). The mechanical and microstructural properties of artificially frozen Sawdust–Ice mixture (PykRetE) and its usability as a retaining structure. *International Journal of Civil Engineering, 21*(1), 119–134. https://doi.org/10.1007/s40999-022-00751-y

Holmes, J. W. (2018). Direct air carbon capture and sequestration using a novel biomass and cold storage process. *Social Science Research Network* [Preprint]. https://doi.org/10.2139/ssrn.3365616

Hunt, A. J., Sin, E. H., Marriott, R., & Clark, J. H. (2010). Generation, capture, and utilization of industrial carbon dioxide. *ChemSusChem: Chemistry & Sustainability Energy & Materials, 3*(3), 306–322.

IPCC. (2021). Climate change 2021: The physical science basis. In Masson-Delmotte, V., Zhai, P., Pirani, A., Connors, S. L., Péan, C., Berger, S., Caud, N., Chen, Y., Goldfarb, L., Gomis, M. I., Huang, M., Leitzell, K., Lonnoy, E., Matthews, J. B. R., Maycock, T. K., Waterfield, T., Yelekçi, O., Yu, R., & Zhou, B. (Eds.), *Contribution of working group I to the sixth assessment report of the intergovernmental panel on climate change*. Cambridge University Press. https://doi.org/10.1017/9781009157896

IPCC. (2023). *AR6 synthesis report: Climate change 2023*. Available from: https://www.ipcc.ch/report/sixth-assessment-report-cycle/. Accessed 24 August 2023.

Keith, D. W., Holmes, G., Angelo, D. S., & Heidel, K. (2018). A process for capturing CO_2 from the atmosphere. *Joule, 2*(8), 1573–1594.

Kimberley, M., Bergin, D., & Beets, P. (2014). Carbon sequestration by planted native trees and shrubs. *Planting and managing native trees: Technical handbook Article 10.5* (p 12). Tane's Tree Trust.

Kirby, K., & Potvin, C. (2007). Variation in carbon storage among tree species: Implications for the management of a small-scale carbon sink project. *Forest Ecology and Management, 246*, 208–221. https://doi.org/10.1016/J.FORECO.2007.03.072. https://www.forbes.com/sites/sergeiklebnikov/2019/10/24/stopping-global-warming-will-cost-50-trillion-morgan-stanley-report/#6b4929e151e2Le

Liu, C. L. C., Kuchma, O., & Krutovsky, K. V. (2018). Mixed-species versus monocultures in plantation forestry: Development, benefits, ecosystem services and perspectives for the future. *Global Ecology and conservation, 15*, e00419. https://doi.org/10.1016/j.gecco.2018.e00419

Lu, C., Bai, H., Wu, B., Su, F., & Hwang, J. F. (2008). Comparative study of CO_2 capture by carbon nanotubes, activated carbons, and zeolites. *Energy & Fuels, 22*(5), 3050–3056.

Maersk. (2022). Sustainability report. Available from: https://www.maersk.com/sustainability/our-esg-priorities/climate-change. Accessed 20 August 2023.

Mathew, M. D. (2022). Nuclear energy: A pathway towards mitigation of global warming. *Progress in Nuclear Energy, 143*, 104080. https://doi.org/10.1016/j.pnucene.2021.104080

McQueen, N., Gomes, K. V., McCormick, C., Blumanthal, K., Pisciotta, M., & Wilcox, J. (2021). A review of direct air capture (DAC): Scaling up commercial technologies and innovating for the future. *Progress in Energy, 3*(3), 032001.

Met Office. (2023b). Antarctic ice hits record low. Official Blog 30 June 2023. Available from: https://blog.metoffice.gov.uk/2023/06/30/antarctic-sea-ice-hits-record-low/. Accessed 9 August 2023.

Met Office. (2023c). Effects of climate change. Available from: https://www.metoffice.gov.uk/weather/climate-change/effects-of-climate-change. Accessed 9 August 2023.

Met Office. (2023a). Spotlight on evidence of a warming world. *Climate News,* 49. Available from: https://content.govdelivery.com/accounts/UKMETOFFICE/bulletins/3649101. Accessed 9 August 2023.

Nanang, D. M., Day, R. J., & Amaligo, J. N. (1997). Growth and yield of neem (*Azadirachta indica* A. Juss.) plantations in Northern Ghana. *The Commonwealth Forestry Review, 76*(2), 103–106.

National Academies of Sciences, Engineering, and Medicine. (2019). *Negative emissions technologies and reliable sequestration: A research agenda.* The National Academies Press. https://doi.org/10.17226/25259. Accessed 31 December 2022.

Neeraj, & Yadav, S. (2020). Carbon storage by mineral carbonation and industrial applications of CO_2. *Materials Science for Energy Technologies, 3*, 494–500.

Oexmann, J., Hensel, C., & Kather, A. (2008). Post-combustion CO_2-capture from coal-fired power plants: Preliminary evaluation of an integrated chemical absorption process with piperazine-promoted potassium carbonate. *International Journal of Greenhouse Gas Control, 2*(4), 539–552.

Okullo, J., & Muhoro, B. (2021). Environmental and social impact assessment: Study report for the proposed commercial plantation forest on Plot LR NO; Kibwezi/Kitengei "B"/137, "B" /1512, "B" /140, "B"/141 and "B"/158 in Kitengei area, Kibwezi, Makueni County. National Environment Management Authority (NEMA). Available from: https://nema.go.ke/images/Docs/EIA_19001909/EIAEIA%201903_SR%202391%20Plantation%20Forest%20(SUBATI)%20kibwezi.pdf. Accessed 18 August 2023.

Olabi, A. G., Obaideen, K., Elsaid, K., Wilberforce, T., Sayed, E. T., Maghrabie, H. M., & Abdelkareem, M. A. (2022). Assessment of the pre-combustion carbon capture contribution into sustainable development goals SDGs using novel indicators. *Renewable & Sustainable Energy Reviews, 153*, 111710. https://doi.org/10.1016/j.rser.2021.111710

Pant, D., Shah, K. K., Sharma, S., Bhatta, M., Tripathi, S., Pandey, H., Tiwari, H., Shrestha, J., & Bhat, A. K. (2023). Soil and ocean carbon sequestration, carbon capture, utilization, and storage as negative emission strategies for global climate change. *Journal of Soil Science and Plant Nutrition 23*, pp.1421–1437. Available from: https://doi.org/10.1007/s42729-023-01215-5.

Phillips, N., Draper, T. C., Richard Mayne, R., Reynolds, D. M., & Adamatzky, A. (2022). Marimo actuated rover systems. *Journal of Biological Engineering, 16*(1). https://doi.org/10.1186/s13036-021-00279-0

Quéré, C., Raupach, M. R., Canadell, J. G., Marland, G., Bopp, L., Ciais, P., Conway, T. J., Doney, S. C., Feely, R. A., Foster, P., Friedlingstein, P., Gurney, K., Houghton, R. A., House, J. I., Huntingford, C., Levy, P. E., Lomas, M. R., Majkut, J., Metzl, N., ... Woodward, F. I. (2009). Trends in the sources and sinks of carbon dioxide. *Nature Geoscience, 2*(12), 831–836.

Renfrew, S. E., Starr, D. E., & Strasser, P. (2020). Electrochemical approaches toward CO_2 capture and concentration. *ACS Catalysis, 10*(21), 13058–13074.

Reuters. (2023). US 'under no circumstances' will pay climate reparations, Kerry says. 13 July 2023. Available from: https://www.reuters.com/world/us/us-under-no-circumstances-will-pay-into-loss-damage-fund-kerry-2023-07-13/. Accessed 24 August 2023.

Sabatino, F., Grimm, A., Gallucci, F., van Sint Annaland, M., Kramer, G. J., & Gazzani, M. (2021). A comparative energy and costs assessment and optimization for direct air capture technologies. *Joule, 5*(8), 2047–2076.

Shah, N. W., Baillie, B. R., Bishop, K., Ferraz, S., Högbom, L., & Nettles, J. (2022). The effects of forest management on water quality. *Forest Ecology and Management, 522*. https://doi.org/10.1016/j.foreco.2022.120397

Sharma, P., Kaur, A., Batish, D. R., Kaur, S., & Chauhan, B. S. (2022). Critical insights into the ecological and invasive attributes of *Leucaena leucocephala*, a tropical agroforestry species. *Frontiers in Agronomy, 4*, 890992.

Shi, X., Xiao, H., Azarabadi, H., Song, J., Wu, J., Chen, X., & Lackner, K. S. (2020). Sorbents for the direct capture of CO_2 from ambient air. *Angewandte Chemie International, 59*(18), 6984–7006.

Smith, P., House, J. I., Bustamante, M., Sobocká, J., Harper, R., Pan, G., West, P. C., Clark, J. M., Adhya, T., Rumpel, C., Paustian, K., Kuikman, P., Cotrufo, M. F., Elliott, J. A., McDowell, R., Griffiths, R. I., Asakawa, S., Bondeau, A., Jain, A. K., ... Pugh, T. A. (2016). Global change pressures on soils from land use and management. *Global Change Biology, 22*(3), 1008–1028.

Snæbjörnsdóttir, S. Ó., Sigfússon, B., Marieni, C., Goldberg, D., Gislason, S. R., & Oelkers, E. H. (2020). Carbon dioxide storage through mineral carbonation. *Nature Reviews Earth & Environment, 1*(2), 90–102.

Solakivi, T., Kiiski, T., & Ojala, L. (2018). The impact of ice class on the economics of wet and dry bulk shipping in the Arctic waters. *Maritime Policy & Management, 45*(4), 530–542. https://doi.org/10.1080/03088839.2018.1443226

Sordello, R., Ratel, O., Flamerie De Lachapelle, F., Leger, C., Dambry, A., & Vanpeene, S. (2020). Evidence of the impact of noise pollution on biodiversity: a systematic map. *Environmental Evidence 9*, 20. https://doi.org/10.1186/s13750-020-00202-y

Sullivan, I., Goryachev, A., Digdaya, I. A., Li, X., Atwater, H. A., Vermaas, D. A., & Xiang, C. (2021). Coupling electrochemical CO_2 conversion with CO_2 capture. *Nature Catalysis, 4*(11), 952–958.

The World Bank. (1991). Environmental assessment sourcebook: volume 2—sectoral guidelines (English). World Bank technical paper, no. WTP 140 Washington DC, World Bank. Available from: http://documents.worldbank.org/curated/en/415971468137388990/Environmental-assessment-sourcebook-volume-2-sectoral-guidelines. Accessed 20 July 2023.

The turning point | Deloitte. (2022). Deloitte [Preprint]. Accessed 12 Sept 2023.

Thonemann, N., Zacharopoulos, L., Fromme, F., & Nühlen, J. (2022) Environmental impacts of carbon capture and utilization by mineral carbonation: A systematic literature review and meta life cycle assessment. *Journal of Cleaner Production, 332*, 130067. https://doi.org/10.1016/j.jclepro.2021.130067

Tripathi, N., Hills, C. D., Singh, R. S., & Atkinson, C. J. (2019). Biomass waste utilisation in low-carbon products: harnessing a major potential resource. *NPJ Climate and Atmospheric Science, 2*(1), 35. https://doi.org/10.1038/s41612-019-0093-5

UN Environment Programme (UNEP). (2020). UNEP and the climate emergency. Briefing paper for the 75th session of the United Nations General Assembly. September 2020. Available from: https://www.unep.org/unep-and-climate-emergency. Accessed 24 August 2023.
United Nations. (2023b). Sustainable Development. Available from: https://sdgs.un.org/. Accessed 9 August 2023.
United Nations. (2023a). Secretary-General's opening remarks at press conference on climate [Press release] 27 July. Available from: https://www.un.org/sg/en/content/sg/speeches/2023-07-27/secretary-generals-opening-remarks-press-conference-climate. Accessed 15 August 2023.
United Nations. (no date a). The Kyoto Protocol. Accessed 24 August 2023.
United Nations. (no date b). Marking the Kyoto Protocol's 25th anniversary | United Nations. Accessed 29 August 2023.
United Nations. (no date c). The Paris Agreement. Available from: https://unfccc.int/process-and-meetings/the-paris-agreement. Accessed 24 August 2023.
United Nations. (no date). https://www.un.org/en/climatechange/marking-kyoto-protocol%E2%80%99s-25th-anniversaryd) COP27 Delivering for People and the Planet. Available from: https://unfccc.int/process-and-meetings/the-kyoto-protocol/history-of-the-kyoto-protocol/text-of-the-kyoto-protocol.Accessed 24 August 2023.
Vanbeveren, S. P., & Ceulemans, R. (2019). Biodiversity in short-rotation coppice. *Renewable and Sustainable Energy Reviews, 111*, 34–43.
Varghese, A. M., & Karanikolos, G. N. (2020). CO_2 capture adsorbents functionalized by amine–bearing polymers: A review. *International Journal of Greenhouse Gas Control, 96*, 103005.
Wang, C., Zhang, W., Li, X., & Wu, J. (2022). A global meta-analysis of the impacts of tree plantations on biodiversity. *Global Ecology and Biogeography, 31*(3), 576–587.
Wang, J., Pu, Q., Ning, P., & Lu, S. (2021). Activated carbon-based composites for capturing CO_2: A review. *Greenhouse Gases: Science and Technology, 11*(2), 377–393.
Wikipedia. (2022). Geoffrey Pyke. Available from: https://en.wikipedia.org/wiki/Geoffrey_Pyke/. Accessed 31 December 2022.
Wildenborg, T., Gale, J., Hendriks, C., Holloway, S., Brandsma, R., Kreft, E., & Lokhorst, A. (2005). Cost curves for CO_2 storage: European sector. In E. S. Rubin, D. W. Keith, C. F. Gilboy, M. Wilson, T. Morris, J. Gale, & K. Thambimuthu (Eds.), *Greenhouse Gas Control Technologies 7* (pp. 603–610). Elsevier Science Ltd.
Wood Recyclers Association. (2023). Available from: https://woodrecyclers.org/. Accessed 20 August 2023.
Woodland Trust. (2023). What Urban Trees do for us. Available from: https://www.woodlandtrust.org.uk/protecting-trees-and-woods/benefits-of-urban-trees/. Accessed 20 August 2023.
Woolf, D., Amonette, J., Street-Perrott, F., Lehmann, J., & Joseph S. (2010). Sustainable biochar to mitigate global climate change. *National Communication, 1*(56). https://doi.org/10.1038/ncomms1053. Accessed 15 August 2023.
World Resources Institute. (2023). Global forest review: Production forests. Available from: https://research.wri.org/gfr/forest-designation-indicators/production-forests. Accessed 21 August 2023.
Xiu, S., & Shahbazi, A. (2012). Bio-oil production and upgrading research: A review. *Renewable & Sustainable Energy Reviews, 16*(7), 4406–4414. https://doi.org/10.1016/j.rser.2012.04.028
Yadav, S., & Mondal, S. (2022). A review on the progress and prospects of oxy-fuel carbon capture and sequestration (CCS) technology. *Fuel, 308*, 122057. https://doi.org/10.1016/j.fuel.2021.122057
Younger, J. L., Clucas, G. V., Kooyman, G., Wienecke, B., Rogers, A. D., Trathan, P. N., Hart, T., & Miller, K. J. (2015). Too much of a good thing: Sea ice extent may have forced emperor penguins into refugia during the last glacial maximum. *Global Change Biology, 21*(6), 2215–2226.
Zarafshar, M., Bazot, S., Matinizadeh, M., Bordbar, S. K., Rousta, M. J., Kooch, Y., Enayati, K., Abbasi, A, & Negahdarsaber, M. (2020). Do tree plantations or cultivated fields have the same ability to maintain soil quality as natural forests? *Applied Soil Ecology, 151*, 103536. https://doi.org/10.1016/j.apsoil.2020.103536

Zeng, N. (2008). Carbon sequestration via wood burial. *Carbon Balance and Management, 3*(1). https://doi.org/10.1186/1750-0680-3-1

Zhou, X., Wen, Y., Goodale, U. M., Zuo, H., Zhu, H., Li, X., You, Y., Yan, L., Su, Y., & Huang, X. (2017). Optimal rotation length for carbon sequestration in Eucalyptus plantations in subtropical China. *New Forests, 48*, 609–627.

Neil Phillips has 20 years of expertise in the implementation of unconventional bioengineering solutions. Co-PI on Horizon 2020 project to commercialise biosensing platform for environmental monitoring. He has worked on the design and manufacture of bioreactors.

Sarah M. Ayling is a plant physiologist with over 40 years of experience. She has studied the impact of environmental conditions on plants, including trees, in the UK, US, and Australia.

Norman M. Ratcliffe has over 30 years of experience in gas/volatile analysis research, specialising in agrifood. He created the first volatile organic compound (VOC) quality control system for storing vegetables.

Ben de Lacy Costello has over 25 years of experience in gas analysis, including production of gas sensors and sensor array systems for volatile detection. He works on projects with sustainable air travel and sustainable farming, where he measures and reduces methane emissions.

William Foord is an MSc environmental management student currently studying at the University of the West of England, with an undergraduate degree in physical geography from the University of Portsmouth, UK.

Thomas C. Draper is a chemist with a wide range of expertise, from nanotechnology to molecular biology to molecular computing. He has over 10 years of experience working on diverse research projects.

Anthony Plumridge is an economist with experience in cost–benefit analyse and a background in social return on investment and carbon accounting. He has expertise in predicting the worldwide economic consequences associated with project implementation.

Jona Razzaque is an expert on environmental laws and policies. She serves as a member of the IUCN World Commission on Environmental Law.

Chapter 13
Developing a National Climate Change Damage Register for Germany

Quirin Oberpriller, Linda Hoelscher, and Caspar Esche

Abstract This paper presents the development of a methodology and concept for a national climate damage register for Germany. Information on past and future climate change-induced damage provides an important basis for a strategically oriented and effective responses to the impacts of climate change. So far, there is no national monitoring program in Germany that systematically records such damage. To begin with, the existing needs and interests of potential users are analysed. In addition, existing guidelines and first practical examples of damage databases are identified. Based on these findings, the structure of a damage register is developed. As the usefulness of a damage register is strongly characterised by the quality of the data used for this purpose, the existing data sources for defined focus areas of damages are examined in detail with regard to their suitability for a register. In terms of timeliness, the focus is on meaningful and reliable loss data, which is more important to most stakeholders than availability as soon as possible. However, a fast track approach is also described for major events, in which initial rough estimates will be published. A special focus is placed on the structure, possible threshold values and the process of data collection and maintenance.

Keywords Extreme weather · Economic damage · Damage registry · Data analysis

Introduction

Climate change is one of the central challenges for society, the economy and politics. Natural hazards, extreme weather events (such as heat waves, storms, floods or heavy rainfall), sea-level rise and other threats can lead to considerable direct and indirect

Q. Oberpriller · C. Esche
INFRAS Forschung und Beratung, Zurich, Switzerland
e-mail: quirin.oberpriller@infras.ch

L. Hoelscher (✉)
Adelphi Research gGmbH, Berlin, Germany
e-mail: hoelscher@adelphi.de

W. Leal Filho et al. (eds.), *Climate Change Mitigation and Adaptation in Practice*, Climate Change Management, https://doi.org/10.1007/978-3-031-85217-6_13

damages in various areas. These are borne by the private sector, the state and the general public, as well as the environment. And these are affected by climate change.

So far, Germany has no national monitoring system to systematically record damage caused by climate change. Yet, information on past and current climate-related damage is essential to align further strategic planning for adaptation in a meaningful and effective way. To provide this information, a national damage register as an accounting and monitoring instrument is of central importance. However, there are hardly any standardised requirements or recording practices, which makes national and international statistics incomplete and difficult to compare. In addition, at the moment, there are only ad hoc studies commissioned after an extreme event, but no systematic way to continuously track damages (see for example PROGNOS, 2022).

Against this background, the German Environment Agency commissioned a research project which INFRAS and adelphi research implemented. This project aimed to prepare the implementation of a user-oriented climate damage register and thus take a first step towards establishing a national monitoring system for climate damage. Therefore, this project's main objective was to develop a concept and methodology for such a register that would run continuously and track damages systematically over time. In doing so, the needs of potential users were examined, an ideal type of register structure was created, the existing data sources for certain types of damages were analysed and, finally, recommendations for further steps were elaborated.

Methodology and Process

The study was divided into the following steps:

- Implementation of a needs analysis of all relevant stakeholders
- Evaluation of templates and guidelines for existing extreme weather damage registers
- Conceptual design of a register for the systematic and consistent accounting of extreme weather and climate change damage at national level.

We conducted various interviews with potential users of a climate damage register as well as potential data providers. In several cases actors have both roles. The needs analysis included interviews with insurances and other financial actors such as the KfW (German public bank), universities, federal authorities, disaster relief agencies, infrastructure providers and municipal authorities. The analysis of potential data sources for different event types and damage categories included desk research as well as interviews and workshops with suppliers of data. Important players in this context are insurance actors, specifically the German Insurance Association (GDV), federal authorities and research bodies, such as the German Statistical Office and the German Meteorological Service, the German Rail and municipal and state authorities. The interviews were an important source of information to gain an understanding of the data, how it was collected and aggregated, and how it could be used and made available for such a register.

During the project two workshops were implemented with data providers and potential users of a register to present and discuss (interim) findings and a concept for the damage register. The workshops also served to reflect on a draft structure for the register, including the different categories of damages included in the prototypical structure, as well as certain methodological aspects of necessary data processing.

This study did not encompass all potential extreme weather events. Instead, a few important event types were selected and analysed in depth. Overall, the focus of the analysis was on monetary damages to buildings and infrastructure due to heavy rainfall and flooding, as well as damage to agriculture caused by drought. For these event and damage types, three case studies were conducted to estimate the damages caused by past events.

In addition, certain types of non-monetary damages from various damage categories were analysed in detail such as loss of or damages to cultural heritage, impacts on human health including loss of life during heat waves, and damages to ecosystems.

For these event types and damage categories, a thorough analysis was conducted of the available data sources and their potential use for the damage register.

Finally, a proposal to integrate extreme event attribution was developed, which—despite the large uncertainties—is a mandatory step for a climate change damage register.

The study developed recommendations for the immediate implementation of a climate damage register, including a prototypical structure of the register and most prevalent steps that should be taken to bring this register to life. In addition, based on the extensive research, advice on further future phases is compiled which will include the continuation, gradual improvement and expansion of the register and the institutionalisation of data collection (Fig. 13.1).

Results

Demands and Needs

The survey of potential users revealed a variety of needs and requirements for a damage register. For most potential users, the main purpose of the damage register is to check the plausibility of their own modelling or business strategies. The insurance industry can use an extensive damage register dataset to compare and perform a plausibility check for its premium policy. Here, details on the total amount of damage per event are just as interesting for the insurance companies as those on the average amount of damage per building unit. For research institutes, damage data on at least the municipal level is desirable in order to be able to guarantee a meaningful level of detail for their modelling. Aggregation of the damage data as late as possible

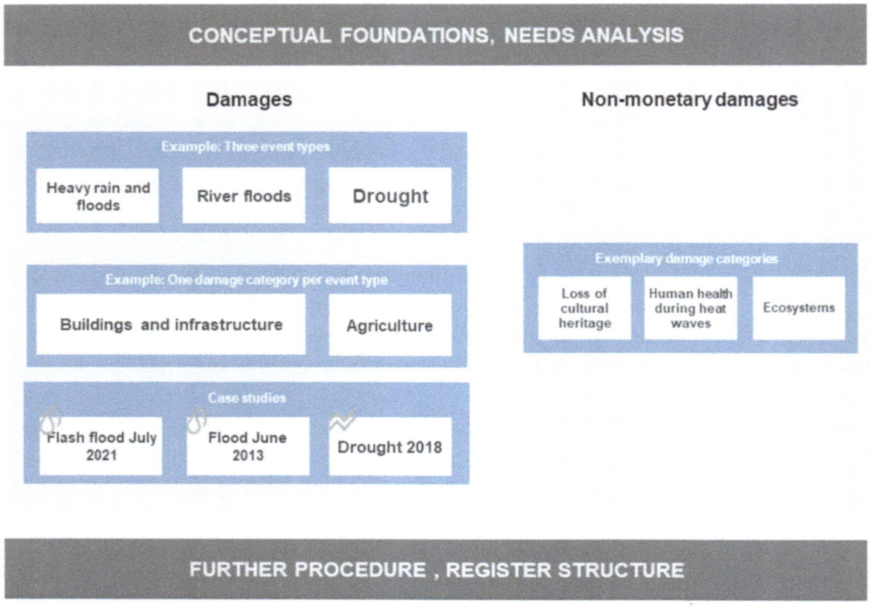

Fig. 13.1 Overview of contents of the study

and correspondingly detailed damage statements in high resolution are therefore important utilisation criteria for several interested parties.

In particular, data on monetary damage and the number of damaged buildings and infrastructure are of great interest. If both elements are known, the average damage value per damaged building or damaged infrastructure unit can be determined. With regard to the timeliness of the data, most users (irrespective of the event under consideration) attach more importance to meaningful and resilient damage values than to availability as close in time to the event as possible. However, especially in the case of major events, there is also a need for rapid availability, e.g., for the federal government.

At the same time, there are several potential users for whom the resolution of damage values is less important than statements about overall damage. This also includes the creation of "situation pictures" regarding the adequacy of political responses and financial expenditures for climate prevention and adaptation. This primarily requires sufficient data coverage, even if the data is not 100% accurate. Therefore, aggregated data should also be made available. It is also important for stakeholders to obtain information on the uncertainties of the entries.

Available Data Sources

Data Sources for Damage to Buildings and Infrastructure

The detailed analysis of the available data shows that a number of potential primary sources would be available for a damage register (Table 13.1). Particularly important is the data from the German Insurance association (GDV), which is of high quality and reliability, but only comprises insured damages. In addition, damage data collected by municipal authorities covering damages to local infrastructure and public buildings, as well as affected stakeholders such as Deutsche Bahn provide information.

In general, however, primary data is rather sparsely available and heterogeneous. Often, available sources are secondary data from non-referenced primary sources or based on rough estimates. In many cases, it is not possible to understand which methodology was used to calculate the damage and which methodological delimitations were chosen. Existing databases use various primary sources in a pragmatic way and sometimes apply a proprietary extrapolation methodology (e.g., Munich RE). Furthermore, damage data is often collected only once in time-limited projects and not updated.

In addition to the very different quality levels of the individual sources of damage, there is hardly consistent data at the national level due to the small-scale data collection of many regional or federal actors. For example, each fire station or community collects local damage data in a different way, as there are no legal requirements.

Data Sources for Damage in Agriculture

Regarding damages in the area of agriculture, the data research was limited to the event type drought. The analysis of data availability for agricultural extreme weather damage shows that there are fewer primary data sources than for damage to infrastructure. The most important primary data source is the "Fachserie 3" (Technical series 3) of the Federal Statistical Office, which includes all data on crop yields and growth in the categories of field crops, vegetables, tree fruit and grape. With the help of the yield changes compared to a normal year, the direct damage or reduced yields of an extreme event can be determined.

Another primary source is data from the GDV. These are primarily determined on the basis of insurance benefits claimed and paid out in the agricultural sector. The availability of data on insured losses varies greatly depending on the type of event under consideration. For drought, data availability is limited due to the low insurance coverage against drought damage. For hail damage, for example, the data situation is much better due to the much higher insurance coverage. A challenge is that damage estimates are often based on indicator-based models such that even insurance data is not collected bottom-up.

In order to improve the available data, remote sensing services such as Copernicus will be increasingly used in the future. Satellite-based remote sensing data

Table 13.1 Overview of data sources for damages to buildings and infrastructure with a high relevance for the damage register

Database/Source	Operator/Publisher	Primary/ Secondary data	Quality	Relevance for damage registers
Damage data of German insures	GDV (insurance association)	Primary data	High	High
RADKLIM/ CATRaRE	German Weather Service	Primary data (meteorological data only)	High	High
Damage data for rail infrastructure	German Rail (Deutsche Bahn AG/DB Netz AG)	Primary data	High	High
Ad-hoc damage assessments by the federal states	Federal Ministry of Finance, Federal Ministry of the Interior	Primary data	High	High
Actual cash outflows, interim/ final reports	Federal Ministry of Finance, Federal Ministry of the Interior, Federal Court of Auditors (Bundesrechnungshof)	Primary data	High	High
Damage data from municipalities	Federal government, federal states	Primary data	Medium	High

enable short-term, detailed and continuous analyses of agricultural damage. It can be assumed that technological advances in the field of satellite-based remote sensing and evaluation methods will be achieved in the future. It is therefore important for the climate damage register to maintain a regular exchange with operators and researchers in this field. Table 13.2 provides an overview of existing databases and sources that (among other things) contain data on damage in agriculture.

Data Sources for Non-monetary Damages

Various types of damage cannot be captured in monetary terms or only individual subcomponents of the damage can be represented in monetary terms.

A significant category of damage that cannot be represented in monetary terms is deaths caused by heat. The Robert Koch Institute provides a suitable data source here. Using a statistical model, the number of heat-related deaths is represented as the "difference between the course of observed mortality and an estimated course of mortality without heat waves" (RKI, 2019). The data can be differentiated according to age groups. It makes sense to set up an annual monitoring that records the number of estimated heat-related deaths in Germany for each calendar year. Deaths caused by other extreme weather events can also be mapped in the register.

Table 13.2 Overview of data sources for damages to agriculture with a medium or high relevance for the damage register

Database/ Source	Operator/ Publisher	Primary/ Secondary data	Quality	Relevance for damage registers	Comment
Claims data from German insurers	GDV (insurance association)	Primary data	High	High	
Data on yields per hectare and sales prices of crops	DESTATIS/ Federal Statistical Office	Primary data	High	High	Important for deviation of yields from the mean value
Copernicus	EU/DLR	Primary data	Medium/High	High	Data quality high if technological progress
EM-DAT International Disaster Database	University of Louvain, Brussels	Secondary data	Medium	Medium	Primarily as a practical example
Copernicus Global Land Service	VITO NV	Primary data	Medium	Medium	Evaluation necessary
Hydrological data, water quality	Federal states	Primary data	Medium	Medium	

Currently, damage to cultural heritage cannot be well integrated into the register. A recording of damages in culturally significant buildings that goes beyond direct property damage is not yet feasible, as a corresponding data basis and methodology is lacking.

A number of data sources are available to integrate a mapping of forest ecosystem damage in the register:

- Logging statistics for recording damaged wood due to drought and storms, available on an annual basis or via direct query at state and municipal level;
- Forest damage registration of the federal research institute "Thünen Institute", based on remote sensing data (currently under development);
- Forest fire statistics, available on an annual basis or via direct query at federal state and municipal level.

Remote sensing data can be used to record changes in the water balance and vitality of critical ecosystems. For example, the Copernicus Land Cover Viewer contains data on wetland degradation, which can be used to monitor wetlands and specifically peatlands. However, remote sensing data cannot be used directly for the damage register. Here, baseline data from reference areas must first be available, followed by an evaluation of the data by experts.

Finally, it should be stressed that in communicating the damage depicted in the register, it is essential to emphasize the importance of non-monetary damage. As it is difficult to express them in monetary terms and include them in the total sum of (economic) damage, or as this is not necessarily advisable, there is a risk that they will occupy less space in public perception. As the following overview of data sources and damage types indicates, for some non-monetary damages it is possible to monetize certain components, such as the cost for treatment of health effects due to heat. Table 13.3 provides an overview of the different data sources and damage types that cannot be fully monetized.

Structure of the Damage Register

One of the results of the research is a prototype for a climate damage register in which entries are made according to a fixed structure. A subdivision into damage categories is provided (e.g. buildings and infrastructure). Damage categories are differentiated by elements (private residential buildings, infrastructure, industry/commerce, agriculture) and sub-elements (e.g. building damage to the building shell and damage to household effects). Entries are optional from a certain level of detail or differentiation (e.g. differentiation by loss carrier and cost unit). The structure should allow data to be easily aggregated from lower to higher levels (from county to state and from state to federal) or to select certain areas (e.g. only damage caused by event type heavy rain). For each event, the register includes the following areas:

- Direct monetary damages (in € and physical units)
- Physical damage
- Non-monetary losses (damages that have no market price).

In principle, each event type causes damage in all categories, but the relevance varies. For example, flash floods mainly cause damage to buildings and infrastructure, while heat waves tend to cause little damage in this category (Kahlenborn et al., 2021). In addition, each event type has its own geographic and temporal scales, which must be taken into account (e.g. droughts tend to be large-scale and have a longer duration; heavy rain falls at small scales and within hours).

The damage register focuses on direct monetary damages. As far as possible, the underlying quantity structure of the physical damage should also be recorded by means of suitable indicators (e.g., number of destroyed residential buildings or kilometres of destroyed railway line). This is important information for many users and relevant for a plausibility or consistency check.

Indirect damage occurs as a result of direct damage, spatially remote or with a time lag (e.g., production losses along the value chain). These are subject to greater uncertainties, as they can usually only be estimated using extensive assumptions. Indirect damages will thus not be considered in the register for the time being and are not the focus of this study. In the medium term, however, indirect damage should be taken into account.

Table 13.3 Overview of data sources and damage types that cannot be fully monetized

Damage category	Subcategory	Indicator, measure	Data origin	To be mapped promptly in the register
Non-monetizable				
Human deaths	Heat-related deaths	Heat-related mortality	Robert Koch Institute	Yes
	Deaths caused by other extreme events		CATDAT	Yes
Human health	Health restrictions due to heat	Hospitalization	Hospital Statistics, GENESIS Database	Yes
Damage to ecosystems	Damage to forests (storms, forest fires or drought)	Crown damage (forest condition), logging, forest fire statistics (forest damage assessment)	Forestry authorities, Destatis, Federal Ministry of Agriculture, Federal Office for Agriculture and Food, Thünen Institute	Yes
	Changes in water bodies	Critical water quality (suspended solids, phytoplankton, pollutants, cyanobacteria)	Federal states, FFH monitoring	Yes
Subcomponents that can be mapped in monetary terms				
Damage to cultural property	Damage to listed buildings		Municipalities	Yes
Damage to ecosystems	Damage to forestry	Economic losses due to damaged wood	Destatis logging statistics	No (calculation necessary)
Human health	Health effects of heat	Treatment costs	Health insurance companies	No
Human health	Reduced productivity due to heat	Economic losses of companies	Lancet countdown	Yes

Definition of Thresholds for Entries into the Register

The focus of the register is on damages caused by extreme events. By definition, only events that exceed certain thresholds are considered extreme. There is therefore a process for the register in terms of the events that are taken into account (for extreme rainfall, for example, this is based on the CatRaRe event database of the

German Weather Service (DWD)). In addition, in the unlikely case that an event causes significant damage according to the insurance data, but is not recorded in the CatRaRe database, such an event would also be considered. The choice of the type and level of the thresholds is a normative process. This should therefore be done transparently and in dialogue with key stakeholders.

It is recommended to use hazard-based thresholds for the register, which are based on meteorological indicators of the respective event type (e.g. precipitation amount). Hazard-based thresholds are easier to verify, as the corresponding indicators are usually collected in a timely manner anyway. In addition, it can be analysed to what extent the damage of similar meteorological events differs over time. This may allow conclusions to be drawn about the effectiveness of adaptation activities on average over many events and can thus serve as a basis for monitoring the German Adaptation Strategy (Die Bundesregierung, 2008) even though it might be expected that the damage data will be quite "noisy". Hazard-based thresholds require a separate threshold to be defined for each event type. Therefore, it must be ensured that the choice of the threshold level is consistent. For droughts and heat, continuous monitoring of the difference to an average year is recommended instead of using mean thresholds. This is a deviation from the event-based entries into the register for the other event types.

Fast-Track and Slow-Track

Overall, the analysis of numerous data sources for the damage-types covered by this study also showed that available data have different maturities: timely data are usually rough estimates by experts with a high degree of uncertainty. They can therefore only be used as a first indication of the order of magnitude. Later, updated data (usually after 0.5–2 years) show a more accurate picture and are then preferable to the first estimates. A number of damages can only be derived using estimation models based on indicators (e.g., changes in labour productivity during heat waves).

The needs analysis also indicated that both quickly available estimates and best possible long-term data for damages are desired. A two-tiered approach is therefore recommended:

1. Fast-track: This path is only implemented for major events. It provides a timely rough estimate with high uncertainty (indicative orders of magnitude). The fast-track must be activated by a special mechanism (e.g. catastrophic event). In the case of major events of national importance, it is important to have estimates available quickly, as these are needed for the short-term implementation of relief and reconstruction programs or for the rapid disbursement of relief funds. In the further course of time, the fast-track data should be regularly updated.
2. Slow-track: This path is always implemented. It provides more reliable and more differentiated entries after all available data are available. It usually takes 0.5–2 years until all data are available. As soon as available, data from the fast-track path are updated with data from the slow-track path.

Attribution

Extreme events occur even without anthropogenic climate change. However, climate change increases their frequency and intensity. Thus, only part of the extent of the damage can be attributed to climate change and extreme weather damage must be corrected with a specific attribution factor.

For a correct calculation of climate damage, the damage with and without climate change must be compared. The damages under climate change occur in reality. The damages without climate change have to be modelled e.g., by means of climatological simulation and a damage function. Therefore, two basic elements are necessary for a damage attribution in the sense of the climate damage register:

- An extreme event attribution that quantifies the increase in intensity of the event due to climate change. For this purpose, data from climate models or observations are used and evaluated through extreme value statistics.
- A damage function, which shows the damage as a function of intensity. The dependency is usually not linear, i.e., a higher intensity causes disproportionately higher damage.

Since the register is a climate damage register, damage attribution is unavoidable. However, as this is only possible with a high degree of uncertainty, both the total damage (without attribution) and the attributed damage should be indicated (with a reference to the uncertainty).

A two-step process should be implemented:

1. For each event type (as well as event severity and region, if applicable), a generic factor for damage attribution is determined in advance. Ideally, the generic factors are based on studies of the corresponding event type in Germany or at least in Europe. If such studies are not available, qualified expert estimates should be made.
2. If a specific study is published for (major) events, specific factors of damage attribution can be derived. The methodology of updating depends on the respective results of the study. Most studies for major events (e.g. World Weather Attribution) will only cover extreme event attribution but not damage attribution.

Implications for Policy

Recommendations for Timely Implementation

The damage register should be centrally managed and operated. Fixed and ongoing responsibilities are necessary to meet the requirements of completeness, consistency and quality.

This requires performance mandates for state actors, the securing of long-term and adequate financial resources as well as a clearly defined institutional framework. A professional IT solution and a regular review of the fulfilment of objectives are also important. Overlaps between different data sources/damage types need to be identified to avoid double counting of damages.

The entries should fulfil specifications for international reporting obligations and enable data exchange with other countries. In order to exploit synergies, existing or planned typologies and monitoring tasks should be taken into account.

It should be clearly communicated that the register currently only records part of the climate damages and thus systematically underestimates the total climate damage in Germany.

Data Collection

All available data must be consistently and systematically included in the register. It is important to involve relevant actors and owners of primary data in the development and maintenance of the damage register. Insurance companies currently play a central role, as they possess detailed data on insured damage. Furthermore, municipalities (for damage to municipal infrastructure) and other infrastructure owners (e.g. Deutsche Bahn) are also important data sources.

In general, data provision by partners must be formalized in a suitable form and the right incentives must be set. Data procurement can also be outsourced (by means of clear specifications), as is the case at EU level with the European Environmental Agency (EEA). In this context, it has proven useful to sign a memorandum of understanding on the content, scope, period, frequency and quality of data deliveries with data providers.

A clearly defined process and responsibilities for the workflow from event identification through data collection to entries in the registry are needed. Data should be collected or transmitted in such a way that it can be integrated into the structure of the register as easily as possible (ideally semi-automated). This requires coordination of the various actors involved in data collection, as well as the establishment of generally accepted standards according to which data are collected and documented. This enables a homogeneous register over the years. The specific data protection requirements of the data providers must be taken into account. For example, the anonymity of the injured parties must be guaranteed.

Only primary sources should be used. These are sources that were collected directly from the damaged parties or can be directly traced back to them. Expert estimates are also primary sources. Secondary sources, on the other hand, should not be used, but the underlying primary source should be used wherever possible. Contact interfaces with authorities, associations and the private sector (e.g., insurance companies) should be defined and maintained so that the same data sources are available in the long term. For state actors, legal requirements, such as reporting obligations, can be used to ensure cooperation. As the available primary sources may change over time (and ideally uncertainty may also decrease), the best available primary sources need to be determined on a regular basis. If new data are available, the data acquisition process must be adjusted and any inconsistencies with older data considered.

With regard to data procurement and preparation, the German Climate Preparedness Portal—KliVO offers an interesting template. There, data owners send relevant information to a responsible person within the German Environment Agency, who checks its quality and processes the data if necessary. In a similar implementation, there would be a network of actors behind the damage register that delivers data to one or two responsible persons. In this case, the damage register offers a central platform that bundles decentralized data and processes it as needed. This network would initially include the most important data providers identified in this study: Insurers, state authorities and municipalities as well as various federal authorities. With the expansion of the damage register, this network would be constantly expanded.

Data Processing

Ideally, raw data from primary data sources can be integrated directly into the register. Often, however, processing is necessary. For example, data from insurance companies only show insured damages. Total damages can be estimated on this basis, assuming the insurance density, which is available in varying degrees of detail for insurance categories. For processing, conversion methods from raw data (often called indicator in this context) to claims have to be defined. The collection of damage data is thus often a combination of primary data and methods. This is especially true for indirect collection methods such as remote sensing. Methods for filling data gaps must also be defined.

It is important, especially in the initial phase of the register, to regularly review the methods of data preparation and adjust them when necessary. Any breaches in the methodology must be clearly communicated so that they can be considered during use. Ideally, the entire time series should be recorded according to a consistent methodology, which may require a new recording of earlier data points or monitoring periods.

Finally, it may be the case that different data sources cover the same damage. In such a case, the most reliable data should be used. The other data should also be included in the register, but with a reference to its higher uncertainty. Such data can be used for plausibility checks, while double counting has to be prevented.

Historical Events

It is recommended to record historical events retrospectively. The same thresholds and methods should be applied that are used for future events, as was done for the three case studies in this report, for example. Thus, entries can be entered immediately without having to wait for the next event. In addition, past climate damage is relevant for various users. For events before the year 2000, the data situation is likely to become increasingly poor, so that no data should be collected before the year 2000 for the time being.

Accessibility

The data of the climate register should be freely available. However, open access can make data acquisition more difficult due to requirements on data protection, confidentiality and terms of use. If data providers do not agree to an open access to their data, such data could still be used for the register if the publicly available data are suitably anonymized (e.g., through aggregation). Alternatively, access restrictions can be provided: Full access is only granted to users who sign a confidentiality agreement and/or can prove a special interest in the detailed data. The appropriate procedure is to be clarified individually with the data providers.

In the initial phase, the spatial resolution of the information in the register can only correspond to the available resolution of the primary data. This is already very high in some cases (e.g., GDV data are in principle available at county level; municipal data are by definition available per municipality).

Continuation, Gradual Improvement and Expansion

The scope, completeness and level of detail of the register should be improved step by step in the next implementation phases. In addition, an institutional anchoring at the level of public authorities in Germany (especially with regard to data collection and aggregation) should be sought, so that the operation of the register becomes permanent.

Table 13.4 shows that several potential improvements can already be identified. However, given the limited financial, human, and technical resources, priorities must be set. A selection of these is described in more detail in the following sections.

Completing the Register

This study only considered a predefined combination of event types and damage categories. For the next phases, further elements should be successively added to the register. In the meantime, it should be clearly communicated that the register only records selected climate change damage, is incomplete in this sense, and thus underestimates the total damage.

The highest priority is to include other important types of extreme events, e.g. heat waves, hail and storms, and storm surges. For this purpose, suitable data providers would have to be identified and, if necessary, methods for data processing would have to be defined. In a first step, damage due to hail and storms should be recorded, as GDV already has suitable data at its disposal.

Table 13.4 Overview of potential improvements of the climate damage register

Potential improvement	Explanation (examples)	Assessment of priority
Improvements in data collection (for event types and damage categories that have already been taken into account)		
Data collection	Collect better primary data (e.g. lower uncertainties) Eliminate data gaps (e.g., more data from municipalities, businesses, agriculture, or uninsured claims) Automate data delivery and quality assurance	High
Process for municipal data collection	Motivating municipalities to collect data Establish uniform processes and assessment bases for claims Establish structures for aggregation of municipal data at the state and federal level	High
Higher regional resolution	Introduce universal collection at NUTS3 level (counties)	Medium
Data preparation	Identify and implement better methods for converting raw data to damage data and quality assurance	High
Completing the register		
Other types of extreme events I	Recording hail and storms (GDV data available)	Very high
Other types of extreme events II	Include other types of events (e.g., heat waves, storm surges)	High
Reduction of threshold values of existing event types	Also include “less extreme” events	Medium
Indirect damage	Consider indirect effects of extreme events (e.g. production losses along the value chain)	Medium
Slow-onset climate damage	Absorbing slow-onset climate damage (e.g. changes in ecosystems, sea-level rise)	Medium
Specify how to deal with non-monetary damage (especially ecosystem damage)	Improve the survey and create an evaluation scheme Better mapping and qualitative assessment of ecosystem damage	Medium
Think about monetization	Consider monetizing non-monetary losses	
Other aspects		

(continued)

Table 13.4 (continued)

Potential improvement	Explanation (examples)	Assessment of priority
User-friendly platform	Replace the Excel template of this study with a professional IT solution in a timely manner	High
Improved attribution	Keep an eye on new attribution research and adjust methodology if appropriate	High
Standard methodologies for fast-track	Standardize processes for initial damage assessments in fast-track	Medium

Events often result in a multitude of indirect damages. For example, an extreme event can interrupt the production of an intermediate product, which in turn leads to production losses along the value chain. Indirect damage can reach a similar magnitude as direct damage and should therefore be considered in the register in the medium term so as not to underestimate climate damage. In a study, Sieg et al. (2019) modelled the ratio of direct to indirect damages as a result of the June flood of 2013. For this purpose, sector-specific ratios (multipliers) of direct to indirect damages were determined for a total of 19 economic sectors on the basis of an input–output model. However, the uncertainties of this method are high and the transferability of an estimate of indirect damages from one event to another is often not possible, as indirect damages are particularly dependent on local conditions, e.g., the presence of industries or infrastructures of supra-regional importance. The effort for surveying indirect damages is therefore high and the uncertainties are considerable. As a first step, a qualitative overview of possible indirect impact pathways should be compiled. Calculation methods can then be successively developed for the individual impact pathways.

Slow-onset climate change damage is currently not recorded in the register. These are climate change damages that are not caused by extreme events but manifest themselves gradually, e.g., damages to ecosystems caused by temperature increases or seasonal shifts. For these damages, too, structures, processes and responsibilities for data collection or data preparation should be established in the long term. This is the only way that the register can represent climate damage as completely as possible. Recording such damage requires continuous long-term monitoring. The same applies to other types of non-monetary damage. Especially for ecosystem damage, a successive development of data sources for various types of damage is appropriate. For this purpose, a system for data processing and evaluation should be created, especially with remote sensing data. The recording of such damage would not be based on events, but on annual entries.

Process for Municipal Data Collection

The analyses within the scope of this study have shown that activities for estimating and aggregating information in municipalities are implemented ad hoc after an event and do not follow any predefined guidelines. An important step forward would therefore be to establish a uniform process for damage assessment, aggregation and dissemination by municipalities and counties. Ideally, data delivery will be largely automated in the future.

Damage estimates in municipalities have so far been implemented based on varying processes and methodologies and the aggregation of data within municipalities or counties does not follow a clear scheme. The same applies to the transfer of data to the state level or beyond. In addition, the categorization of damage types also does not follow a uniform system. The development of such a system, which is followed by all municipal actors in the categorization of damages, would therefore represent a major step forward, not least in terms of data consistency. Repeated use of this systematization would also improve the comparability of damage data across different events. The same applies to the processes for data collection and dissemination: there are no clear structures, processes or responsibilities according to which action is taken when an event occurs. Thus, collecting consistent data is currently very difficult.

An important further development in the context of the climate damage register is therefore the establishment of a clear structure and uniform processes for the recording of extreme weather damage by municipal actors. A central location where data can be filed and collected can be made available nationwide in the future. At the same time, a uniform input mask with specifications on which information should be estimated for which damage categories promotes a uniform procedure in the municipalities. A practical approach could be the establishment of an online platform, possibly linked to the damage register, where municipalities can enter their recorded damage. On this municipal data platform, all data would be collected centrally and could be processed automatically. It would be helpful if this platform were available both in a browser and as a mobile app. In this way, municipal employees could enter their collected information on their mobile devices directly during the damage recording process. The provision of such a platform by the federal government would be very conducive to uniform data collection. The data recorded should include a quantitative description of the extent of the damage (length of the affected watercourse, etc.), a damage assessment and a rough estimate for the repair cost. By geographically mapping the recorded damage, it would also be possible see graphically where damage has already been entered and recorded and in which areas this is still outstanding. Such mapping can be of great help in organizing work in the immediate aftermath of an extreme weather event.

The processes for delivering and aggregating data also need to be clearly structured. These vary from state to state and can only be standardized in dialogue with the states. Such a dialogue process should be initiated and coordinated by the federal government but requires cooperation across the federal-state-local levels.

The processes described so far in this chapter focus primarily on the short-term estimation of damage in the aftermath of an extreme weather event and thus as a basis for the fast-track procedure of the damage register. For the slow-track process, the retrieved relief and subsidy funds are particularly relevant as a data basis to obtain accurate and reliable information on the amount of damage. In order to establish a process for obtaining this data that can be implemented repeatedly, an overview of the responsible administrative unit in each case is first needed. These vary depending on the federal states.

Conclusions

A national register for damages caused by extreme weather events and climate change will provide valuable information for strategic decisions in adaptation policy. It can help identify priority areas for action by indicating sectors or geographic areas with a high occurrence of damages. The establishment of such a register is also mentioned in the draft federal adaptation law.

Extreme weather events occur even without anthropogenic climate change. However, climate change increases their frequency and intensity. Thus, only part of the extent of damage can be attributed to climate change and extreme weather damage must be corrected with a specific attribution factor. An attribution should be carried out in any case, even if it is fraught with high uncertainties. Given the high uncertainties of attribution, it is important to show the extreme weather damage and the associated share of climate change damage separately for each event, so that users can choose which value to use depending on the application.

It can also make a significant contribution to the monitoring of impact of adaptation policy and of climate change. By systematically tracking changes in the amount of damage caused by extreme weather events over time, this information can provide a basis for measuring the success of adaptation measures in different areas. This will however be challenging given the many factors that influence damages and would a require long, consistent time series. A climate change damage register can support two types of cost–benefit analysis:

- A comparison of adaptation costs versus climate damages
- A comparison of the costs of mitigation of climate change versus the costs of inaction (the calculation of which requires climate damages as a key input).

Our analysis shows that there are already several data sources which can form the basis of such a climate change damage register. The most significant data sources are insurances, federal public authorities and research institutes, local and municipal authorities and affected parties. However, there is a clear need for further research and improvements in the data basis. An important step forward would therefore be to establish a uniform process for damage recording, aggregation and dissemination by municipalities and counties.

The climate change damage register will start now. The scope, completeness and level of detail should be gradually improved in the next implementation phases.

References

DESTATIS. (n.D.). Deutschlands Indikatoren der VN Sustainable Development Goals. SDG 11—Städte und Siedlungen inklusiv, sicher, widerstandsfähig und nachhaltig gestalten. Available at: https://sdg-indikatoren.de/11/. Accessed October 4, 2023.

Die Bundesregierung. (2008). Deutsche Anpassungsstrategie an den Klimawandel (DAS). Vom Bundeskabinett am 17.12.2008 beschlossen. Available at: https://www.bmuv.de/fileadmin/Daten_BMU/Download_PDF/Klimaanpassung/das_gesamt_bf.pdf. Accessed October 4, 2023.

Kahlenborn, W., Porst, L., Voß, M., Fritsch, U., Renner, K., Zebisch, M., Wolf, M., Schönthaler, K., & Schauser, I. (2021). *Klimawirkungs- und Risikoanalyse für Deutschland 2021* (26/2021; Climate Change). Umweltbundesamtes. Available at: https://www.umweltbundesamt.de/sites/default/files/medien/479/publikationen/kwra2021_teilbericht_zusammenfassung_bf_211027_0.pdf. Accessed October 4, 2023.

PROGNOS. (2022). Extremwetterschäden in Deutschland seit 2018. Available at: https://www.prognos.com/sites/default/files/2022-07/Prognos_KlimawandelfolgenDeutschland_Kurzzusammenfassung_Extremwettersch%C3%A4den%20seit%202018_AP2_3d_.pdf. Accessed October 4, 2023.

Robert-Koch Institut (RKI). (2019). Indikatoren für die Deutsche Anpassungsstrategie. Indikator-Factsheets zum Handlungsfeld Menschliche Gesundheit, GE-I-2. Available at: https://www.umweltbundesamt.de/sites/default/files/medien/380/dokumente/ge-i-2_indikator_hitzebedingte_todesfaelle_2019.pdf. Accessed 4th October 2023.

Sieg, T., Schinko, T., Vogel, K., Mechler, R., Merz, B., & Kreibich, H. (2019). Integrated assessment of short-term direct and indirect economic flood impacts including uncertainty quantification. *PLoS ONE, 14*(4), e0212932.

Quirin Oberpriller is Senior Project Manager at INFRAS. He works in the field environment, climate and energy. He primarily focuses on national and international climate policy, climate mitigation instruments as well as economic evaluations.

Linda Hoelscher is Senior Advisor at adelphi research in the area of adaptation to the consequences of climate change and climate risk analyses. Her primary focus is the financing of climate adaptation and the further development of adaptation policy in Germany.

Caspar Esche is Consultant at INFRAS. He works in the thematic areas of transportation as well as in environment, climate, and energy, with focus on economic analyses and environmental economics.

Chapter 14
Navigating the Yanomami Health Crisis in the Era of Climate Change: Multidisciplinary Team Coordination and Governance Implications for Policymakers

Heloisa H. S. De Pontes, Lucas C. Stocco, and Luciana O. Cezarino

Abstract In the initial months of 2023, many news reports emerged, shedding light on and denouncing the profound humanitarian crisis endured by the Yanomami People in Brazil. Indigenous communities, serving as stewards of the forest, are essential conservation agents in the fight against climate change. The profound connection these communities maintain with the forest holds substantial implications for policymakers, both in the public and private spheres. This underscores the need to formulate policies that address their fundamental needs and respond effectively to urgent health interventions. To mitigate this issue, the Brazilian Ministry of Health has established a Public Health Emergency Operations Centre to respond and offer primary healthcare more effectively. The following question emerges within this context: "How can public policymakers establish multidisciplinary primary care teams to enhance the health of indigenous populations and mitigate the impacts of climate change?" This study aims to contribute to comprehending policymakers and other political and organisational stakeholders when dealing with emergencies affecting indigenous communities and structuring multidisciplinary primary health teams. The research is underpinned by a theoretical framework centred on Stakeholder Management and Quality of Work Life (QWL), offering a robust foundation and equipping the study with appropriate analytical tools. We gathered information from diverse sources, including company and governmental reports, newspapers, recordings, and photographs. The findings underscore the critical necessity of upskilling internal stakeholders to respond proficiently to humanitarian crises impacting ethnic groups. Furthermore, adopting a quality work-life approach proves instrumental in discerning

H. H. S. De Pontes · L. C. Stocco (✉) · L. O. Cezarino
University of Sao Paulo (USP), Sao Paulo, Brazil
e-mail: lucascstocco@gmail.com

H. H. S. De Pontes
e-mail: heloisapsico@yahoo.com.br

L. O. Cezarino
e-mail: lcezarino@usp.br

W. Leal Filho et al. (eds.), *Climate Change Mitigation and Adaptation in Practice*, Climate Change Management, https://doi.org/10.1007/978-3-031-85217-6_14

the damaging physical and psychological repercussions resulting from stimuli in the work environment for multidisciplinary teams.

Keywords Yanomami · Indigenous populations · Health crisis · Health care · Indigenous Governance · Quality of work-life · Stakeholders management · Climate change · Indigenous public policy · Sustainability: Governance · Multidisciplinary

Introduction

The Bulletin of the Organization for the Amazon Cooperation Treaty (OTCA, 2004) highlighted that "The Amazon Rainforest is the largest tropical forest in the world, but there is still no scientific consensus regarding its physical boundaries. According to OTCA (2004), the Amazon covers approximately 7.5 million km^2, based on the political-administrative criteria used by Amazonian countries, with approximately 68% of the total Amazon Forest territory belonging to Brazil." The Amazon rainforest harbours the highest biodiversity on the planet (Ab'sáber, 1996). Still, it has been losing its original area since the boom of cultivar production, such as rubber, cocoa, açai, and Brazilian nuts, in the nineteenth century (Homma, 2014).

Deforestation is a primary factor contributing to the degradation of the Amazon rainforest, mainly through activities such as road construction, hydroelectric dams, agriculture, extensive livestock farming, logging, mining, and urbanisation (Cunha, 2008). According to Becker (2005), the economic expansion of the Amazon region between the 1970s and 1990s promoted significant deforestation and regional conflicts. Indigenous populations are essential for understanding the history of the Amazon region (Ab'sáber, 1996). According to the UN, indigenous peoples are vital stewards of the environment. Approximately "28% of the world's land surface, including some of the most ecologically intact and biodiverse forest areas, is primarily managed by indigenous peoples, families, smallholders, and local communities" (UN News, 2019). The UN further emphasises that "[...] the lifestyles of indigenous peoples and their livelihood can teach the world a great deal about preserving natural resources, sustaining and cultivating food sustainably, and living in harmony with nature" (UN News, 2019). Also, disseminating the knowledge stemming from this heritage and historical legacy is crucial for addressing the challenges that agriculture and food face today and in the future.

Given their importance for the continuity of the Amazon rainforest, the Yanomami and Ye'kwana peoples, together through the Hutukara Yanomami Association, have developed the Yanomami Indigenous Land Territorial and Environmental Management Plan (PGTA). The plan is a management tool based on the National Policy for Environmental and Territorial Management of Indigenous Lands (PNGATI), created by Decree No. 7,747/12 (Brazil, 2012). This document presents ideas about what is necessary for all Yanomami and Ye'kwana people to live well today and in the future. The paper comprises seven themes: governance, territory surveillance, income generation and management, traditional knowledge, natural resources, health, and education (Vieira & Lima, 2019).

Regarding the health theme, it is fundamental to note that the increased contact with non-indigenous people, e.g. illegal miners through the invasion of indigenous lands (Watts, 2023), has brought diseases previously unknown to the Yanomami people (Paniz-Mondolfi et al., 2019). Furthermore, the problems associated with the consumption of industrialised foods introduced into the communities must be addressed, as dietary habits have led to diabetes, hypertension, malnutrition in children, and other health issues (Moraes et al., 2023).

Those responsible for the health of the Yanomami include the Yanomami and Ye'kwana Special Indigenous Health District (DSEI-YY), the Secretariat of Indigenous Health (SESAI), CONDISI (the council overseeing the actions of DSEI-YY), associations, and communities. The National Indigenous People Foundation (FUNAI) and partners assist in seeking support to improve indigenous health care, including researching diseases and contamination (Vieira & Lima, 2019). In the development of the PGTA, the Yanomami prioritise their health care needs as follows: taking preventive measures to reduce the number of displacements from Yanomami indigenous land, improving patient care at CASAI (Special Indigenous Health District), reducing the time spent in the city; ensuring efficient control of diseases such as malaria, onchocerciasis, pneumonia, and tuberculosis, which have become frequent in their communities; improving specific health care for women, emphasising traditional care practices during pregnancy and childbirth; promoting women's meetings in assemblies to discuss issues specific to their health (Vieira & Lima, 2019), respecting their traditional knowledge and territory governance system (Freitas et al., 2021). The proposals for Yanomami health outline the Brazilian No. 7,747/12 legal decree guidelines: ensuring the quality of care in communities with minimal relocation; promoting autonomy in decision-making and participatory health management; ensuring a workforce with appropriate training; strengthening indigenous health care practices; improving infrastructure and services in communities; and caring for indigenous environments and lives (Brazil, 2012).

In the early days of 2023, the world became aware of the severe humanitarian crisis faced by the Yanomami people residing in the states of Roraima and Amazonas due to a lack of healthcare assistance (Scherf & da Silva, 2023). The absence of aid stemmed from a political agenda aimed at undermining governmental entities responsible for indigenous populations' welfare, beginning in 2019 with the inauguration of President Bolsonaro (The Lancet, 2019). The then-president, Jair Bolsonaro, endorsed a campaign promoting illegal activities in the Amazon region and indigenous territories. He weakened surveillance agencies, supported mining, and proposed a bill permitting gold prospecting in indigenous territories (Watts, 2023). Following a change in government and Lula's reelection, community assistance systems were reinstated, shedding light on the state of emergency in which these populations found themselves (Watts, 2023).

The Ministry of Health established a Public Health Emergency Operations Centre (COE-Yanomami) as a national mechanism for coordinated response management under the responsibility of the SESAI (Brazil, 2023) with support from partners in this emergency action, including UNICEF, which has provided specialised personnel for healthcare services aimed at the entire Yanomami people (UNICEF, 2023).

In this study, the teams of specialised professionals assigned to essential healthcare services will be referred to as stakeholders. According to Macke and Carrion (2006), this is a widely used term to denote any individual or group that can influence an organisation or be affected by it, including internal staff, suppliers, consumers and local communities. Complementary, the QWL proposed by Timossi et al. (2010) is a concept that defines QWL as the perception held in a work environment regarding factors such as pressures, work-life balance, signs of stress, dietary habits, lifestyle, and technological impacts, contributes as an integrative approach to understand how these specialised professionals experienced working in such a challenging environment.

Building upon this relationship between the stakeholder approach and QWL, this study aims to contribute to comprehending policymakers and other political and organisational actors in addressing emergencies involving indigenous communities. It seeks to contribute to formulating public policies that address these populations' specific needs and enhance the teams' preparedness for effective response. Beyond the challenges faced by the Yanomami people, there is a necessity for in-depth discussions on proposals and initiatives to combat climate change. Indigenous communities, acting as stewards of the forest, play a crucial role due to their connection with the surrounding environment and their cultural and social aspects. This connection has a direct implication for improving both local and global climatic conditions. Therefore, initiating new discussions stemming from indigenous themes to ensure primary access to health and quality nutrition contributes to enabling these communities to sustain their daily activities intricately linked to environmental preservation and sustainability.

Nevertheless, the event highlighted managing specialised stakeholders, such as health professionals, when ethnic populations need urgent assistance. The ex-post analysis is also essential to improve governance mechanisms for multiple disciplinary stakeholders to tackle this complex assistance and actively contribute to the prospective formulation of public policies addressing fundamental healthcare provision while elucidating the multifaceted implications of this challenge for policymakers and governmental and non-governmental entities. Drawing on a qualitative approach, this documentary research retrieves information from documents, such as reports, newspaper articles, magazines, photographs, and other promotional materials. Therefore, this study addresses the following research question: "How can public policymakers establish multidisciplinary primary care teams to improve indigenous populations' health and fight climate change?". This study aims to contribute to comprehending policymakers and other political and organisational stakeholders when dealing with emergencies affecting indigenous communities and structuring multidisciplinary primary health teams.

Theoretical Background

Stakeholders Management Approach

Stakeholder management can be summarised as a process of engaging and overseeing the interested parties of a project. Typically, individuals with various roles will be involved in this effort, each having different levels of influence on the task. Hence, maintaining a solid rapport among stakeholders is crucial to fostering greater harmony.

Before commencing a project, it is essential to identify which stakeholders wield greater or lesser influence over the project. This determination helps prioritise specific stakeholders over others (Clarkson, 1995). When managing a business model or a particular project, it is essential to consider how these stakeholders align with the proposal and how they can contribute to social and environmental benefits, for example (Fooks et al., 2012).

Proper stakeholder management makes it feasible to exert more control over project constraints and the risks inherent in decision-making. Furthermore, it becomes easier to recognise and address the interests of these groups, thereby ensuring the successful execution of the project (Donaldson & Preston, 2022).

Through stakeholder management, it becomes possible to negotiate with parties that exhibit resistance to specific actions or changes that the organisation wishes to implement (Boesso & Kumar, 2009). The significance of stakeholder management also lies in the organisation gaining a broader understanding of the impacts of its decisions at various levels (Parmar et al., 2010). Therefore, stakeholder management is believed to be a process of engaging and overseeing the parties involved in a project. Typically, individuals with distinct roles are part of this kind of work, each with varying levels of influence on each project, making it essential to maintain harmonious relationships among all involved parties (Kaur, 2023).

This management becomes relevant for this project when managing stakeholder engagement, which involves developing strategies for resolving conflicts that may arise during the project's progression. It is necessary to ensure the positive participation of all parties throughout the project's entire lifecycle.

Quality of Work Life (QWL) and Multidisciplinary Teams

Quality of work life has received notable attention over the past two decades; however, several uncertainties still need to be addressed regarding the exact meaning of the term (Dechawatanapaisal, 2017; Rodrigues, 1995). The evolutionary conception of QWL began between 1959 and 1972 when the quality of work life was considered merely as a variable. However, new analytical perspectives emerged from the need to investigate how to improve living conditions in the workplace and how individuals within an organisation could attain better work-life conditions (Nayak et al., 2018).

The coefficient of Quality of Work Life arises from three main variables: task dimensions, critical psychological states, and personal and work outcomes (Pedroso et al., 2014). It affects the individual, service provision, and quality, necessitating preventive efforts (Prado, 2016). 'Autonomy' pertains to the extent of responsibility that a worker feels regarding their task. Autonomy leads the worker to experience increased accountability for their work (Rosa et al., 2018). 'Interrelationship' relates to an individual's contact with others in the workplace or how they interact with other organisation members (Deon, 2021). Additionally, 'feedback' refers to the information a worker needs to influence their performance within the process (Pedroso et al., 2014; Rodrigues, 1995).

The task dimension variable and its sub-variables are vital factors in achieving the three critical psychological states: 'perceived task significance', which is the degree to which a worker smells or feels how meaningful, important, or valuable their task is within the organisation (Rosa et al., 2018).

Perceived job responsibility refers to the extent to which an individual understands or personally feels responsible for the outcomes of the task they perform. Regarding knowledge of work outcomes, this critical psychological state refers to the degree to which the worker knows and understands how they effectively complete their task (Nayak et al., 2016).

These three critical psychological states, in turn, take into account personal and work outcomes, which include overall job satisfaction, which represents the worker's general level of well-being about their job; internal motivation for employment, which is the degree of internal reason an individual experiences with positive inner feelings when performing their tasks with quality and negative internal feelings when performing them with low quality (Rosa et al., 2018).

Specific satisfaction represents the worker's well-being concerning Supervision, Job Security, Compensation, Social Environment, and Growth. High-quality work production is how a worker perceives the difference between high-quality and low-quality performance. Additionally, low absenteeism and turnover reflect the degree to which a worker feels the need to change the organisational environment (Rodrigues, 1995, p. 23).

Organisations must be aware that they must participate in transforming this reality, finding a balance between work and improving the quality of life, as emotional difficulties arising from personal life can impact performance. Several necessary workplace conditions highlighted by Rodrigues (1995, p. 125) are described:

- Working conditions: cleanliness, organisation, safety, and health hazards.
- Health conditions: assistance, education, and operational health.
- Morality relationships: task identity, interpersonal relationships, recognition/feedback, people orientation, and job security.
- Compensation: salaries, variable compensation, and benefits.
- Participation: creativity, personal expression, and the impact of given ideas.

Among several factors influencing human behaviour, we are particularly interested in motivation. To understand people's behaviour, it is essential to comprehend human motivation. According to Zaman et al. (2021), motivation is the drive to achieve a specific goal for which an individual expends energy. People exhibit varied attitudes towards motivation because their needs differ from one person to another, resulting in different behaviour patterns. Social values also vary, as do strengths and weaknesses. The definition of motivation, whether in an individual context or within institutional relationships, contributes to the organisational climate. Individuals are deeply engaged in adapting to various situations to satisfy their needs and naturally maintain emotional equilibrium. Emotional issues in one's personal life can also interfere with the work environment and even assume similar forms (Zaman et al., 2021).

Healthcare services fail to address health and social inequities as integral components of healthcare delivery (Browne et al., 2016). It is imperative to approach indigenous health issues through a multidisciplinary framework. This approach is advantageous due to the collaborative exchange of experiences among healthcare professionals, fostering a culture of respect for cultural, ethnic, social, and ideological diversity. This, in turn, facilitates a more extensive sharing of knowledge for effective problem-solving (Sillitoe, 2004). Within the indigenous context, the efficacy of healthcare, spanning from primary care to high-complexity interventions, must integrate indigenous idiosyncrasies while respecting their traditions and beliefs (Taylor et al., 2020).

Consequently, utilising a multifunctional team is one of the most compelling and impactful approaches to delivering respectful and efficient healthcare to indigenous communities (Gray, 2010; Topp et al., 2021). Ensuring universal access to proactive first-contact care and public health interventions is paramount (Macinko & Harris, 2015). The multifunctional team's strategy enhances healthcare quality and contributes to the broader objective of providing comprehensive and inclusive health services to indigenous populations. This integrated approach aligns with cultural competence and evidence-based practice principles, fostering a healthcare environment responsive to indigenous communities' unique needs and perspectives.

Research Design

Drawing on a qualitative approach (Corbin & Strauss, 2015), this research sought to gather information to assist in characterising the phenomenon under investigation. Qualitative analysis and its central aspect relate to how the researcher can access an abundance of descriptive data acquired through the direct engagement of the researcher with the studied context (Denzin & Lincoln, 2018) once it seeks to capture the nuanced perspectives of participants (Flick, 2017). The research design is intentionally open and flexible, focusing on a complex and contextualised understanding of reality (Corbin & Strauss, 2015).

The selected method for data collection is documentary research (Denzin & Lincoln, 2018). The rationale for adopting this methodology stems from its capacity to unveil a wealth of information, thereby enhancing comprehension of entities requiring sociocultural and historical contextualisation (Corbin & Strauss, 2015).

Moreover, documentary analysis enables observing complex systems involving individuals, groups, concepts, knowledge, behaviours, mentalities, and practices (Cellard, 2008).

Websites and journalistic articles addressing the health emergency theme, particularly from January 2023 onward, recordings, and photographs were utilised as data sources for the research, and this period witnessed an increased proliferation of online materials. Additionally, government websites from Brazil, prevailing legislation about indigenous matters in the country, publications disseminated by non-governmental organisations (NGOs), and regulatory and monitoring government agencies such as the Pan American Health Organization (PAHO) and the United Nations Children's Fund (UNICEF) were also incorporated into the research methodology.

A content analysis was conducted to analyse the data based on the material's coding and categorisation. Following the criteria outlined by Bardin (2016), the acquired material underwent the stages of pre-analysis, material exploration, and results treatment. Beyond comprehending the analytical process, some questions were formulated to guide the research from data collection through analysis to presentation, as follows:

- How did the Brazilian government structure the emergency healthcare response for the Yanomami?
- Which professionals were selected to form the multidisciplinary team for emergency healthcare for the Yanomami?
- What is the management approach for the various stakeholders involved in the project?

Results and Discussion

Yanomami People's Territorial and Health Aspects

In Roraima, the highest percentage of indigenous people lives in Uiramutã. In the region of the Raposa Serra, the Sol indigenous land, 88.1% of the population is indigenous. The city of Normandia (56.9%) ranks second in the percentage of indigenous people, followed by Pacaraima (55.4%) and Amajarí (53.8%). The Yanomami land, located in both Amazonas and Roraima, has the largest indigenous population, with 25,700 indigenous individuals, accounting for 5% of the total indigenous population in the country.

Primary healthcare in the indigenous context, operated by 34 special indigenous sanitary districts, faces the challenge of ensuring access and the comprehensive nature of care based on intercultural dialogue and the appreciation of the specific traditional knowledge of each indigenous group. This encompasses a range of services, including health promotion (e.g., guidance on better nutrition) and prevention (such as vaccination and family planning) to treat acute and infectious diseases, control chronic conditions, provide palliative care, and rehabilitation.

Primary Healthcare for the Yanomami people is the responsibility of the SESAI, a branch of the Ministry of Health. Under its supervision is the SASISUS, the indigenous healthcare subsystem. Through this subsystem, indigenous peoples are supposed to have their quality of life ensured. Mobile teams established in each FUNAI's regional office carry out sporadic health actions. The Ministry of Health collaborates with the control of major endemic diseases (Brazil, 2021). Over the years, FUNAI has signed agreements with governmental and non-governmental entities to address the need for proper infrastructure.

The Principal Domains of the Basic Healthcare Initiative for the Yanomami People

According to the Ministry of Health (Brazil, 2021), the DSEI-YY was the first DSEI created within indigenous health, as per Interministerial Ordinance No. 316 of April 11, 1991, issued by the Ministries of Health and Justice. Its headquarters are in Boa Vista—RR, covering an area of 9,664,975 ha, including regions in the states of Roraima and Amazonas, on the border with Venezuela.

The DSEI-YY comprises a CASAI, 37 Base Units, and 78 Indigenous Basic Health Units (UBSI), providing healthcare to 27,723 indigenous individuals distributed across 366 villages inhabited by five distinct indigenous groups: Sanumã, Ninan, Yawari/Xamathari, Ye'kuana, and Yanomami, the most numerous among them (Brazil, 2021).

DSEI's managers acknowledge that certain areas are affected by conflicts, leading to interruptions in entering multidisciplinary indigenous health teams for service at specific points. These conflicts predominantly arise due to the encroachment upon indigenous territories by invaders, such as those engaged in mining and logging activities. This results in an escalation of violence and tension between the invaders and indigenous leadership. Such tension constrains the effectiveness of healthcare teams, considering that basic healthcare facilities must be closed until the situation is normalised and the security of health agents can be ensured. By the time of this research, of the 37 base units in the DSEI, two were temporarily closed: Kataroa and Lahaka.

Table 14.1 below provides a concise overview of healthcare units that necessitated closure during periods of heightened conflict escalation, along with the allocation of professionals tasked with preventive interventions against afflictions such as malaria and malnutrition affecting the Yanomami people in the region. Notable base units where conflicts have occurred include, as presented in Table 14.1.

Given the existing conflicts that directly impact the entry of multidisciplinary indigenous health teams, there is a direct impact on the quality of work these teams provide. Additionally, Rodrigues (1995) relates the responsibility for employment

Table 14.1 Base units in Yanomami territory and their intercurrences and events

Base unit	Intercurrences and events
Kataroa	Closed on October 7, 2021, due to a lack of security for professionals. DSEI has been taking specific actions
Lahaka	Closed in August 2020 due to a lack of security for professionals. The District has been taking specific actions
Parima	Closed on several occasions
Surucucu	A team composed of a nurse, nursing technicians, a physician, and a pharmacist was sent to provide care and fulfil a 30-day rotation
Paapiu	Features a team of nursing technicians and professionals from the Brazilian Onchocerciasis Elimination Program who fulfilled a 30-day rotation
Uxiú	Features a team of nursing technicians who fulfilled a 30-day rotation
Kayanau	Features a team of nurses, nursing technicians, and disease control agents who fulfilled a 30-day rotation
Alto Catrimani	Features a team of physicians and nursing technicians who fulfilled a 30-day rotation
Baixo Mucajaí	Features a nurse and nursing technicians' team who fulfilled a 30-day rotation
Alto Mucajaí	Features a team of nursing technicians who fulfilled a 30-day rotation
Pewaú	Features a team of nurses, nursing technicians, and disease control agents who fulfilled a 30-day rotation
Xaruna	A team composed of a nurse, nursing technicians, a physician, and a pharmacist was sent to provide care
Hewethe-ú	A team composed of a nurse, nursing technicians, a physician, and a pharmacist was sent to provide care
Palimiú	The DSEI Yanomami resumed routine care at the Palimiú Base Unit, with support from the National Force and Funai, enabling the maintenance of 780 indigenous individuals

Source Brazil—Ministry of Health (2021)

(RP) to understanding the quality of life at work and how workers feel about the potential outcomes of their activities. Thus, the closure of specific base units and the significant number of base units where conflicts have occurred create instability in continuing actions. In these conflicts, threats from indigenous leaders are present, causing fear for their lives.

Over the years, various potential projects have been developed to alleviate the hardships experienced by the Yanomami people since their first contact with Westerners. For instance, the Urihi-Health Yanomami project constituted a significant initiative that operated from the 1990s to the 2000s, providing essential healthcare services to the indigenous population in the territory (Delgado, 2023). According to Delgado (2023) in a Sumaúma online article, the model proposed by the NGO is regarded as ideal by indigenous leaders who have experienced this type of healthcare, as mentioned in Delgado's report (2023, paragraph 2): "It is necessary to bring back to the villages the Urihi model." These and other initiatives that have taken shape over the years aimed to implement projects for essential healthcare services for indigenous peoples in Brazil, addressing institutional gaps resulting from resource limitations (such as changes in government and regulatory systems for public service provision), difficulties in accessing territories, and awareness of social and cultural differences that may impact healthcare delivery.

Additional fronts for healthcare and Yanomami health maintenance were established, starting from January 20th, 2023, under a Public Health Emergency of National Importance (PHENI) (PAHO, 2023a). For example, the Pan American Health Organization (PAHO) established a knowledge hub that provides various reports and guidelines for professionals in multiple fields, organised into thematic sections such as infant and adult malnutrition, malaria, and mercury exposure in children and adults (PAHO, 2023a). Another initiative was the collaborative development between the Ministry of Health of Brazil, the Professor Fernando Figueira Comprehensive Medicine Institute (IMIP), and PAHO for training professionals attending to malnutrition cases (PAHO, 2023b).

Furthermore, as indicated by the Ministry of Health, by February 20th, 2023, 5000 consultations had already been conducted for the Yanomami. Simultaneously, 5500 emergency food baskets were delivered, with plans to increase the number to alleviate food insecurity in the territory (Carta Capital, 2023).

This demonstrates how the federal government, in collaboration with public entities, international regulatory bodies, NGOs, and other institutions, sought to develop a system to control the impacts experienced by the Yanomami people. This led to the involvement of multiple professionals in diverse teams aiming to address various aspects, whether in welcoming indigenous families, understanding their cultural and social systems and their relationship with the forest, or providing primary healthcare, humanitarian aid, and financial assistance to ensure human dignity for those living in the territory.

Promoting Quality of Life at Work to Multidisciplinary Teams

In a remote point in the middle of the Amazon jungle at the Surucucu Base Pole, in the extreme north of the country, a few kilometres (km) from the border between the state of Roraima and Venezuela, the only way of access between different communities is by air.

In this context, it is possible to understand the working conditions of health teams that work in the region where the Yanomami people live, as highlighted in the statement below:

> Here in Surucucu, most of the villages are difficult to access. It is an immense challenge for all those responsible for helicopter logistics and some employees of the DSEI, who tirelessly strive to rescue patients with the most diverse problems, such as malaria, malnutrition and dehydration. These diseases have had a significant increase, directly related to the increase in the number of invaders of the territory, in addition to incidents inherent to life in the forest,

> such as snake bites, falls from heights, among others (quote from Gabriela Mafra interview, who has been treating the Yanomami) (Vilela for Agência Brasil online, 2023).

The work carried out is challenging due to the conditions in which it is carried out, access to the site, including some dangers typical of regions where the forest is dense, the logistics necessary for the operation of the centres, in short, even distance from family are factors that can be considered coefficients of quality of life at work (Zaman et al., 2021). All existing tensions can also impact your relationship with labour; it can have a physical aspect even if present from emotional damage, as they are inserted in the process of elements that affect the individual psychically and with this, harm is established physically (Margis et al., 2003).

The Emergency Operations Centre (COE), created by the government to face the humanitarian crisis, carried out 111 removals within its territory, transporting patients from villages to be treated in local health centres in the Yanomami indigenous land. The priority for rescues and transportation of supplies makes monitoring primary care with community visits challenging (Vilela, 2023).

> "I didn't think twice about coming here. It's a joy to know that I can contribute to reducing their suffering", he says. But joy is mixed with a feeling of helplessness. "Because we're going to leave, and they will continue. I have a family, and I can't stay for long", he gets emotional (quote from Marcos Fonseca interview who is a member of the National Force of the Unified Health System (SUS), which is reinforcing services in the indigenous land) (Vilela for Agência Brasil online, 2023).

In the case mentioned above, an emotional bond with indigenous populations, which these professionals deal with daily, can be seen in his speech. This refers to the limited time to carry out the activities necessary for care; this condition that gives rise to the sensation of being unable to complete the work is one of the biggest enemies of the worker's health and productivity (Zaman et al., 2021). Thus, it is believed that man's relationship with his work is influenced by external threats from different origins, including human perception, which denotes both personal vulnerability and the ability to adapt. Then, it is understood that the stimuli of the work environment demand responses; thus, a specific stress phenomenon is directly associated with the individual's perception of considering the events as stressful or not, including the cognitive aspect, which plays a role of great importance in the process in which stimuli occur.

The approach to quality of life at work also considers the stressors in the work environment. Genuíno et al. (2009) use the term occupational stressor to designate stimuli generated in the work environment and have negative physical and psychological consequences for many people exposed to these stimuli.

In the biological aspect, stress is characterised by the body's exhaustion level. At the same time, the person's affective, emotional and intellectual processes denote the psychological aspect of how the individual relates to other people and the environment. In the external world, the sociological element also consists of understanding the variables established in the world around you (Deon, 2021; Prado, 2016).

Key Actors Shaping the Contextual Dimensions of Quality of Work-Life

The Ministry of Health has declared a Public Health Emergency of National Importance due to the need to combat the lack of healthcare for the people living in the Yanomami territory (Brazil, 2023). In addition to the emergency declaration, the Ministry has also established the Emergency Public Health Operations Centre (COE—Yanomami) as a national mechanism for the coordinated management of the emergency response at the national level. The administration of COE will be under the responsibility of SESAI, considering the typology of the emergency.

Regarding the selection of stakeholders who make up the Multidisciplinary Emergency Healthcare Team for the Yanomami, primarily, these were voluntary professional groups who arrived in Roraima. In total, 40 professionals, including nutritionists, pharmacists, social workers, doctors, and nurses, trained in the Ministry of Health's program. They form nine multidisciplinary teams focused on providing care at CASAI. They travel to three healthcare hubs in the indigenous territories of Auaris, Surucucu, and Missão Catrimani to actively seek out patients. To do this, the professionals undergo specific training to address cases of malnutrition and malaria (Melo, 2023).

Managing the various stakeholders involved in the project involves coordinating and planning the Emergency Operations Centre (COE). Established by the Ministry of Health and coordinated by SESAI, it is composed of the Civil Defense, the Office of the Chief of Staff of the Presidency of the Republic, the FUNAI, the Pan American Health Organization (PAHO), the Ministry of Defense, the Ministry of Justice and Public Security, the Ministry of Social Development, and the Oswaldo Cruz Foundation (Fiocruz). In addition to the volunteers, COE coordinates food distribution and replenishment of health supplies and deploys mobile internet structures to improve communication with healthcare hubs. Patients in more critical conditions are referred to reference hospitals during the care provided.

Regarding measures to ensure the quality of work life for stakeholders (Zaman et al., 2021) directly involved in healthcare services for the Yanomami, the training of human resources for indigenous health should be prioritised as a fundamental tool for aligning the actions of SUS health professionals and assistance with the specificities of indigenous healthcare and the new technical, legal, political, and organisational realities of the services. Courses for updating, refining, and specialising managers, healthcare professionals, and technical advisors (indigenous and non-indigenous) from various institutions within the system should be promoted (Brazil, 2017).

Work is an essential element of a productive individual's life, occupying a significant portion of their daily routine, forming their identity and subjectivity, and facilitating their integration into social life with a unique perspective on society and themselves. Occupational risk factors are identified in organisational environments, including chemical, biological, ergonomic, psychosocial, mechanical, and accidental risks. It is essential to ensure that internal stakeholders have a safe working environment. A secure environment is crucial (Boesso & Kumar, 2009; Kaur, 2023). It is essential to take all necessary measures to ensure occupational safety, especially for professionals working in such unique environments as the Yanomami territory. Therefore, it is fundamental to establish actions that minimise or eliminate the risks in this environment, acknowledging that there are many challenges to address.

Therefore, internal stakeholders (Boesso & Kumar, 2009; Kaur, 2023) are often susceptible to occupational illnesses. It is crucial in the workplace to recognise existing risks for their prevention. Consequently, internal stakeholders carry out their professional duties in the face of threats, including occupational accidents, which pose a significant problem for public health and a country's economy. They result in costs not only for the injured party but also for the state. However, understanding the actual working conditions, such as the reality of healthcare delivery, the scope of occupational hazards, the outcomes of intervention actions, the consequences of workplace accidents, or the training requirements, makes it possible to "implement" preventive policies.

For Yanomami healthcare services, teams are accompanied by one indigenous health agent and one local professional, who will serve as interpreters. These professionals received anthropological and cultural guidance about the indigenous people before entering the region. They were also trained to combat malaria, as well as in cases of severe malnutrition in children. Each group sent to the area consists of 1 doctor and two nurses. Servers must go on foot, by boat or helicopter to the villages where the Yanomami live. The means of travel depends on the location where the service will be provided.

Furthermore, as presented in the Urihi NGO proposal, as reported by Delgado (2023), the perspective of increased productivity for healthcare professionals can only be achieved, according to the interviewee Deise Alves, by ensuring favourable working conditions for the professionals to keep the system operational (Delgado, 2023). However, these projects require a substantial amount of work and resources for their operation, given the operational and logistical challenges often imposed by the movement of these teams to the territories. In cases such as the humanitarian crisis highlighted in this research, the difficulty is further compounded by the need to ensure the proper protection of these teams and mitigate issues arising from the confrontation between the Yanomami people and invaders of the territory, primarily associated with illegal mining.

The federal government is investigating alleged threats made to healthcare professionals and federal public servants deployed to Roraima to aid the Yanomami indigenous people. According to members of the Ministry of Human Rights and Citizenship delegation who were in Boa Vista, ensuring the physical safety of healthcare and rescue teams is an additional challenge in addressing the healthcare crisis affecting the largest exclusive indigenous reserve in the country. There is growing concern about the human rights violations and insecurity within this scenario.

Stakeholder Mapping and Improvement Proposals

Stakeholder mapping contributes to identifying those involved in the project and organising contact between the team and all interested parties to maintain alignment between all proposed objectives, as shown in Fig. 14.1.

Figure 14.1 illustrates the interconnection among various governmental instances and institutions for essential healthcare provision to the Yanomami people. On the one hand, public organisations such as the Unified Health System (SUS) and the National Health Foundation (FUNASA) connect with other indigenous healthcare institutions such as SESAI and CASAI, in a relationship mediated by the overarching instance, the FUNAI. This relationship led to the Brazilian government's structuring of the National Coordination Committee, with the initiative and operational proposal to provide emergency services for primary Yanomami healthcare to mitigate the impacts resulting from the illegal occupation of the territory and abandonment by governmental spheres. These factors have contributed to increased health problems such as food insecurity, severe malnutrition, acute diarrheal diseases, malaria, tuberculosis, and mercury contamination (PAHO, 2023b). In other words, the collaboration among these entities has enabled the circumvention and control of sanitary, environmental, sociocultural, and economic impacts on the communities arising from the illegal occupation of the Yanomami territory.

It is essential to highlight the participation of government ministries in sharing financial and personnel resources to establish healthcare mechanisms. Several ministries actively participated in this effort, including but not limited to the Ministry of Indigenous Peoples, Ministry of Health, Ministry of Environment, and Ministry of Human Rights. The active involvement of ministries creates an opportunity for

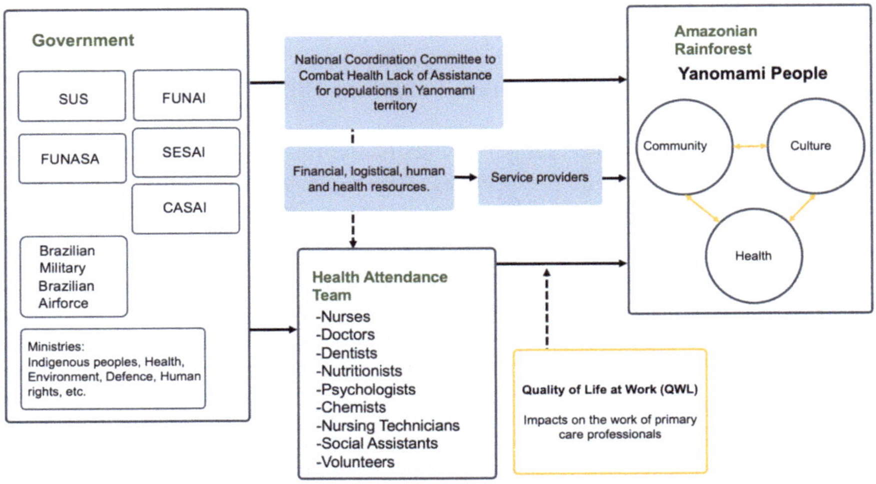

Fig. 14.1 Identified stakeholders. *Source* Authors

sharing expertise across multiple healthcare fronts to minimise impacts and control the situation the Yanomami people face.

Furthermore, sharing financial, logistical, human, and health resources determines how these resources are allocated throughout the process and within a specified time frame for healthcare interventions and service providers. In other words, the formed multi-professional teams have different operations and methods for controlling these sanitary, environmental, sociocultural, and economic impacts.

It was possible to understand that Stakeholder Management focuses on comprehending the relevance and level of influence of all those involved in the project. Stakeholder mapping aids in identifying who wields positive, negative, and moderate power. In this context, the concept of quality of life at work stands out, which encompasses various interpretations ranging from the medical focus on the absence of diseases in workers to the requirements for resources, objects, and procedures that meet collective demands in a particular situation. Safety is one of the primary foundations for preserving the workforce, as they interact to safeguard both personal and material working conditions that provide a certain level of employee health. To protect workers, companies should strive to provide employees with a clean, healthy, and pleasant environment, which can lead to increased productivity.

Indigenous Territories as Vital Frontlines in Combating Climate Change: An Urgent Call for Policymakers

Addressing climate change issues has been the subject of numerous discussions aimed at developing more efficient production and consumption systems and preserving forests and natural resources. Beyond reducing carbon emission rates in the atmosphere to achieve a net-zero target, directly impacting natural ecosystems (Low et al., 2024), fighting climate change involves a deeper and more systemic understanding of the intricate interactions among different and complementary causes.

For example, studies have sought to understand this phenomenon from social, environmental, and economic perspectives. Delving deeper, several key issues have emerged from these discussions, such as the productive efficiency of food and more sustainable food systems (Ran et al., 2024), initiatives to phase out coal use for energy based on compensations and policy orientation (Nacke et al., 2024), the development of production systems grounded in the Circular Economy to mitigate climate change (Gallego-Schmid et al., 2020), and the pursuit of a zero waste framework (Greyson, 2007).

The energy transition has emerged as an essential international front, linked to reducing fossil fuel use and adopting cleaner energy sources, such as bioenergy (Low et al., 2024), wind, and solar energy (Tan et al., 2021). Adapting energy systems has prompted economies to explore new ways to compensate for potential imbalances this transition might cause, directly affecting groups economically dependent on activities like coal mining and related power plants (Nacke et al., 2024).

Complementary, various initiatives have been established and need to be effectively implemented to achieve significant results in this fight against the harm done to nature and life preservation. Public policy development has proven to be an efficient means of encouraging the adoption of sustainable practices that can significantly reduce the impact of human activity on the climate. This can involve organising around social issues related to ecosystem adaptation, such as those faced by riverine communities (Eneh, 2024), or changing economic perspectives to promote the use of renewable resources in energy generation (Low et al., 2024; Nacke et al., 2024). The climate issue has driven global economies due to the urgent need to meet CO_2 emission levels that can sustain life, preventing the reach of a tipping point that would directly impact environmental and human systems (Hoegh-Guldberg et al., 2018).

A preview of these tipping points has already shown its potential through the increased number of extreme climate events observed, especially in recent years. Events such as floods in East Africa and droughts in Southern Africa (WMO, 2024), the increased destructive potential of cyclones and tornadoes (Berardelli, 2019), or the historic drought that devastated rivers in the Amazon region in 2023 (WWF Brazil, 2023), are some examples of why combating climate change must be a common goal among nations.

This highlights the importance of forest regions worldwide and raises concerns about the capacity to preserve and maintain these ecosystems to combat climate change. Despite some efforts, certain areas still see increasing activities such as deforestation, wildfires, and the loss of native vegetation to cattle ranching or mining. Understanding the fight against climate change based on the role forests play for indigenous populations is crucial, as these communities act as stewards of these ecosystems. Considering that approximately 36% of the remaining intact forest zones are located in indigenous territories (World Bank, 2023), we can see the importance of these communities in maintaining ecosystems vital for mitigating the potential effects of climate change. This work delves deeper into the Brazilian context, examining how an indigenous community navigates issues related to preserving their territory and maintaining their cultural, social, and health aspects, which are crucial for life preservation.

Based on the research question that was foundational for this study, "How can public policymakers establish multidisciplinary primary care teams to improve indigenous populations' health and fight climate change?" we sought to understand how the case experienced by the Yanomami people and the healthcare provided by the Brazilian government can establish means to comprehend the reality experienced by indigenous populations and their needs.

However, healthcare must be a connected factor that runs parallel to the environmental element and the protection/preservation of forests and indigenous territories. Indigenous *weltanschauung* is intrinsically linked to their territory, as this space directs the daily life of indigenous existence. Building on the idea that indigenous peoples are responsible for preserving rivers, fauna, and flora and that their relationship with the forest is the structuring axis of their culture, social system, and expression of daily life, it opens space for reflection on the need for public and private entities to establish mechanisms aimed at minimising the impacts resulting from human activity in the territories of these peoples, as a pathway to fight climate change. The case experienced by the Yanomami people is primarily characterised by the illegal occupation of their territory for gold extraction. However, activities such as logging and pasture for cattle also create room for processes of degradation of indigenous territory and directly impact the economic and social system of these peoples, with direct repercussions on their health.

The occupation of the territory by illegal agents and the consequent deforestation have practical implications for indigenous health, as rivers and forests are a source of food and income for this population. An example is the mercury contamination experienced by the Yanomami. The use of mercury by illegal prospectors in the search for gold on riverbanks causes the water, when in contact with riverbanks and streams, to carry this contaminant to the communities (Jacques et al., 2023; Pestana et al., 2022). Moreover, this activity has a series of other implications, such as increased violence in the region, the use and formation of routes for drug trafficking, and abuse of indigenous children, adolescents, and women by prospectors (Lobo & Lúcia, 2023).

These factors demonstrate how the preservation of indigenous territory cannot be viewed from a simplistic perspective but from a systemic standpoint that starts from the environmental aspect to preserve their environment without disconnecting it from social and economic factors.

Practical Implications for Policy Makers

In practical terms, policymakers must be attentive to the macro-environment and the multiple facets of resources that must be allocated. Considering the various stakeholders involved in healthcare, for example, in the case of an indigenous community, it is necessary for the public authorities and public regulatory agencies to pay attention to the needs of both the indigenous population and the professionals involved. Relevant issues for policymakers include:

- The connection of the indigenous community with the forest as a means of constructing their culture and social factors that are knottily linked to everyday life in the forest.
- Financial, logistics, human, and health resources to structure basic indigenous healthcare systems still need to be available.
- There is an urgent call for integrated efforts among multiple government agencies to guide these resources, organise activities, and establish control and monitoring mechanisms.
- Healthcare systems need to respect the cultural aspects of the population, considering these are complex social systems permeated with traditional knowledge, including the qualification of health professionals.

- Implementing the QWL approach minimises psychological and psych-emotional impacts on these agents and seeks to reduce team stress and fatigue.

The Yanomami health emergency has established a new paradigm for indigenous health care and overcoming health and humanitarian crises. Because it involves a series of activities provided by multifunctional teams, attention needs to be given to the healthcare agents in the field conducting these activities, as they are vulnerable to the emotional impacts resulting from the precarious humanitarian situation.

Establishing the Emergency Operations Centre (COE) and integrating different ministries and other public agents for emergency care proved to be an effective tool for reducing the impacts resulting from the neglect experienced by the Yanomami people in their territory. However, it is worth noting that this study sought to establish a discussion about factors that may contribute to increasing the impact on the physical and emotional health of healthcare agents providing these services.

Conclusions

In the early days of 2023, the world became aware of the severe humanitarian crisis faced by the Yanomami people residing in the states of Roraima and Amazonas due to a lack of healthcare assistance. In response, this study sought to understand the perception of internal stakeholders regarding their QWL in a humanitarian crisis context, aiming to identify stakeholder management strategies and their impact on the healthcare services provided to the Yanomami community.

There are areas affected by conflicts that have led to interruptions in the entry of multidisciplinary indigenous health teams to provide healthcare in certain areas. Two of the 37 Base DSEIs, Kataroa and Lahaka, are temporarily closed. The absence of local infrastructure and the scarcity of resources are noteworthy, highlighting the need for a structured healthcare service. Even though the Centre for Emergency Public Health Operations (COE-Yanomami) is working in this regard, as previously mentioned, access is complex, and the number of professionals available to work in the area needs to be increased. Not all internal stakeholders have a compatible profile to work under the provided conditions, resulting in a significant turnover of professionals in certain areas.

Internal stakeholders require specific preparation for interaction with different ethnic groups, considering language, culture, and customs. Adaptation time on both sides is necessary to strengthen healthcare services within the indigenous community. Therefore, it is understood that primary healthcare is an integral and inseparable part of an individual, and to ensure effective and quality healthcare for indigenous populations, it is highly relevant to respect their ethnocultural conditions without interfering in their daily lives in the community.

In conclusion, the imperative for providing primary healthcare assistance to indigenous populations, such as the Yanomami people in the Amazonia Rainforest, must be supported if we aim to fight climate change. The recent humanitarian crisis faced by the Yanomami due to inadequate healthcare underscores the urgency of prioritising and enhancing primary healthcare services tailored to the ethnocultural conditions of these communities. Establishing a Public Health Emergency Operations Centre by the Brazilian Ministry of Health represents a commendable step towards more effective healthcare responses. Policymakers must continually improve the governance process of multidisciplinary primary care teams by focusing on stakeholder management and the quality of work life. This approach ensures not only the qualification of internal stakeholders but also addresses the physical and psychological consequences faced by healthcare teams working in challenging environments.

By respecting the legacy and unique needs of indigenous populations, policymakers can contribute to improved health outcomes and play a crucial role in mitigating the impact of climate change on vulnerable communities. Recognising that effective primary healthcare is a humanitarian imperative and a key strategy in building resilience and preventing new disasters among indigenous populations is essential.

References

Ab'sáber, A. N. (1996). *Amazônia: Do discurso à práxis.* Edusp.

Bardin, L. (2016). *Análise de conteúdo.* Edições 70.

Becker, B. K. (2005). Geopolítica da Amazônia. *Estudos Avançados, 19*(53), 71–86.

Berardelli, J. (2019). How climate change is making hurricanes more dangerous. *Yale Climate Connections.* [online]. Available at: https://yaleclimateconnections.org/2019/07/how-climate-change-is-making-hurricanes-more-dangerous/

Boesso, G., & Kumar, K. (2009). An investigation of stakeholder prioritisation and engagement: Who or what counts. *Journal of Accounting & Organizational Change, 5*(1), 62–80. https://doi.org/10.1108/18325910910932214

Brazil. (2012). *Decreto nº 7.747, de 5 de Junho de 2012.* Institui a Política Nacional de Gestão Territorial e Ambiental de Terras Indígenas—PNGATI. Available at https://www.planalto.gov.br/ccivil_03/_ato2011-2014/2012/decreto/d7747.htm. Accessed April 17, 2023.

Brazil. (2017). *Política Nacional de Atenção à Saúde dos Povos Indígenas.* Available at: https://bvsms.saude.gov.br/bvs/saudelegis/gm/2017/MatrizesConsolidacao/comum/37488.html. Accessed May 10, 2023.

Brazil. (2021). *Ministério de Saúde. Ações de Saúde no DSEI Yanomami.* Available at: https://www.gov.br/saude/pt-br/assuntos/noticias/2021-1/novembro/acoes-de-saude-no-dsei-yanomami. Accessed May 15, 2023.

Brazil. (2023). *Ministério da Saúde. Ministério da Saúde declara emergência em saúde pública em território Yanomami.* Available at: https://www.gov.br/saude/pt-br/assuntos/noticias/2023/janeiro/ministerio-da-saude-declara-emergencia-em-saude-publica-em-territorio-yanomami. Accessed May 10, 2023.

Browne, A. J., Varcoe, C., Lavoie, J., Smye, V., Wong, S. T., Krause, M., Tu, D., Godwin, O., Khan, K., & Fridkin, A. (2016). Enhancing healthcare equity with Indigenous populations: Evidence-based strategies from an ethnographic study. *BMC Health Services Research, 16*(1), 544. https://doi.org/10.1186/s12913-016-1707-9

Carta Capital. (2023). *Yanomami: mais de 5 mil atendimentos médicos foram feitos em um mês*. Available at: https://www.cartacapital.com.br/saude/yanomami-mais-de-5-mil-atendimentos-medicos-foram-feitos-em-um-mes/. Accessed July 15, 2023.

Cellard, A. (2008). A Análise documental. In J. Poupart et al. (Orgs.), *A pesquisa qualitativa: Enfoques epistemológicos e metodológicos* (pp. 295–316). Vozes.

Clarkson, M. E. (1995). A stakeholder framework for analyzing and evaluating corporate social performance. *Academy of Management Review, 20*(1), 92–117. https://doi.org/10.5465/amr.1995.9503271994

Corbin, J., & Strauss, A. (2015). *Basics of qualitative research: Techniques and procedures for developing grounded theory* (4th ed.). Sage.

Cunha, H. B. (2008). *O mundo das águas. Scientific American Brasil. Amazônia: tesouros* (pp. 6–13). Duetto Editorial.

Dechawatanapaisal, D. (2017). The mediating role of organisational embeddedness on the relationship between quality of work life and turnover. *International Journal of Manpower, 38*(5), 696–711. https://doi.org/10.1108/ijm-12-2015-0205

Delgado, M. (2023). De Copérnico a Kafka. Ou como o Estado puniu os médicos que revolucionaram a saúde indígena no Brasil: A história da Urihi. *Sumaúma*. Avalilable at: https://sumauma.com/copernico-kafka-estado-puniu-medicos-revolucionaram-saude-indigena-brasil-historia-urihi-yanomami/. Accessed August 20, 2023.

Denzin, N. K., & Lincoln, Y. S. (2018). *The Sage handbook of qualitative research* (5th ed.). Sage.

Deon, A. P. R. (2021). *Estresse ocupacional e estratégias de coping no trabalho de psicólogos*. PhD Thesis, Universidade Federal de Santa Catarina.

Donaldson, T., & Preston, L. E. (2022). The stakeholder theory of the corporation: Concepts, evidence, and implications. *Academy of Management Review, 20*(1), 65–91. https://doi.org/10.5465/amr.1995.9503271992

Eneh, O. C. (2024). Livelihood adaptation to changes in River Ekulu depth, channel width and riverbank vegetation cover type by household breadwinners in selected communities in Enugu, Nigeria. *Discover Sustainability, 5*(1). https://doi.org/10.1007/s43621-024-00268-5

Flick, U. (2017). *Introduction to qualitative research* (5th ed.). SAGE Publications.

Fooks, G., Gilmore, A., Collin, J., Holden, C., & Lee, K. (2012). The limits of corporate social responsibility: Techniques of neutralization, stakeholder management and political CSR. *Journal of Business Ethics, 112*(2), 283–299. https://doi.org/10.1007/s10551-012-1250-5

Freitas, F. P. P., Luna, W. F., Bastos, L. O. A., & Ávila, B. T. (2021). Experiências de médicos brasileiros em seus primeiros meses na Atenção Primária à Saúde na Terra Indígena Yanomami. *Interface-Comunicação, Saúde, Educação, 25*.https://doi.org/10.1590/interface.200212

Gallego-Schmid, A., Chen, H.-M., Sharmina, M., & Mendoza, J. M. F. (2020). Links between circular economy and climate change mitigation in the built environment. *Journal of Cleaner Production, 260*, 121115. https://doi.org/10.1016/j.jclepro.2020.121115

Genuíno, S. L. V. P., Gomes, M. S., Moraes, E. M. (2009). O Estresse Ocupacional e a Síndrome de Burnout no Ambiente de Trabalho: Suas Influências no Comportamento dos Professores da Rede Privada do Ensino Médio de João Pessoa. *Anagrama, 3*(2), 1–9. https://doi.org/10.11606/issn.1982-1689.anagrama.2009.35426

Gray, R. (2010). Is accounting for sustainability actually accounting for sustainability, and how would we know? An exploration of narratives of organisations and the planet. *Accounting, Organizations and Society, 35*(1), 47–62. https://doi.org/10.1016/j.aos.2009.04.006

Greyson, J. (2007). An economic instrument for zero waste, economic growth and sustainability. *Journal of Cleaner Production, 15*(13–14), 1382–1390. https://doi.org/10.1016/j.jclepro.2006.07.019

Hoegh-Guldberg, O., Jacob, D., Taylor, M., Bindi, M., Brown, S., Camilloni, I., Diedhiou, A., Djalante, R., Ebi, K. L., Engelbrecht, F., Guiot, J., Hijioka, Y., Mehrotra, S., Payne, A., Seneviratne, S. I., Thomas, A., Warren, R., & Zhou, G. (2018). Impacts of 1.5 °C global warming on natural and human systems. In V. Masson-Delmotte, P. Zhai, H.-O. Pörtner, D. Roberts, J. Skea, P. R. Shukla, A. Pirani, W. Moufouma-Okia, C. Péan, R. Pidcock, S. Connors, J. B. R. Matthews,

Y. Chen, X. Zhou, M. I. Gomis, E. Lonnoy, T. Maycock, M. Tignor, & T. Waterfield (Eds.), *Global warming of 1.5 °C. An IPCC special report on the impacts of global warming of 1.5 °C above pre-industrial levels and related global greenhouse gas emission pathways, in the context of strengthening the global response to the threat of climate change, sustainable development, and efforts to eradicate poverty*, (pp. 175–312). Cambridge University Press. https://doi.org/10.1017/9781009157940.005

Homma, A. K. O. (2014). Extrativismo ou Plantio: qual a opção para a Amazônia? In A. K. O. Homma (Ed.), *Extrativismo Vegetal na Amazônia: história, ecologia, economia e domesticação* (pp. 17–45). Centro de Pesquisa Agroflorestal da Amazônia Oriental.

Jacques, P. D., Viglio, E. P., & Pinto, D. (2023). Environmental geochemical analysis in the Yanomami Indigenous Land, Mucajaí River Basin, State of Roraima, Brazil. *Toxics, 11*(10), 861–861. https://doi.org/10.3390/toxics11100861

Kaur, A. (2023). Stakeholder management challenges in a health-care medical device project in India—An SME perspective. *Emerald Emerging Markets Case Studies, 13*(2), 1–19. https://doi.org/10.1108/eemcs-09-2022-0318

Lacombe, F. (2015). *Recursos Humanos*. Saraiva Educação S.A.

Lobo, M., & Lúcia, M. (2023). Lessons from urgent times: The experience of Yanomami health care then and now. *Cadernos De Saude Publica, 39*(4), 1–6. https://doi.org/10.1590/0102-311xen065623

Low, S., Fritz, L., Baum, C. M., & Sovacool, B. K. (2024). Public perceptions on carbon removal from focus groups in 22 countries. *Nature Communications, 15*(1). https://doi.org/10.1038/s41467-024-47853-w

Macinko, J., & Harris, M. J. (2015). Brazil's family health strategy—Delivering community-based primary care in a universal health system. *New England Journal of Medicine, 372*(23), 2177–2181. https://doi.org/10.1056/NEJMp1501140

Macke, J., & Carrion, R. M. (2006). Planejamento, implementação e avaliação de programas sociais: uma proposta de inovação. *REAd–Edição 53, 12*(5), 320–344.

Margis, R., Picon, P., Cosner, A. F., & Silveira, R. O. (2003). Relação entre estressores, estresse e ansiedade. *Revista De Psiquiatria Do Rio Grande Do Sul, 25*, 65–74. https://doi.org/10.1590/S0101-81082003000400008

Melo, K. (2023). *Governo envia novas equipes de socorro à Terra Indígena Yanomami*. Available at: https://agenciabrasil.ebc.com.br/direitos-humanos/noticia/2023-02/governo-envia-novas-equipes-de-socorro-terra-indigena-yanomami. Accessed June 2023.

Moraes, A. O. S., Magalhães, E. I. S., Orellana, J. D. Y., Gatica-Domínguez, G., Neves, P. A. R., Basta, P. C., & Vaz, J. S. (2023). Food profile of Yanomami indigenous children aged 6 to 59 months from the Brazilian Amazon, according to the degree of food processing: A cross-sectional study. *Public Health Nutrition, 26*(1), 208–218.https://doi.org/10.1017/S1368980022001306

Nacke, L., Vinichenko, V., Cherp, A., Jakhmola, A., & Jewell, J. (2024). Compensating affected parties necessary for rapid coal phase-out but expensive if extended to major emitters. *Nature Communications, 15*(1). https://doi.org/10.1038/s41467-024-47667-w

Nayak, T., Sahoo, C. K., & Mohanty, P. K. (2018). Workplace empowerment, quality of work life and employee commitment: A study on the Indian healthcare sector. *Journal of Asia Business Studies, 12*(2), 117–136. https://doi.org/10.1108/jabs-03-2016-0045

Nayak, T., Sahoo, C. K., Mohanty, P. K., & Sundaray, B. K. (2016). HR interventions and quality of work life of healthcare employees: An investigation. *Industrial and Commercial Training, 48*(5), 234–240. https://doi.org/10.1108/ict-02-2015-0019

OTCA. (2004). Bulletin, Brasília: OTCA, n° 1, jun./ago. Available at http://www.otca.info/br/institucional/index.php?id=1306. Accessed May 10, 2023.

PAHO - Pan American Health Organization. (2023a). *OPAS trabalha em cooperação técnica com Brasil para melhorar saúde dos povos indígenas yanomami*. Available at. https://www.paho.org/pt/historias/opas-trabalha-em-cooperacao-tecnica-com-brasil-para-melhorar-saude-dos-povos-indigenas. Accessed August 15, 2023.

PAHO - Pan American Health Organization. (2023b). *Contribuição da BIREME à Emergência Sanitária no Território Yanomami*. Available at https://www.paho.org/pt/noticias/23-2-2023-contribuicao-da-bireme-emergencia-sanitaria-no-territorio-yanomami. Accessed August 15, 2023.
Paniz-Mondolfi, A. E., Grillet, M. E., Tami, A., Oliveira-Miranda, M. A., Noguera, L. D., Hotez, P., & Oletta, J. F. (2019). Venezuela's upheaval threatens Yanomami. *Science, 365*(6455), 766–767. https://doi.org/10.1126/science.aay6003
Parmar, B. L., Freeman, R. E., Harrison, J. S., Wicks, A. C., Purnell, L., & de Colle, S. (2010). Stakeholder theory: The state of the art. *The Academy of Management Annals, 4*(1), 403–445. https://doi.org/10.1080/19416520.2010.495581
Pedroso, B., Kovaleski, A., Ferreira, C. L., Pilatti, L. A., Gutierrez, G. L., & Picinin, C. T. (2014). Desenvolvimento e validação da versão brasileira do Diagnóstico do Trabalho (Job Diagnostic Survey) de Hackman e Oldham. *Gestão & Produção, 21*(2), 285–301. https://doi.org/10.1590/0104-530x198
Pestana, I. A., de Rezende, C. E., Almeida, R., de Lacerda, L. D., & Bastos, W. R. (2022). Let's talk about mercury contamination in the Amazon (again): The case of the floating gold miners' village on the Madeira River. *The Extractive Industries and Society, 11*, 101–122. https://doi.org/10.1016/j.exis.2022.101122
Prado, C. E. P. (2016). Estresse ocupacional: Causas e consequências. *Revista Brasileira De Medicina Do Trabalho, 14*(3), 285–289. https://doi.org/10.5327/z1679-443520163515
Ran, Y., Van Rysselberge, P., Macura, B., Martin Persson, U., Abu Hatab, A., Jonell, M., Lindahl, T., & Röös, E. (2024). Effects of public policy interventions for environmentally sustainable food consumption: A systematic map of available evidence. *Environmental Evidence, 13*(1). https://doi.org/10.1186/s13750-024-00333-6
Rodrigues, M. V. (1995). *Qualidade de vida no trabalho evolução e análise no nível gerencial*. Petrópolis Vozes.
Rosa, J. S., Wilhelm, F. A., Deon, A. P. R., & Said, F. L. S. (2018). *Estratégias de enfrentamento: uma pesquisa com servidores estaduais na Amazônia Legal*. ISMA. Available at: http://www.ismabrasil.com.br/congressos/congresso-2018/trabalho. Accessed July 10, 2023.
Scherf, E. L., & da Silva, M. V. V. (2023). Brazil's Yanomami health disaster: Addressing the public health emergency requires advancing criminal accountability. *Frontiers Public Health, 11*, 1–5. https://doi.org/10.3389/fpubh.2023.1166167
Sillitoe, P. (2004). Interdisciplinary experiences: Working with indigenous knowledge in development. *Interdisciplinary Science Reviews, 29*(1), 6–23. https://doi.org/10.1179/030801804225012428
Tan, X., Zhu, K., Meng, X., Gu, B., Wang, Y., Meng, F., Liu, G., Tu, T., & Li, H. (2021). Research on the status and priority needs of developing countries to address climate change. *Journal of Cleaner Production, 289*, 125669. https://doi.org/10.1016/j.jclepro.2020.125669
Taylor, E. V., Lyford, M., Parsons, L., Mason, T., Sabesan, S., & Thompson, S. C. (2020). "We're very much part of the team here": A culture of respect for Indigenous health workforce transforms Indigenous health care. *PLOS ONE, 15*(9), e0239207. https://doi.org/10.1371/journal.pone.0239207
Timossi, L. S., De Francisco, A. C., Santos Junior, G., & Xavier, A. A. P.(2010). Análise da qualidade de vida no trabalho de colaboradores com diferentes níveis de instrução através de uma análise de correlações. *Production 20*(3), 471–480. https://doi.org/10.1590/S0103-65132010005000031
The Lancet. (2019). Bolsonaro threatens the survival of Brazil's indigenous population. *Lancet, 394*, 444. Available at https://www.thelancet.com/pdfs/journals/lancet/PIIS0140-6736(19)31801-X.pdf. Accessed September 10, 2023.
Topp, S. M., Tully, J., Cummins, R., Graham, V., Yashadhana, A., Elliott, L., & Taylor, S. (2021). Unique knowledge, unique skills, unique role: Aboriginal and Torres Strait Islander Health Workers in Queensland, Australia. *BMJ Global Health, 6*(7). https://doi.org/10.1136/bmjgh-2021-006028

UNICEF. (2023). *I feel a deep connection with the Yanomami people as if they were my own.* Available at https://www.unicef.org/brazil/en/stories/i-feel-deep-connection-yanomami-people-if-they-were-my-own. Accessed June 12, 2023.
UN News. (2019). *5 maneiras que os povos indígenas estão ajudando o mundo a alcançar a #FomeZero.* Available at: https://www.news.un.org/pt/story/2019/08/1683741&gt. Accessed August 10, 2023.
Vieira, M. A. R. M., & Lima, L. P. N. S. (2019). *Plano de Gestão Territorial e Ambiental: terra indígena Yanomami.* Hutukara Associação Yanomami (HAY). Available at https://acervo.socioambiental.org/acervo/livros/terra-indigena-yanomami-plano-de-gestao-territorial-e-ambiental-com-protocolo-de. Accessed September 12, 2023.
Vilela, R. (2023). *Profissionais de saúde superam desafios para salvar vidas yanomami.* Available at: https://agenciabrasil.ebc.com.br/saude/noticia/2023-02/profissionais-de-saude-superam-desafios-para-salvar-vidas-yanomami. Accessed June 12, 2023.
Watts, J. (2023). Health emergency over Brazil's Yanomami people. *The Lancet, 401*(10377), 631. https://doi.org/10.1016/s0140-6736(23)00384-7
World Bank. (2023). Climate stories|Indigenous peoples. [online] Available at: https://www.worldbank.org/en/news/feature/2023/08/09/empowering-indigenous-peoples-to-protect-forests
World Meteorological Organization. (2024). *Widespread parts of Asia and Africa reel under extreme weather.* [online] Available at: https://wmo.int/media/news/widespread-parts-of-asia-and-africa-reel-under-extreme-weather. Accessed May 12, 2024.
WWF Brazil. (2023). *Seca extrema na Amazônia pode se agravar e bater recorde histórico.* [online] Available at: https://www.wwf.org.br/?86961/Seca-extrema-na-Amazonia-pode-se-agravar-e-bater-recorde-historico
Zaman, S., Ansari, A. H., & Chaturvedi, S. (2021). Work-life enablers for job satisfaction in healthcare: The moderating role of organisation type. *Industrial and Commercial Training, 54*(1), 95–122. https://doi.org/10.1108/ict-05-2020-0060

Heloisa H. S. De Pontes Holds a master's degree in administration, specialising in Social Management Strategies, from the University Foundation of Blumenau (PPGAD/FURB). Has an MBA in Project Management from the University of São Paulo (FEA-RP/USP) and a degree in Psychology. Engaged in health missions in the indigenous areas of Surucucu and Auaris as a Brazilian Air Force Health Team member. Provides Hospital Psychology services to indigenous communities at the General Hospital of Roraima, Brazil, to the present day.

Lucas C. Stocco PhD in Management, University of São Paulo (FEA-RP/USP), with a split doctorate program at Imperial College London, Business School. Research Assistant at Queen's University Belfast, and Assistant Editor for the Latin American Journal of Management for Sustainable Development (LAJMSD). Engaged in research projects related to indigenous themes in Brazil and productive systems of Amazonian biodiversity. My research interests encompass sustainable operations, sustainable supply chains, business models, indigenous entrepreneurship, and paradoxes and tensions in supply chains.

Luciana O. Cezarino Associate Professor of Business Management at School of Economics, Business Administration and Accounting at Ribeirão Preto (FEARP) at University of Sao Paulo (USP), Brazil. Research themes include the systemic connection between corporate strategy, socio-emotional competencies, and cleaner operations. Regarding professional education, she completed her post-doctoral studies at Politecnico di Milano. She holds a doctorate and master's degree in Business Administration from University of São Paulo (USP) with a split doctoral program at the University Luigi Bocconi. Coordinator of Latin2s research group, focusing on sustainability issues in Latin America. Member of the Golden for Sustainability, hosted by

Bocconi University and ISSS (Brazilian Chapter)—at FEARP-USP. Editor-in-chief of the Latin American Journal of Management for Sustainable Development.

Chapter 15
Climate Variability and Asthma Attacks in Two Climatic Zones in Togo: The Case of the Savanes Region in a Unimodal Sahelian Area and the Case of the Grand-Lome Region in a Bimodal Coastal Area

Essoninam Passike Pokona, Essohanam Boko, Pascal Yaka, and Brama Kone

Abstract Climate change and climate variability trigger certain respiratory conditions, but the extent of its effect is still poorly understood, especially from one geographical area to another. The present study was conducted on Asthma using five (05) years of data from 2018 to 2022 in two climatic areas in Togo, namely the Savanes region in a unimodal Sahelian zone in the north and the Grand-Lome region in a bimodal coastal zone in the south. The methodology employed consisted of an analysis of the correlation coefficient and the significance at 5% of the p-values; identifying environmental parameters, in particular the Normalized Difference Vegetation Index (NDVI) and climatological parameters, namely rainfall (RR), maximum (TMAX) and minimum (TMIN) temperatures, maximum (HMAX) and minimum (HMIN) relative humidity, wind speed (Windspeed) and insolation (Inso), which are strongly correlated with asthma in each region of the study area. The results indicated that NDVI, TMIN and Windspeed are strongly correlated with Asthma in the Savanes region with correlation coefficients and p-values of 0.063 at 2.6%, − 0.53 at 4.2% and − 0.60 at 3.5% respectively. In contrast to the Grand-Lome region, the correlation between UMAX and Inso for Asthma was established, with (r) and

E. Passike Pokona (✉)
West African Science Service Centre on Climate Change and Adapted Land Use (WASCAL), University of Lome, Lome, Togo
e-mail: essoninam38@gmail.com

E. Boko
Faculty of Health Sciences, University of Lome, Lome, Togo

P. Yaka
Climate Services, World Meteorological Organization (WMO) West Africa Sub-Regional Office, Dakar, Senegal

B. Kone
Department of Research Programs, Ministry of Higher Education and Scientific Research, Abidjan, Côte d'Ivoire

W. Leal Filho et al. (eds.), *Climate Change Mitigation and Adaptation in Practice*, Climate Change Management, https://doi.org/10.1007/978-3-031-85217-6_15

p-values of 0.75–0.5% and 0.66 to 1.7% respectively. Asthma prevalence was very high in the highly industrialized coastal region of Grand-Lome (274.8 per 100,000) in the south of the country, in contrast to the Savanes region (64.88 per 100,000) in the north. Finally, October to December seems to be the month's most prone to asthma attacks in both regions, with January and February being particularly prone. This situation might probably be explained by the fact that during these months, the weather is cold, the air is dry and his quality is not proper for breath because of dust and mold. Therefore, actions such as establishing early warning systems for climate-sensitive respiratory diseases, implement multisectoral approach, conduct training and awareness campaigns for both communities and healthcare professionals on the health impacts of climate change, appear necessary to reduce the prevalence of asthma However, this study has been constrained by the limited availability of long-term data, necessitating a study period restricted to five years, potentially limiting the ability to observe longer-term trends in the effects of climate on asthma prevalence.

Keywords Climate variability · Asthma · Grand-Lome and Savanes regions · Togo

Introduction

Climate change has a significant impact on various factors related to public health, including the availability of clean drinking water, air quality, secure housing, and food supply. According to the 2020 World Health Statistics Report from the World Health Organization (WHO), it is anticipated to result in approximately 250,000 additional annual fatalities between 2030 and 2050. The primary causes of these additional deaths are expected to be heat stress, malaria, malnutrition, and diarrhea. The consequences of climate change encompass both direct effects, such as extreme weather events, droughts, floods, landslides, and rising sea levels, particularly affecting densely populated megacities and coastal slums. Additionally, there are indirect effects, including poor water quality and malnutrition due to crop failures, which can lead to an increase in communicable diseases like diarrheal diseases and dengue fever, as well as mental disorders and non-communicable diseases like cardiovascular and respiratory illnesses (WHO, 2018).

Recent years have seen a notable increase in obstructive respiratory diseases, with asthma being a prominent example. This can be attributed, in part, to environmental changes characterized by the rising presence of chemical triggers such as particulate matter and gaseous components like nitrogen dioxide and ozone, as well as biological triggers like aeroallergens in the atmosphere. Aeroallergens can sensitize the airways in individuals with allergies, leading to symptoms of bronchial asthma (WHO, 2020).

Asthma, a condition characterized by reversible airway obstruction and bronchospasm that affects the lungs, results from a combination of genetic and environmental factors. It manifests with common symptoms like wheezing, coughing, and shortness of breath and primarily affects genetically predisposed individuals exposed to environmental risk factors. The onset of asthma is influenced by genetic predisposition and environmental factors, including climate conditions, air pollution, allergens, and airborne chemical irritants (Kristeen, 2021). With a global impact, asthma affects approximately 300 million people and leads to 250,000 deaths annually, as reported by Global Initiative for Asthma (GINA) in 2011. In 2019, an estimated 262 million individuals suffered from asthma, with 455,000 related deaths (Lancet, 2020). In addition, projections a suggest that the number of asthma patients may reach 400 million by 2025, with an annual death toll of 250,000 people worldwide. The prevalence of asthma has been steadily increasing on a global scale in recent decades, posing a significant public health challenge, especially in the context of a growing world population and the challenges posed by adverse climate changes (Kassebaum et al., 2016).

Respiratory diseases, including asthma, rank as the second leading cause of death in Togo National Health Development Plan established by the Ministry of Health and Public Hygiene (MSHP) (MSHP, 2023). The Grand-Lome and Savanes regions, like other parts of the world, are also susceptible to the effects of climate change, significantly contributing to increased respiratory illness and mortality among asthma patients. This results in elevated rates of asthma attacks, hospitalizations, and mortality, all of which negatively impact the well-being of communities, productivity, and the economies of countries.

While there is currently no cure for asthma, it can be effectively controlled and managed, reducing the risk of asthma attacks and related fatalities. Although genetics play a crucial role in the development of asthma, the observed increase over the past few decades cannot be solely attributed to genetic factors Americans with Disabilities Act (ADA, 2016). Identifying the risk factors for asthma is essential in implementing effective strategies for its management.

Considering these circumstances, it is imperative to identify the climatic and environmental factors that trigger asthma and its adverse effects, and to gain insights into how these factors may evolve in the future within the study areas. This study aims to contribute to a deeper understanding of the relationship between climate and asthma in two distinct regions of Togo.

Methodology

Presentation of the Study Area

This research was conducted in two distinct regions of Togo: the Grand-Lome region and the Savanes region. The Grand-Lome region is situated in the far southwestern part of the country, along the Gulf of Guinea coastline, and serves as the location for the nation's capital (Fig. 15.1a). As of 2022, the estimated population of Grand-Lome stands at 2,188,376 according to National Institute of Statistics and Economic and Demographic Studies (INSEED, 2023). This region experiences a

Guinean climate characterized by two distinct rainy seasons. The primary rainy season typically spans from April to July, followed by a secondary, less prominent season starting in September and concluding in late November. Average maximum temperatures under shelter reach approximately 30 °C in the afternoon, while the average minimum temperature hovers around 23 °C in the morning according to the National Meteorology Directorate (Afidégnon, 2003). During December to February or March, the Harmattan, a dry wind from the Sahara, can occasionally reduce morning temperatures to 19 °C. Annual rainfall in this region averages around 900 mm.

The vegetation in the coastal plain exhibits a diverse mosaic. It includes seagrass areas dominated by key species such as Remirea Maritima, Sporobolus Virginicus, and Canavalia Roseus. Additionally, coastal grassy savannahs host species such as Aristida Sieberiana and Aristida Adscensionis, forming pure stands in specific areas, while scattered bushes throughout these coastal savannahs showcase a co-dominance of species with Guinean-Congolese affinities and sub-Guinean species. Fallow regions under coconut plantations feature plant species like Cleome Viscosa, Phyllanthus Amarus, Spermacoce Radiata, Passiflora Foetida, Portulaca Oleracea, Andropogon Gayanus, Indigofera Arrecta, and Indigofera Pilosa.

Moving to the Savanes region (Fig. 15.1b), this area is situated between 0°10 and 1°00 longitude east and 9°55 and 11°05 latitude north, with an estimated population of 114,352 inhabitants in 2022 (INSEED, 2023). It represents the southernmost region of Togo and covers an expanse of 8533 km^2, equivalent to 15% of the national territory according to the Ministry of Agriculture, Livestock, and Fisheries (MAEP, 2014). The region experiences a Sudanian climate characterized by a rainy season from June to September, followed by a dry season extending from October to May (Inoue et al., 2018).

Annual rainfall in this region averages around 1050.0 mm, with an annual maximum temperature of approximately 34 °C and an annual minimum temperature of around 23 °C. Relative humidity levels fluctuate between 50 and 70% on average (Inoue et al., 2018). The region experiences an average wind speed of approximately 2 m/s, and the daily sunshine duration typically lasts for around 7 h (Inoue et al., 2018).

The predominant soil types found in the area are primarily sedimentary, consisting of clays and sandstones according to Food and Agriculture Organization of the United Nations (Afidégnon, 2003). The region is traversed by two major rivers: the Koumangou, which flows in an East–West direction, and the Oti, which flows in an East-South-South-West direction. These two rivers, both of which are tributaries of the Volta, are the main water sources in the region.

As for vegetation, the Savanes region is marked by Sudanian wooded savannah vegetation, with a few remnants of gallery forests along the banks of the River Oti. In the northern sectors of this zone, on eroded, armored soils, a "steppe savannah" emerges, characterized by thorny shrubs (MAEP, 2014). Common species in the area include Adansonia Digitata (Baobab), Parkia Biglobosa (Néré), and Butyrospermum Parkii (Shea).

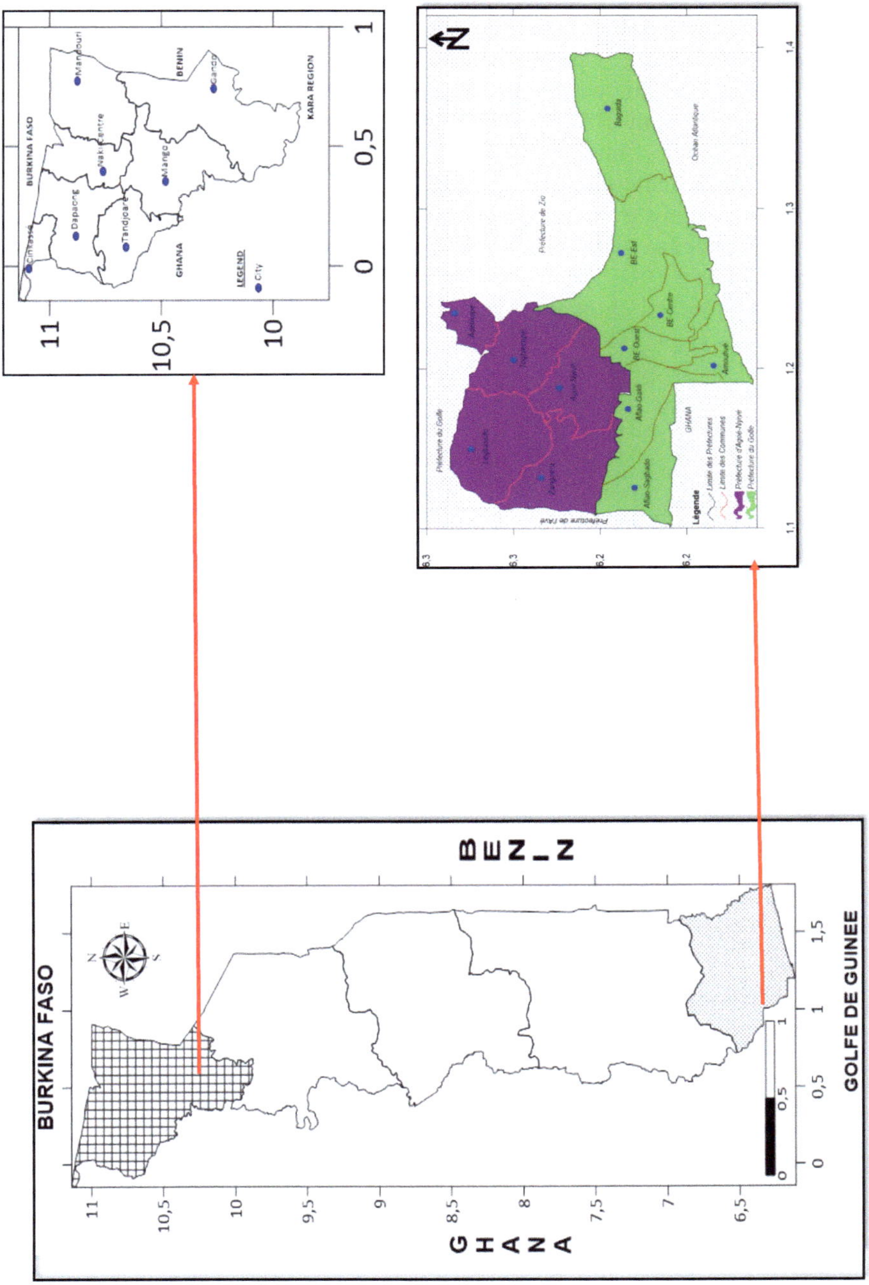

Fig. 15.1 Study area

Data Collection

This study was a retrospective, descriptive, and analytical investigation spanning the period from 2018 to 2022. Climatic data were gathered by accessing the database of the Direction Générale de la Météorologie Nationale, encompassing parameters such as daily rainfall (RR), minimum and maximum temperatures (TMIN and TMAX), maximum and minimum relative humidity (UMAX and UMIN), wind speed (Windspeed), and insolation (Inso). Health data were centered around asthma cases recorded monthly across all age groups, sourced from the Ministry of Health's District Health Information Software (DIHS2) databases. Environmental data relied on vegetation indices (NDVI) and were collected decennially from the Meteorological Operational Satellite Program (METOP) databases at the Institut de Recherche Agronomique du Togo (ITRA).

Data Analysis

Analysis of Seasonal Variability in Climatological and Sanitary Parameters

This phase involved monthly graphical representations of the seasonal trends in climatological parameters from 2018 to 2022. Additionally, it entailed monthly graphical depictions of the seasonal patterns of asthma attacks reported during the same period to discern temporal variations.

Analysis of Asthma Prevalence

Asthma prevalence by region was determined by normalizing the data using Eq. 15.1. It's important to note that standardizing the data is crucial to account for variations in the population sizes across regions, ensuring fair comparisons of asthma prevalence.

$$\text{Prevalence} = ((\text{Number of existing cases of the disease in the population}) / (\text{Total size of the study population})) \times 100{,}000 \tag{15.1}$$

Analysis of the Correlation Coefficient (R)

The correlation coefficient (r) quantifies the strength of the linear relationship between two variables, as described by Eq. 15.2. This coefficient ranges between − 1 and 1, with values closer to the extremes (− 1 or 1) indicating a stronger correlation (either inverse or direct). Conversely, a correlation of 0 suggests that the variables being studied are linearly independent.

$$r = \frac{\mathrm{Cov(X, Y)}}{\partial X \partial Y} \tag{15.2}$$

where

Cov(X,Y) denotes the covariance of the variables X and Y; and
∂X∂Y are their standard deviations.

Results

Seasonal Trends of Asthma Cases from 2018 to 2022 in Grand-Lome Region

An examination of the seasonal pattern in asthma occurrences within the Grand-Lome region unveils substantial fluctuations throughout the year (Fig. 15.2a). The peak of asthma cases is observed during the months from October to December, with these three months collectively constituting nearly 29.3% of the annual total, highlighted by a peak of 550 cases in October. Conversely, March and April register the lowest occurrences of asthma, with 378 and 395 cases, respectively. It is evident that the period spanning from October to December stands out as the most conducive time for the onset of asthma attacks.

Seasonal Trends of Asthma Cases from 2018 to 2022 in Savanes Region

The examination of the seasonal pattern in asthma occurrences within the Savanes region reveals a marked variation over the course of the year, as illustrated (Fig. 15.2b). The highest number of asthma cases is documented in February, August, and the period from October to December. During these three months, they collectively account for a substantial 48.72% of the annual total, with October and November being the months of peak prevalence, each recording 68 cases. Conversely, the months of March and April report the lowest occurrences of asthma, with 40 and 45 cases, respectively. It becomes evident that the months spanning from October to December, along with August and February, are the periods most conducive to the onset of asthma attacks.

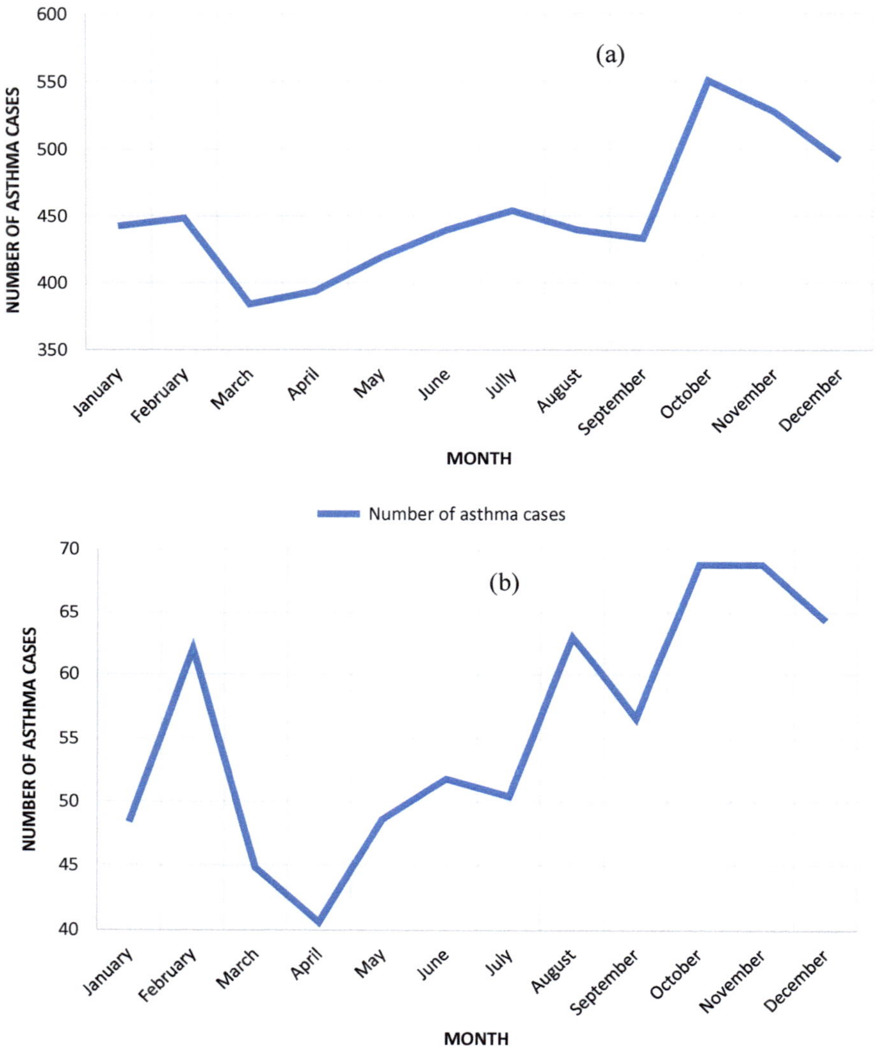

Fig. 15.2 Seasonal trends in asthma cases from 2018 to 2022 in the Grand-Lome region (**a**) and the Savanes region (**b**)

Regional Prevalence of Asthma in Togo

The Savannah and Grand-Lome regions illustrate the two opposite ends of asthma prevalence within the nation's six regions (Fig. 15.3). Notably, prevalence exhibits an ascending gradient from the southern (Grand-Lome region) to the northern (Savanes region) regions, demonstrating a declining trend as one moves away from the Guinean

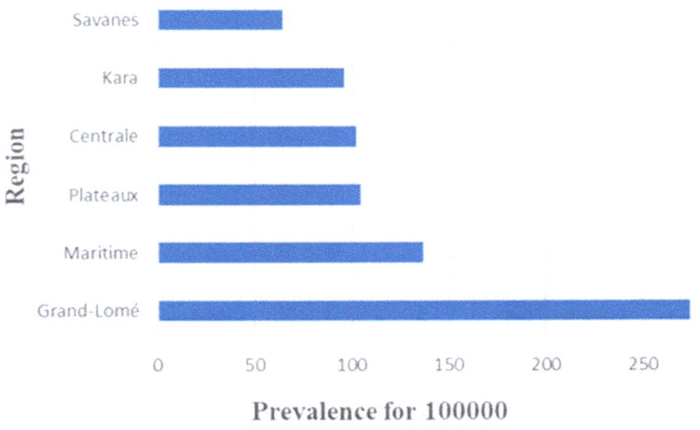

Fig. 15.3 Asthma prevalence by region in 2022

climate zone towards the Sudanian climate zone. Specifically, prevalence diminishes from 274.26 per 100,000 in the Grand-Lome region to 64.88 per 100,000 in the Savanes region. These regions also exhibit distinctive climatic characteristics, with the Grand-Lome region experiencing a relatively humid Guinean climate featuring two rainy seasons, in contrast to the Savanes region characterized by a dry Sudanian climate, with a single rainy season.

Seasonal Trends of Climatic and Environmental Parameters from 2018 to 2022 in the Grand-Lome Region

Seasonal Rainfall Trends

The analysis of seasonal trends in average rainfall within the Grand-Lome region reveals a bimodal rainfall pattern (Fig. 15.4a). The monthly average rainfall exhibits significant variability throughout the year, consistently remaining below 160 mm. The initial rainy season, notably the lengthiest, encompasses the period from April to July, characterized by monthly rainfall exceeding 100 mm, reaching its zenith at 157 mm.

The subsequent season, which is relatively brief in comparison, spans from September to October, featuring a peak of 145 mm in October. Consequently, two dry seasons interpose between the two rainy periods. The shorter dry season extends from the conclusion of July to the end of August, while the more extended dry season covers the months from November to February. It is worth noting that these periods do not necessarily equate to "favorable weather conditions." The atmosphere can periodically be highly humid, and the skies may remain hazy for several days,

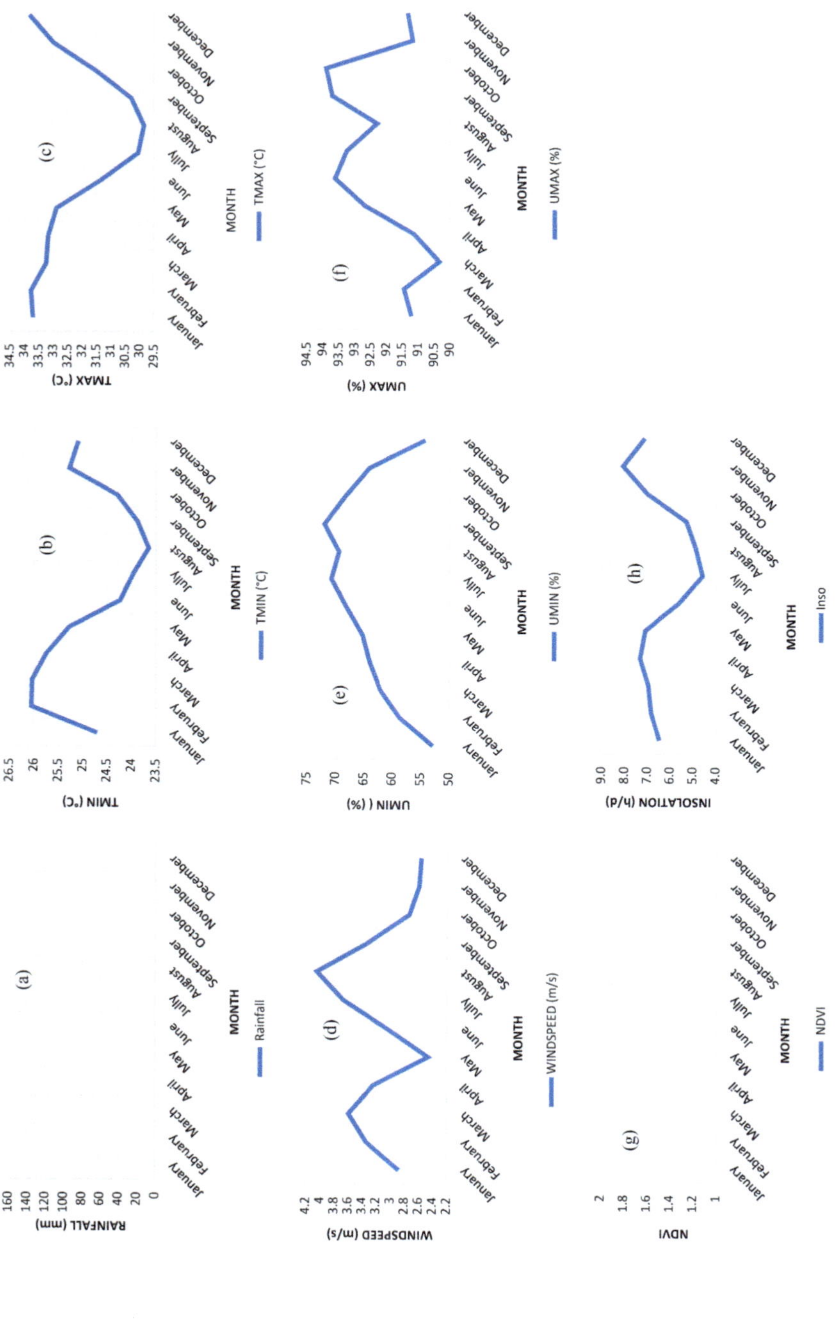

Fig. 15.4 Seasonal trends of climatic and environmental parameters from 2018 to 2022 in Grand-Lome region

particularly during the occurrence of the Harmattan, characterized by continental trade winds, haziness, and a decrease in temperature.

Seasonal Trends of Maximum and Minimum Temperatures

The average monthly temperature fluctuations span from 23.6 to 34 °C (Fig. 15.4b, c). Average monthly maximum temperatures (TMAX) exhibit a range from 29.7 °C in August to 33.8 °C in March, as illustrated in Fig. 15.4c. Meanwhile, the variations in minimum temperatures (TMIN) fluctuate between 23.6 °C in August and 26 °C in February, as indicated in Fig. 15.4b. It is notable that the warmest months appear to be December to February, while July to September are seemingly associated with the lowest recorded temperatures.

Seasonal Changes of Wind Speed

Wind speed demonstrates seasonal variations (Fig. 15.4d) with typically ranging values between 2.5 and 4 m/s. The months of May, October to December exhibit the lowest wind speeds, falling below 2.5 m/s. In contrast, the months of February to April and July to September register the highest wind speeds, all surpassing 3.2 m/s, with the peak, reaching 4 m/s, notably observed in August.

Seasonal Trends in Maximum and Minimum Relative Humidity

Monthly relative humidities, both maximum and minimum, exhibit significant fluctuations throughout the year, typically ranging between 52 and 94% (Fig. 15.4e, f). The maximum relative humidity (UMAX) varies from 90% in March to 94% in October, as depicted in Fig. 15.4e. On the other hand, the minimum humidity (UMIN) shows a range from 52% in January to 73% in September, as shown in Fig. 15.4f. Consequently, humidity levels tend to be elevated during the rainy season and more frequently decrease during the dry season.

Seasonal Evolution of the Vegetation Index

The NDVI exhibits values ranging between 1 in February and 1.9 in May (Fig. 15.4g). Specifically, during the period from January to March, which represents the relatively arid months of the year, the vegetation index indicates relatively modest values, ranging from 1 to 1.4. This decline is attributed to cold climatic conditions and reduced sunshine, inducing plant dormancy in numerous regions, resulting in a decrease in vegetative growth. Starting from April onward, the initial rains act as a stimulus for plant growth, which is reflected in a gradual elevation of the vegetation index.

From May to June, corresponding to the winter months, the vegetation index attains its pinnacle at 1.9. Favorable weather conditions, characterized by ample sunlight and rainfall, foster robust vegetative growth. Subsequently, during August to September, as the principal rainy season approaches its conclusion, the vegetation index recedes, subsequently surging once more in October with the advent of the second rainy season, only to taper off as the second dry season ensues.

Seasonal Variation of Insolation

The average monthly duration of sunshine (insolation) across the year falls within the range of 4.5–8 h per day, as presented in Fig. 15.4h. Notably, the peak duration occurs in November, contrasting with the lowest values recorded from June to September. These months coincide with periods characterized by substantial cloud cover, acting as an effective shield against the sun's rays.

Seasonal Trends of Climatic and Environmental Parameters from 2018 to 2022 in the Savanes Region

Seasonal Rainfall Patterns

The examination of seasonal rainfall trends within the Savanes region reveals an unimodal rainfall pattern (Fig. 15.5a). Average monthly rainfall exhibits considerable variability throughout the year, consistently remaining below 250 mm. The period spanning from May to September emerges as the wettest, marked by average monthly rainfall exceeding 100 mm. July stands out as the month when the Savanes region records the most substantial rainfall, with average monthly amounts reaching 250 mm. Cumulative rainfall during the rainy season, extending from May to October, accounts for more than 80% of the average annual total, translating to 932 mm out of 1100 mm.

Seasonal Temperature Variations

Average monthly temperature variations range from 22.5 to 35.2 °C, as represented (Fig. 15.5b, c). Average monthly minimum temperatures (TMIN) fluctuate from 22.5 °C in August to 23.7 °C in June, as shown in Fig. 15.5b. Meanwhile, variations in maximum temperatures (TMAX) span from 33.7 °C in August to 35.2 °C in June, as illustrated in Fig. 15.5c. Notably, the hottest months appear to be April to June, while the lowest temperatures are registered from July to September.

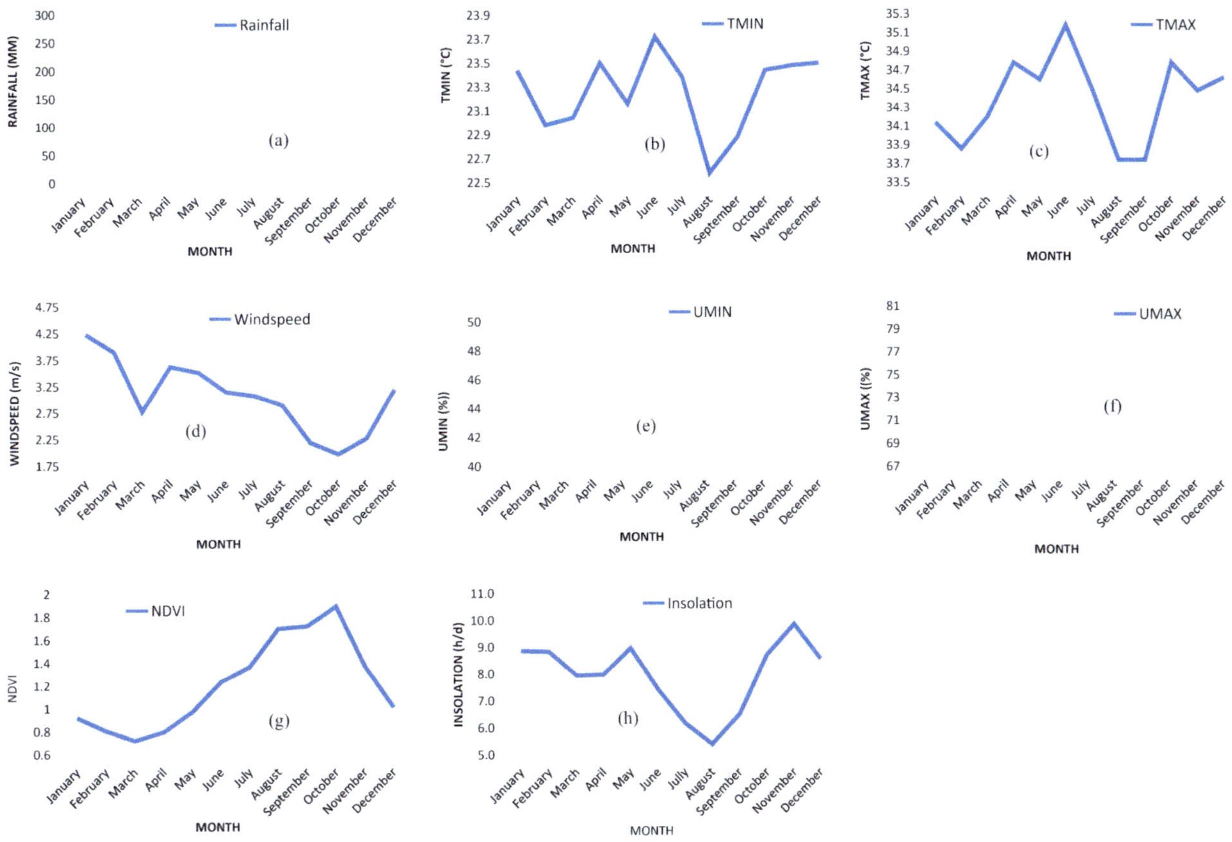

Fig. 15.5 Seasonal trends in climatic and environmental parameters from 2018 to 2022 in the Savanes region

Seasonal Wind Speed Variations

Wind speed displays seasonal variations throughout the year (Fig. 15.5d), typically ranging between 1.9 and 4.25 m/s. The months from September to November record reduced wind speeds, falling below 2.25 m/s, whereas January and February report the highest values, all exceeding 3.75 m/s, with a pinnacle of 4.25 m/s documented in January.

Seasonal Relative Humidity Patterns

Monthly relative humidities, encompassing both maximum and minimum values, undergo significant fluctuations over the course of the year, generally oscillating between 40 and 80% (Fig. 15.5e, f). The minimum relative humidity (UMIN) demonstrates a range from 40% in October to 50% in November, as depicted in Fig. 15.5e. Meanwhile, the maximum humidity (UMAX) fluctuates from 67% in July to 80% in June, as portrayed in Fig. 15.5f.

Seasonal Changes in Vegetation Index

The Normalized Difference Vegetation Index (NDVI) exhibits substantial variations throughout the year, typically oscillating between 0.7 and 1.9 (Fig. 15.5g). The lowest indices manifest during the arid months of January to April, all falling below 1. Vegetation growth commences from May onwards, corresponding to the onset of the initial rains of the season. Elevated NDVI values, ranging from 1.4 to 1.9, are observed during the rainy season, culminating in a peak in October, reaching 1.9.

Seasonal Insolation Variations

The average monthly duration of sunshine throughout the year fluctuates between 5.3 and 10 h per day (Fig. 15.5h). The highest duration of 10 h is observed in November, whereas the lowest values, less than 6 h (specifically 5.3 h), are recorded in July and August. These periods align with extensive cloud cover, effectively acting as a screen against the sun's rays.

Analysis of Correlations Between Climatic and Environmental Parameters and Asthma

The findings underscore a robust correlation between specific climatic and environmental parameters and asthma incidence in both the Savanes and Grand-Lome regions (Table 15.1).

Within the Savanes region, three parameters, namely the Normalized Difference Vegetation Index (NDVI), minimum temperature (TMIN), and wind speed (Windspeed), exhibited statistically significant correlations with asthma at the 5% significance level. NDVI displayed a positive correlation, with correlation coefficients (r) and p-values of 0.63 and 0.02, respectively. In contrast, TMIN and Windspeed demonstrated negative correlations with asthma, characterized by (r) values of − 0.53 and 0.60, along with corresponding p-values of 0.04275 and 0.035.

In the Grand-Lome region, only two parameters, Maximum Relative Humidity (UMAX) and insolation (Inso), exhibited significant correlations with asthma at the 5% significance level. UMAX displayed a positive correlation of 0.75, with a p-value of 0.005, while Inso exhibited an (r) of 0.66, accompanied by a p-value of 0.01.

Correlation Between Asthma and Maximum Relative Humidity in the Grand-Lome Region

An augmentation in maximum relative humidity corresponds to an escalation in the incidence of asthma cases (Fig. 15.6a). These trends both culminate in their zeniths during October and November, respectively, before experiencing a subsequent decline.

Correlation Between Asthma and Insolation in the Grand-Lome Region

The variation in the number of asthma cases mirrors a proportional relationship with the duration of insolation, except during the period from June to August when the trends deviate (Fig. 15.6b). From January to May and from September to December, the duration of sunshine parallels that of asthma cases. In contrast, from June to August, the number of asthma cases increases while the duration of sunshine appears to decrease.

Correlation Between Asthma and Vegetation Index in the Savanes Region

It has been observed that an increase in the vegetation index is paralleled by an uptick in the number of asthma cases from March to October, reaching their peaks during this period, and subsequently declining until December. Notably, a peculiar anomaly

Table 15.1 Correlation between climatological and environmental parameters and asthma

Région	Pearson's correlation	TMAX	NDVI	TMIN	Windspeed	UMAX	UMIN	RR	Inso
Savanes	r	− 0.31	0.63	− 0.53	− 0.60	− 0.02	0.07	− 0.04	0.14
	p-value	0.318	0.026*	0.042*	0.035*	0.943	0.805	0.886	0.656
Grand-Lome	R	− 0.02	0.30	− 0.20	− 0.36	0.75*	− 0.03	− 0.051	0.66*
	p-value	0.943	0.332	0.526	0.242	0.005	0.912	0.874	0.017

*Significance at 5% confidence level

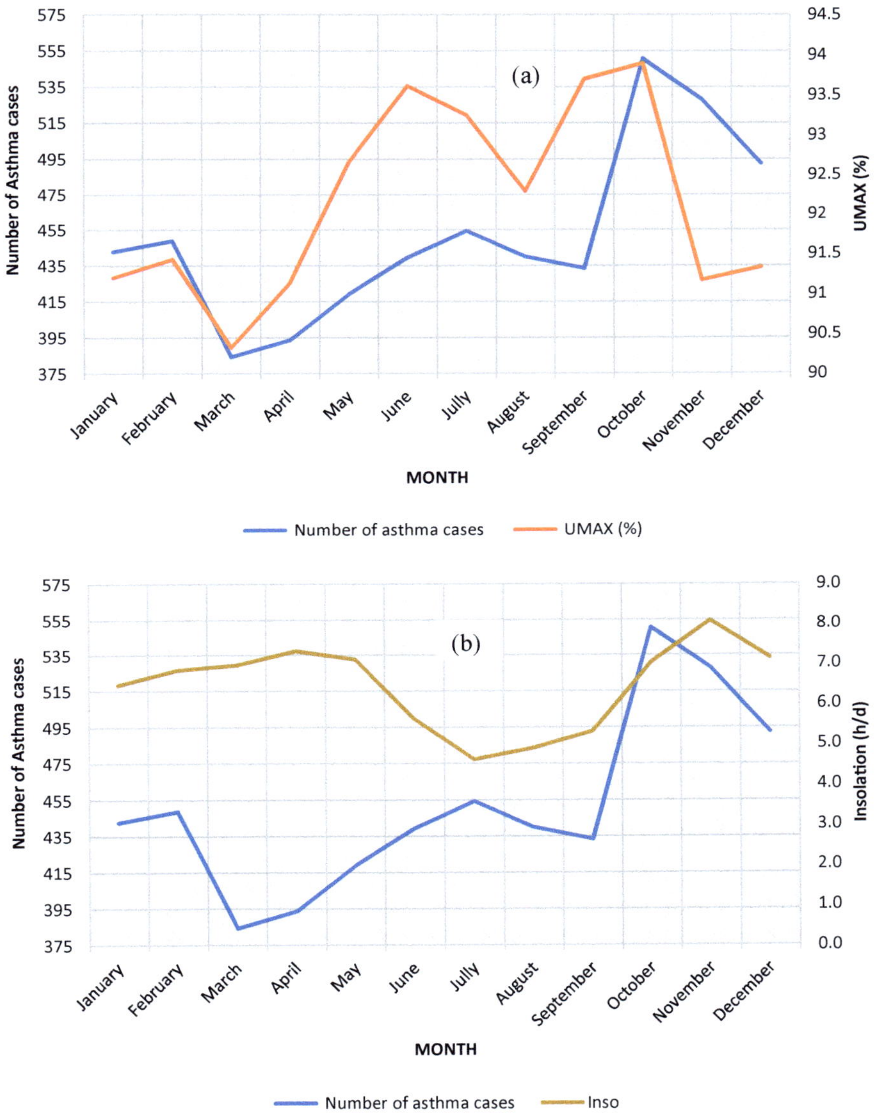

Fig. 15.6 Correlation between asthma and climatic parameters (maximum relative humidity (**a**) and insolation (**b**) in the Grand-Lome region)

surfaces in January and February, where the incidence of asthma cases increases despite a decline in the vegetation index (Fig. 15.7a).

Correlation Between Asthma and Wind Speed in the Savanes Region

Examination of the association between asthma cases and wind speed unveils a contrasting pattern (Fig. 15.7b). The trajectory of asthma cases appears to oppose that of wind speed throughout the year. When wind speed is relatively high, cases of asthma tend to decrease, while a decrease in wind speed coincides with an increase in asthma cases.

Correlation Between Asthma and Minimum Temperature in the Savanes Region

Analysis of the correlation between asthma and minimum temperature reveals a divergent trend (Fig. 15.7c). Specifically, during the period from February to July, when minimum temperatures are elevated, the number of asthma cases appears to be notably low. In contrast, from August to December, the period of relatively lower minimum temperatures, there is an increase in asthma cases.

Discussion

Climatic and Environmental Parameters Influencing Asthma

The findings from this investigation revealed notable associations between asthma and various climatic and environmental factors, including vegetation, wind speed, minimum temperature, maximum relative humidity, and sunshine duration. These findings align with the research conducted by several authors, including Epstein (2005), who have emphasized the climate-related nature of numerous common human ailments, particularly respiratory diseases. These associations are attributed to variations in air quality and reduced air temperatures.

Climatic and Environmental Parameters Influencing Asthma in the Savanes Region

The analysis demonstrates a positive correlation between the incidence of asthma cases in the Savanes region and vegetation, while it reveals negative correlations with minimum temperature and wind speed. These results signify that heightened

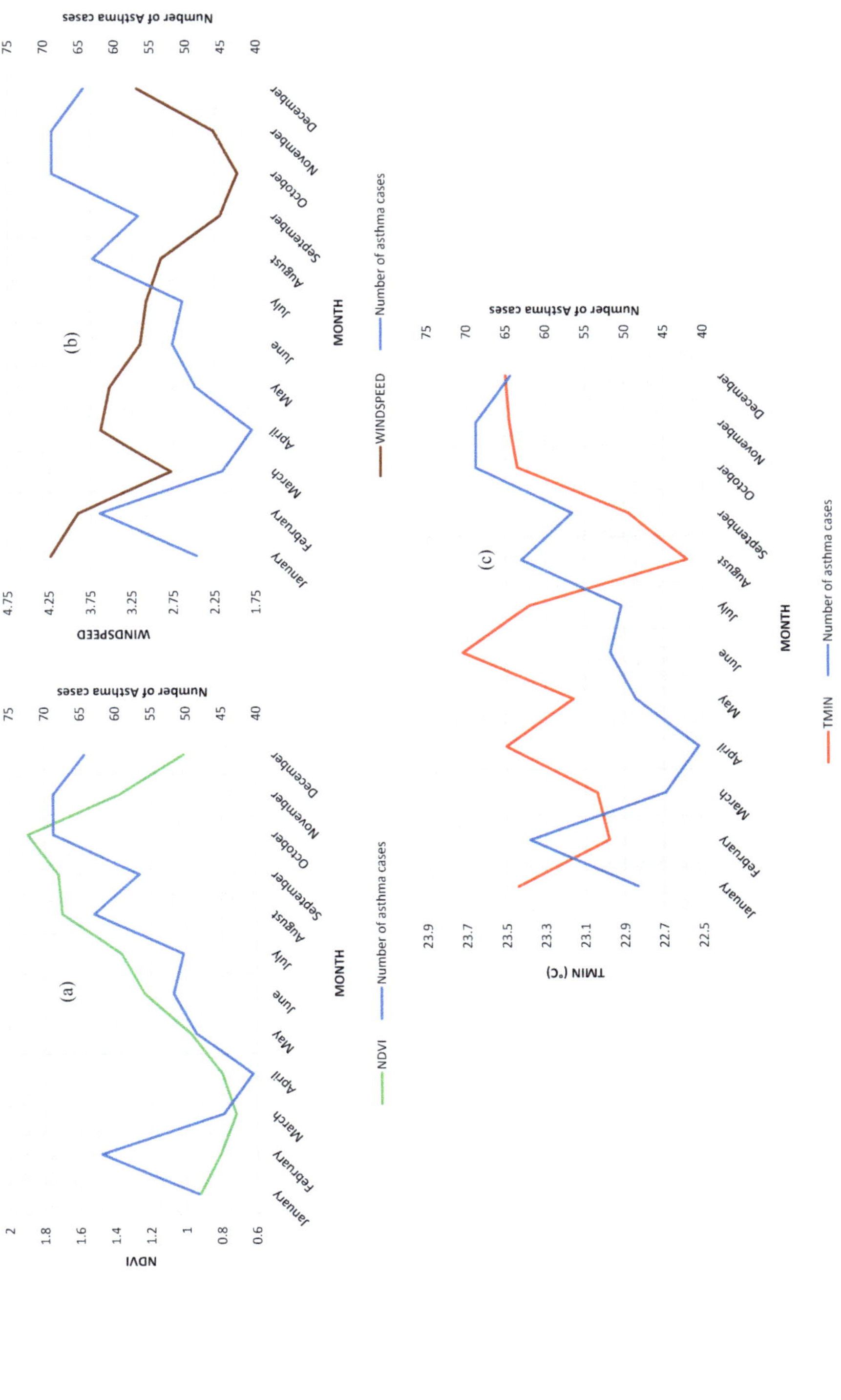

Fig. 15.7 Correlation between asthma and climatic parameters in the Savannah region

vegetation, reduced wind speed, and lower temperatures create favorable climatic conditions conducive to the occurrence of asthma attacks.

- ***Link between vegetation and Asthma***

The findings of our study demonstrate a parallel increase in asthma cases with the vegetation index. This phenomenon can be attributed to the capacity of pollen from flowers, trees, and grasses to induce allergic reactions in susceptible individuals, thereby exacerbating asthma symptoms. Furthermore, certain plant species emit volatile chemical substances, such as essential oils, which can irritate the airways of asthma patients, leading to asthma attacks. Annesi-Maesano et al. (2010) observed that specific plant species, notably deciduous trees, were associated with a heightened prevalence of asthma in children. Biel et al. (2007) investigated the impact of allergenic plants like ragweed on the prevalence of asthma in farmers and found an association between ragweed exposure and an increased prevalence of asthma in this group.

The work by Zock et al. (2005) supported the adverse effects of vegetation on asthmatics by studying the influence of airborne allergens, including those originating from plants, on the exacerbation of asthma symptoms. Their study revealed that exposure to certain allergens like pollen could contribute to the worsening of asthma symptoms in susceptible individuals.

Furthermore, other researchers, such as Heinrich et al. (2005), confirmed the detrimental impact of vegetation on individuals with asthma. Heinrich et al. (2005) explored the relationship between exposure to vegetation and the development of asthma in children, and their results indicated that certain microbial components within vegetation, such as endotoxins, could be associated with an elevated risk of asthma.

Nevertheless, it is essential to acknowledge that some studies present results that contradict our findings. For instance, research by Reese et al. (2007) examined the presence of vegetation around dwellings and its impact on asthma prevalence in adults. They found that adults residing in areas with higher vegetation coverage had a lower prevalence of asthma compared to those in less green areas. Similarly, other studies conducted in 2009 by the same authors, focusing on the role of urban parks in alleviating asthma symptoms in children, indicated that children living near urban parks experienced a reduction in asthma symptoms. These results suggest that the presence of vegetation in green spaces may have beneficial effects on the health of individuals with asthma by enhancing air quality and reducing pollution levels.

- ***Link between minimum temperature and Asthma***

Our study reveals that a decrease in minimum temperatures is associated with an increased likelihood of the onset or exacerbation of asthma. These findings align with those of Song et al. (2021), who observed that a decrease in temperature correlated with a rise in emergency department visits for asthma among children in South Korea. Similarly, Bloomfield et al. (2018) discovered that periods of severe cold were linked to an upsurge in emergency department visits for asthma in children in Atlanta, USA, indicating that colder temperatures may exacerbate asthma symptoms.

Furthermore, other authors investigating respiratory infections, such as Ferrari et al. (2012), have reported a 1% increase in outpatient consultations for Chronic Obstructive Pulmonary Disease (COPD) for every degree Fahrenheit decrease in Bavaria. Tseng et al. (2019) also noted an increase of 0.8% in the number of outpatient consultations for COPD for each 1 °C drop in temperature during their studies in Taiwan, accompanied by a rise in emergency room consultations in colder regions.

The exacerbation of asthma symptoms in cold weather can be attributed to several factors. Inhaling cold air has the potential to irritate airways, inducing bronchial constriction in some individuals with asthma, leading to symptoms like coughing, shortness of breath, and wheezing. Additionally, Monto (2000) has noted that acute respiratory infections, including respiratory viruses like rhinoviruses and influenza viruses, are more prevalent during colder periods. These infections can trigger asthma symptoms in susceptible individuals, as individuals with asthma may be more susceptible to respiratory infections when exposed to colder temperatures, thus worsening their asthma symptoms.

It is also noteworthy that during the colder months, people tend to spend more time indoors, where the concentration of specific asthma triggers, such as dust mites, mold, and animal dander, can be higher. These allergens and irritants have the potential to provoke asthma symptoms in sensitive individuals, as evidenced by Dharmage (2019), who analyzed data from a large cohort of Australian children and found an association between early exposure to pet hair and an increased risk of asthma.

Moreover, Tischer et al. (2018) examined the association between exposure to mold in homes and children's respiratory health, suggesting that exposure to mold and its compounds may contribute to respiratory problems in children. This finding was consistent with the observations of Sordillo et al. (2018), who explored the relationship between the gut microbiome of infants and the risk of asthma. Their results indicated that factors such as exposure to environmental allergens, including mold and animal hair, can influence the composition of the gut microbiome and contribute to the development of asthma. Additionally, cold weather is a meteorological factor that can act as a trigger for or an aggravator of asthma attacks.

- ***Link between wind speed and Asthma***

Our data analysis has unveiled a compelling association between low wind speed and an upswing in asthma attacks. These findings are in consonance with the research of several other authors, including Sunyer et al. (2000), who investigated the relationship between wind speed and respiratory symptoms in elderly individuals. They observed that days characterized by low wind speed were linked to an increased incidence of respiratory symptoms, including asthma. The plausible explanation for this association lies in the phenomenon that, as wind speed diminishes, the air tends to stagnate. Such stagnation can foster the accumulation of various pollutants, including fine particles, allergens, chemical irritants, or toxic gases in the atmosphere. Inhaling these substances can incite asthma symptoms in susceptible individuals. Similarly, Gavett et al. (2003) noted in their study that days featuring low wind speeds were correlated with an escalation in respiratory symptoms, including asthma. They pointed out that fine particles of air pollution are typically less dispersed on low-wind-speed days, exacerbating respiratory symptoms.

In line with these findings, Gupta et al. (2017) conducted an evaluation of the impact of weather conditions on asthma emergency visits in India. They established that low wind speed was linked to an increase in asthma emergency visits, suggesting a connection between limited pollutant dispersion and the exacerbation of asthma symptoms. Consistent observations were also made in China by Xu et al. (2014), who identified that low wind speed was associated with a heightened risk of asthma attacks in children, particularly in the presence of elevated air pollution levels.

However, it is imperative to acknowledge that, despite our results showcasing a positive correlation between low wind speeds and an increase in asthma attacks, some studies paradoxically reveal that high wind speeds can also induce the aggravation or initiation of asthma attacks. In this context, Halonen et al. (2010), following a study involving asthmatic children, disclosed that strong winds were linked to an augmentation of asthma symptoms. This suggests that high wind speeds can result in a greater dispersion of allergens in the air, thereby triggering asthma attacks in susceptible individuals. For Rönmark et al. (2011), robust winds might lead to an amplification in the dispersion of atmospheric pollutants, thereby intensifying asthma symptoms. According to Herrera-Van Oostdam et al. (2016), strong winds could facilitate the dispersion of pollutants such as vehicle exhaust fumes, dust storms, or sandstorms. These conditions can irritate airways and incite asthma symptoms in certain individuals with asthma.

Climatic Parameters Influencing Asthma in Grand-Lome Region

As in the Savanes region, asthma exhibits correlations with specific climatic parameters. Asthma displays a positive correlation with both maximum relative humidity and insolation, implying that heightened levels of both maximum relative humidity and insolation are climatic conditions conducive to the occurrence of asthma attacks.

- ***Correlation between Maximum Relative Humidity and Asthma***

Our study findings suggest that increased relative humidity can serve as a triggering factor for asthma symptoms or exacerbate existing ones. High relative humidity tends to foster the proliferation of common asthma triggers, such as molds and dust mites. Furthermore, mold spores and dust mite allergens can permeate indoor air, resulting in elevated exposure and a potential worsening of asthma symptoms. Chen et al. (2018) also noted a connection between high relative humidity and increased hospitalizations for asthma in adults. They proposed that relative humidity might influence indoor air quality by promoting mold growth and the release of volatile organic compounds, which can worsen asthma symptoms. A study in Tunisia by Denguezli et al. (2020) uncovered a correlation between high relative humidity and the prevalence of asthma in children, suggesting that relative humidity may favor the growth of house dust mites a common source of allergens for asthma sufferers that also thrive in damp environments. These mites are often found in household dust and can trigger asthma symptoms upon inhalation.

A meta-analysis conducted by Zhang et al. (2019) substantiates a significant association between high relative humidity and an upsurge in asthma symptoms. The same authors propose that this link can be attributed to the role of relative humidity in promoting the proliferation of allergens and irritants within indoor environments, thereby constituting factors that either trigger or exacerbate asthma attacks. Research carried out in Japan by Kusaka et al. (2017) unveiled an association between high relative humidity and increased airway inflammation in asthma patients. Some individuals with asthma possess airways that are particularly sensitive to changes in relative humidity. Elevated relative humidity levels can result in airway swelling, heightened mucus production, and increased bronchial reactivity, culminating in asthma symptoms like coughing, shortness of breath, and wheezing. This suggests that relative humidity can act as a stressor on the airways, aggravating asthma symptoms. It is worth considering that individuals residing in regions characterized by low relative humidity may exhibit increased sensitivity to sudden spikes in relative humidity. These abrupt shifts in atmospheric relative humidity can lead to airway constriction in people with asthma, culminating in heightened symptoms such as wheezing, shortness of breath, and coughing.

- ***Link between Insolation and asthma***

Extended exposure to sunlight appears to induce or exacerbate asthma attacks. Prolonged exposure to high temperatures can elevate respiratory rates and lead to shallow breathing, resulting in breathing difficulties for individuals with asthma. Additionally, as indicated by Silverman et al. (2014), excessive sun exposure may contribute to airway inflammation, thereby increasing the risk of asthma attacks. It is worth noting that prolonged exposure to sunlight can heighten the risk of dehydration. Dehydration, in turn, can lead to the thickening of bronchial secretions, which has the potential to worsen asthma symptoms.

For other researchers, such as Castro-Rodriguez et al. (2019), the onset or exacerbation of asthma attacks due to prolonged sun exposure can be attributed to bronchial reactivity. According to their findings, ultraviolet rays may enhance airway sensitivity to various stimuli, including allergens or physical exertion, potentially aggravating asthma symptoms in susceptible individuals. Hajat et al. (2014) uncovered that prolonged exposure to intense sunlight can be associated with increased air pollution, particularly elevated ozone levels. These heightened ozone concentrations can irritate the airways and trigger asthma attacks in individuals with heightened sensitivity.

Comparison of Asthma Prevalence Between the Grand-Lomé Coastal Region and the Savanes Region

Numerous factors may contribute to variations in asthma prevalence across different geographic regions, particularly when comparing coastal and savannah areas.

Coastal environments typically exhibit higher humidity levels, fostering the proliferation of molds. This phenomenon may elucidate the elevated concentration of common asthma triggers, such as pollens, molds, and dust mites, in coastal areas. Furthermore, coastal regions often host major economic centers, rendering them more susceptible to air pollution stemming from industrial activities, vehicular traffic, and maritime operations. This air pollution, characterized by fine particulates and noxious gases, has the potential to exacerbate asthma symptoms and elevate the risk of asthma development. Distinct climatic patterns between coastal areas and savannahs can also influence asthma prevalence. Coastal regions tend to experience greater relative humidity, milder temperatures, and variable weather conditions, all of which can impact the frequency and severity of asthma cases.

Socioeconomic disparities between these regions may additionally contribute to variations in asthma prevalence. Limited access to healthcare in rural or savannah areas can result in delayed diagnosis and treatment of asthma, thus affecting the overall prevalence of the disease.

Recommendations

Based on the findings of the study, here could be some recommendations to better mitigate the impact of climate variability on asthma prevalence and improve public health outcomes:

- Enhanced Long-Term Data Collection: Given the constraints of limited data availability over only five years, there is a need to establish and maintain long-term data collection systems. This would provide a more comprehensive understanding of how climate variability influences asthma prevalence over extended periods in different climatic zones.
- Regional Specificity in Monitoring: Recognizing the significant regional differences in asthma prevalence and climate correlations between the Savanes and Grand-Lome regions, it is advisable to tailor monitoring efforts to specific environmental parameters identified as influential. This approach can enhance targeted interventions and resource allocation.
- Integration of Environmental and Climatic Data: Integrate environmental and climatic parameters into ongoing health monitoring systems. This holistic approach can provide early warning system and proactive measures against asthma exacerbations linked to climate variations.

- Public Health Awareness and Preparedness: Given the seasonal peaks in asthma cases during the colder and drier months (October–February), public health campaigns should focus on raising awareness about asthma triggers exacerbated by weather conditions (cold, dry air, dust, and mold). This includes educating vulnerable populations and healthcare providers on preventive measures and timely management strategies.
- Policy and Urban Planning Considerations: In highly industrialized and densely populated coastal areas like Grand-Lome, where asthma prevalence is notably higher, urban planning policies should prioritize green spaces, air quality improvements, and measures to mitigate industrial emissions. This can help reduce environmental triggers contributing to asthma outbreaks.
- Research and Collaboration: Encourage further research collaborations between climatologists, epidemiologists, and public health officials to deepen the understanding of complex interactions between climate variables and respiratory health outcomes. This interdisciplinary approach can lead to more effective mitigation strategies and policy recommendations.

Limitations of the Study

The health data spanning only a five-year period may offer an incomplete portrayal of Asthma's progression within the examined regions. A more comprehensive understanding of prolonged trends and seasonal fluctuations in the disease necessitates protracted investigations. Furthermore, while the study has identified correlations between climatic factors and Asthma, it is imperative to acknowledge that correlation does not inherently imply causation. Numerous other elements, including genetic distinctions, specific environmental exposures, individual conduct, and lifestyle choices, may also exert influence over the emergence of asthma attacks.

Moreover, the study primarily concentrated on climatic parameters, yet various unaccounted variables such as air pollution, region-specific allergens, and additional environmental factors might also contribute to the prevalence and clinical manifestations of asthma. It is essential to recognize that the study's scope was geographically confined to the savannah and Grand-Lome regions of Togo. Consequently, the findings should not be indiscriminately extended to other parts of the country. Subsequent investigations conducted in diverse geographical and cultural contexts are indispensable for a more comprehensive comprehension of the interplay between climatic factors and asthma in Togo.

Conclusion

The findings underscore a notably robust association between climatic parameters and Asthma, with temperature, relative humidity, insolation, vegetation density, and wind speed emerging as the principal contributing factors to the ailment. Lower temperatures can incite heightened airway inflammation, exacerbating Asthma symptoms. Elevated relative humidity fosters the proliferation of common Asthma triggers, including dust mites and mold. Intense solar radiation can lead to the generation of air pollutants, intensifying the severity of respiratory issues. The density of vegetation can elevate pollen concentrations, a significant allergen for individuals afflicted by Asthma. Additionally, wind speed can shape the dispersion of atmospheric pollutants, thereby exacerbating Asthma symptoms.

Moreover, this investigation has unveiled an incongruity in the prevalence of Asthma between the Grand-Lome region, characterized by its coastal climate, and the Savanes region, marked by its Sahelian climate. The Grand-Lome region exhibits a higher Asthma prevalence, attributable to a constellation of factors including heightened population density, greater exposure to air pollutants stemming from urban industrial and transport activities, disparate living conditions, and discrepancies in healthcare accessibility. Consequently, these findings underscore the imperative of integrating climatic and environmental parameters into the management of Asthma, particularly within regions where these factors wield significant influence.

References

Afidégnon, D. (2003). *Carte de la végétation du Togo: Notice explicative.* Laboratoire de Botanique et Écologie Végétale, Université Paul Sabatier, Laboratoire d'Écologie Terrestre.

Annesi-Maesano, I., Krieger, J., & Pollitt, R. (2010). *The complex interplay of genetic and environmental factors in the etiology of asthma.* Springer Science & Business Media.

Asthma, allergies, and the Americans with Disabilities Act (ADA). (2016). https://aafa.org/asthma/living-with-asthma/asthma-allergies-and-the-american-with-disabilities-act/

Biel, M. A., Fanta, C. H., & Zibrak, J. D. (2007). *Asthma: Comorbidities, coexisting conditions, and differential diagnosis.* Springer.

Bloomfield, M. G., Whittington, G., Vyse, A., Hanna, G. B., Athanasiou, T., & Winstanley, M. (2018). Clinical prediction models for mortality in acute pancreatitis: A systematic review and meta-analysis. *Annals of Surgery, 267*(5), 769–780.

Castro-Rodriguez, J. A., Saglani, S., Rodriguez-Martinez, C. E., Oyarzun, M. A., Fleming, L., & Bush, A. (2019). The relationship between inflammation and remodeling in childhood asthma: A systematic review. *Pediatric Pulmonology, 54*(7), 973–983.

Chen, Z., Salam, M. T., Eckel, S. P., Breton, C. V., Gilliland, F. D., & Berhane, K. (2018). Chronic effects of air pollution on respiratory health in Southern California children: Findings from the Southern California Children's Health Study. *Journal of Thoracic Disease, 10*(Suppl 19), S2260–S2271.

Denguezli, Y., Nouira, S., Zaouali, M., Bougmiza, I., Feki, I., Elloumi, N., & Ben Cheikh, A. (2020). Association between obesity phenotypes and asthma severity, spirometric values, and control level. *International Archives of Allergy and Immunology, 181*(9), 699–706.

Dharmage, S. C. (2019). *Asthma: Epidemiology, etiology and risk factors.* Academic Press.

Ferrari, R., Ballabio, A., Todeschini, R., & Vichi, M. (2012). *Asthma: Advances in diagnosis and treatment*. Publisher.

Gavett, S. H., Haykal-Coates, N., Highfill, J. W., Ledbetter, A. D., Chen, L. C., Cohen, M. D., & Costa, D. L. (2003). World Trade Center fine particulate matter causes respiratory tract hyperresponsiveness in mice. *Environmental Health Perspectives, 111*(7), 981–991.

Gupta, R., Sheikh, A., Strachan, D. P., & Anderson, H. R. (2017). Time trends in allergic disorders in the UK. *Thorax, 72*(4), 391–393.

Hajat, S., Haines, A., & Goubet, S. A. (2014). Associations of air pollution with indicators of acute respiratory health in a panel of young adults in London, UK. *Occupational and Environmental Medicine, 71*(7), 472–479.

Halonen, M., Stern, D. A., Wright, A. L., Taussig, L. M., & Martinez, F. D. (2010). Alternaria as a major allergen for asthma in children raised in a desert environment. *American Journal of Respiratory and Critical Care Medicine, 162*(5), 1738–1744.

Heinrich, J., Holscher, B., Douwes, J., Richter, K., Koch, A., Bischof, W., & Wichmann, H. E. (2005). Respiratory symptoms and infections in schoolchildren: Associations with fungal and endotoxin levels in classrooms. *Journal of Allergy and Clinical Immunology, 115*(4), 810–817.

Herrera-Van Oostdam, A. S., Páez-Hernández, M. E., Hernández-Ramírez, R. U., Morales-Hernández, R. M., & Hernández-Cadena, L. (2016). Respiratory health effects of indoor air pollution exposure in Mexican women and children: A review. *Environmental Pollution, 212*, 678–686.

Inoue, M., Matsui, E., & Kodama, T. (2018). Strategies for the prevention and management of asthma: The importance of avoiding allergens and environmental irritants. *Journal of Allergy and Asthma Management, 2018*(3), 45–56.

Kassebaum, N. J., Arora, M., Barber, R. M., Bhutta, Z. A., Brown, J., Carter, A., Casey, D. C., Charlson, F. J., Coates, M. M., Coggeshall, M., Cornaby, L., Dandona, L., Dicker, D. J., Erskine, H. E., Ferrari, A. J., Fitzmaurice, C., Foreman, K., Forouzanfar, M. H., Fullman, N., ... Murray, C. J. L. (2016). Global, regional, and national disability-adjusted life-years (DALYs) for 315 diseases and injuries and healthy life expectancy (HALE), 1990–2015: A systematic analysis for the Global Burden of Disease Study 2015. *The Lancet, 388*(10053), 1603–1658.

Kristeen, C. (2021). Is asthma genetic? https://www.healthline.com/health/asthma/is-asthma-genetic.

Kusaka, H., Shimojo, M., & Yamanaka, T. (2017). The relationship between weather-related factors and outpatient visits for allergic rhinitis in Fukuoka, Japan. *International Journal of Environmental Research and Public Health, 14*(5), 486.

Lancet. (2020). Global burden of 369 diseases and injuries in 204 countries and territories, 1990–2019: A systematic analysis for the Global Burden of Disease Study 2019. *The Lancet, 396*(10258), 1204–1222.https://doi.org/10.1016/S0140-6736(20)30925-9

Ministry of Agriculture, Livestock, and Fisheries (MAEP). (2014). Programme national d'investissement agricole et de sécurité alimentaire. MAEP/SG, Plan d'opération, 28p.

Monto, A. S. (2000). Epidemiology of viral respiratory infections. *American Journal of Medicine, 112*(6), 4-12S.

National Institute of Statistics and Economic and Demographic Studies (INSEED). (2023). Résultat du cinquième recensement général de la population et de l'habitat (RGPH-5).

Reese, G., Ayuso, R., Leong-Kee, S., Plante, M., & Lehrer, S. B. (2007). Identification of major allergens of Bermuda grass pollen. *Clinical and Experimental Allergy, 37*(10), 1514–1523.

Rönmark, E. P., Ekerljung, L., Lötvall, J. O., Wennergren, G., Torén, K., Rönmark, E., & Lundbäck, B. (2011). Large-scale questionnaire survey on respiratory health in Sweden: Effects of late- and non-response. *Respiratory Medicine, 105*(12), 1704–1712.

Silverman, R. A., Boudreaux, E. D., Woodruff, P. G., Clark, S., Camargo Jr, C. A., & National Emergency Department Asthma Surveillance (NEDAS) Investigators. (2014). Cigarette smoking among asthmatic adults presenting to 64 emergency departments. *Chest, 145*(4), 917–926.

Song, H., Li, X., Zhu, T., Guo, X., & Xia, H. (2021). Impact of climate change on crop yield in China: A meta-analysis. *Science of the Total Environment, 764*, 144498.

Sordillo, J. E., Hoffman, E. B., Celedón, J. C., Litonjua, A. A., Milton, D. K., Gold, D. R., & O'Connor, G. T. (2018). Multiple microbial exposures in the home may protect against asthma or allergy in childhood. *Clinical & Experimental Allergy, 48*(12), 1536–1546.

Sunyer, J., Antó, J. M., Kogevinas, M., & Roca, J. (2000). Asthma is associated with urinary excretion of 8-isoprostane, a marker of oxidative stress. *American Journal of Respiratory and Critical Care Medicine, 162*(6), 2097–2100.

Tischer, C. G., Hohmann, C., Thiering, E., Herbarth, O., Müller, A., Henderson, J., & Heinrich, J. (2018). Meta-analysis of mould and dampness exposure on asthma and allergy in eight European birth cohorts: An ENRIECO initiative. *Allergy, 73*(12), 2320–2331.

Tseng, E., Wu, C., Liao, Y., Su, M., Huang, C., Chen, L., & Lin, W. (2019). Association of air pollution with childhood allergic rhinitis: A case-control study. *Allergy, Asthma & Immunology Research, 11*(4), 509–519.

World Health Organization (WHO). (2018). *Climate change and health*. Available from: https://www.who.int/news-room/fact-sheets/detail/climate-change-and-health. Consulted October 23, 2022.

World Health Organization (WHO). (2020). *Urgent health challenges for the next decade*. Available from: https://www.who.int/news-room/photo-story/photo-story-detail/urgent-health-challenges-for-the-next-decade. Consulted October 23, 2022.

Xu, H., Qu, F., Xu, H., Lai, W., Andrew, J., Wang, Y., et al. (2014). Environmental exposure to cadmium: Health risk assessment and its associations with hypertension and impaired kidney function. *Science of the Total Environment, 470–471*, 875–881.

Zhang, Y., Xu, S., Huo, X., Wang, Y., Wang, S., Li, B., et al. (2019). The effects of air pollution on allergic rhinitis: A multi-center study in northern China. *Environmental Research, 171*, 365–374.

Zock, J. P., Plana, E., Antó, J. M., Benke, G., Blanc, P. D., Carosso, A., et al. (2005). Domestic use of hypochlorite bleach, atopic sensitization, and respiratory symptoms in adults. *Journal of Allergy and Clinical Immunology, 116*(3), 648–654.

Chapter 16
Entrepreneurship for Climate Change Mitigation and Adaptation in Nigeria and Ghana: Motivations, Challenges and Policy Implications

Agu Godswill Agu, Dike Onwubiko Ngozi, and Robert Ebo Hinson

Abstract Sustainable entrepreneurship has gained growing research attention, but studies that dig deep into the emerging area of climate entrepreneurship remain scarce. There are currently very few studies that explored the motivations and challenges of climate entrepreneurship, and the African context has not received reasonable scholarly attention. Yet, this knowledge is relevant in the formulation and implementation of policies for climate change mitigation and adaptation in the region given the inevitable role of entrepreneurs. The aim of this study is to contribute to the scarce literature on the motivations and challenges of climate entrepreneurship in the context of emerging African economies by exploring the lived experiences of technology-driven climate entrepreneurs in the solar energy sector. A qualitative study involving 24 climate entrepreneurs in Nigeria and Ghana was undertaken by means of semi-structural online interviews. Inductive thematic analysis was performed on the generated data. Findings indicate that climate entrepreneurs are motivated by passion, opportunity, technology, empowerment, and training/education (POTENT), while they face financial, environmental, and weak policy formulation/implementation (FEW) as major challenges. By proposing the POTENT-FEW-Policy Model, the study provides directions that stimulate policies for enhancing the involvement of entrepreneurs in climate change mitigating and adaptation in Africa.

A. G. Agu (✉)
Department of Marketing, Abia State University, Uturu, Nigeria
e-mail: godswill.agu@abiastateuniversity.edu.ng

D. O. Ngozi
Department of Management Sciences (Marketing Program), Rhema University, Aba, Nigeria
e-mail: dike@rhemauniversity.edu.ng

R. E. Hinson
Ghana Communication Technology University, Accra, Ghana

University of Johannesburg, Johannesburg, South Africa

R. E. Hinson
e-mail: rehinson@gctu.edu.gh

W. Leal Filho et al. (eds.), *Climate Change Mitigation and Adaptation in Practice*,
Climate Change Management, https://doi.org/10.1007/978-3-031-85217-6_16

Keywords Climate change · Climate entrepreneurship · Adaptation · Mitigation · POTENT-FEW-Policy model

Introduction

As the world grapples with the impacts of climate change, emphasis has been on mitigation and adaptation strategies. While mitigation strategies aim at avoiding or lowering the atmospheric emissions of greenhouse gases, adaptation strategies are adjustments deployed to limit the present and future negative effects of climate change and utilizing the opportunities it presents. According to the United Nations (2023: 1), climate change refers to long-term shifts in temperatures and weather patterns which can be natural, due to changes in the sun's activity or large volcanic eruptions. However, the nature-induced climate change has been overtaken by the human activity-induced climate change which results from burning of fossil fuels like coal, oil and gas, and leading to greenhouse effect and global warming (Ani et al., 2022; United Nation, 2023).

Abbass et al. (2022) explored the effects of climate change on different sectors including agriculture, biodiversity, human health, forestry, tourism, and the economic sector, and conclude that changes in these systems resulting from climate change will have catastrophic global effects. Since no country is immune to climate change impacts (The World Bank, 2023), governments across the globe and other stakeholders, including private enterprises endeavour to deploy innovative approaches and policies towards climate change mitigation and adaptation. More specifically, the essential role that entrepreneurs play in creating climate technologies and services has been highlighted (Schaer & Kuruppu, 2018; United Nations Framework Convention on Climate Change, UNFCCC, 2018). According to UNFCCC (2018), entrepreneurs actively contribute to the creation of technologies, business models, and services that society may employ to achieve low-emission and climate-resilient sustainable development. Through developing and deploying new technologies or facilitating the local adoption of existing solutions, entrepreneurs advance mitigation of or adaptation to climate change (Oraftik, 2021).

The global demand for energy remains on the increase and more than 80% of the energy produced worldwide is still derived from fossil fuels (United Nation, 2023). Due to their finite, depleting supply and negative effects on the environment, fossil fuels are still not a sustainable source of energy (Krishnan et al., 2021). According to the United Nations Climate Action (2023: 1), "Renewable energy sources—which are available in abundance, provided by the sun, wind, water, waste, and heat from the Earth—are replenished by nature and emit little to no greenhouse gases or pollutants into the air". Thus, solar and bioenergy are examples of renewable energy sources that are crucial for reducing the environmental issues brought on by greenhouse gas emissions. In Nigeria, Ghana and other African countries, some entrepreneurs have deployed entrepreneurial efforts and technology to mitigate the impacts of climate change by engaging in climate entrepreneurship that aims at producing and marketing climate-friendly energy products. This is in a bid to accelerate the actualization of the United Nations' sustainable development goals (SDGs) and specifically, SDG 7 "Ensure access to affordable, reliable, sustainable and modern energy for all."

The emerging field of climate entrepreneurship in Africa deserves more scholarly attention. Climate entrepreneurship has been defined as sensing, seeking and integrating climate change issues in business (Lee & Ahn, 2019). While growing research attention has been given to sustainable entrepreneurship in general, very few studies have focused on the emerging area of climate entrepreneurship (Akinbami et al., 2019; Lee & Ahn, 2019; Oraftik, 2021). Yet, the literature is still limited as research that digs deep into the motivations and challenges of climate entrepreneurship, particularly, renewable energy remains scarce (Kaesehage et al., 2019). But this knowledge is relevant in the formulation and implementation of policies for climate change mitigation and adaptation. Climate entrepreneurs, especially, technology-driven renewable energy entrepreneurs are growing in their numbers in Africa and their activities in various sectors of the African economy are helpful in meeting the United Nations' SDGs. Africa has unique climate change issues which entrepreneurial actions can help to address. However, climate entrepreneurs in the region face distinct challenges and opportunities (Akinbami et al., 2019; Oraftik, 2021) which call for special research attention.

Therefore, given this research gap, this chapter makes critical contributions by addressing the following three research questions:

RQ1: What factors drive climate entrepreneurship in Nigeria and Ghana?
RQ2: What challenges do climate entrepreneurs face in Nigeria and Ghana?
RQ3: What policy-measures can be deployed to help climate entrepreneurs in Africa to tackle the challenges and scale up?

Literature Review

Entrepreneurship in Renewable Energy: Nigeria and Ghana

Entrepreneurship in renewable energy is a key component of climate entrepreneurship. Although the continent of Africa does not emit a lot of greenhouse gases as it accounts for 4% of global greenhouse gas emissions, the continent still faces the greatest risks from the effects of climate change (UNFCCC, 2023; World Economic Forum, 2022). In order to tackle these impacts, adaptation and mitigation strategies for climate change have received improved attention in policy agendas at the global, national and local levels (Karki et al., 2022). As relevant stakeholders in the pursuit of the United Nations' SDGs, entrepreneurs in Africa have made noticeable efforts at developing and marketing products and services that aim at addressing climate change challenges. Thus, a growing number of businesses in Africa are developing novel climate change solutions to enhance mitigation and adaptation (World Economic Forum, 2022). The Agricultural sector for instance has witnessed influx of micro, small and medium-scale enterprises (MSMEs) investing in organic farming, precision agriculture, agroforestry, digital technology-enabled agribusiness, and other sustainable agricultural practices which can increase productivity while reducing adverse effects on ecosystems (United Nations' Environmental Programme, UNEP, 2023).

Renewable energy sources are constantly replenished by nature and are obtained either directly or indirectly from the sun or other environmental factors (Kaartemo & Gonzalez-Perez, 2020). Renewable energy has the potential to be a key enabler of socioeconomic growth in Africa and industrialization (IRENA & AfDB, 2022). According to UNEP (2023: 1), "Africa can become a trailblazer in renewable energy solutions, with abundant solar, wind, hydro, biomass, and geothermal resources that may contribute to a 6.4% increase in GDP from 2021 to 2050".

Africa has a growing number of MSMEs providing renewable energy solutions to climate change. To enable these entrepreneurs scale up, the World Bank in partnership with various African governments (e.g., federal government of Nigeria) and other stakeholders (e.g., Ashesi University, Ghana) have established centers that provide venture development and capacity-building support to MSMEs operating in different areas of climate entrepreneurship. For instance, the Nigeria Climate Innovation Center (NCIC) and the Ghana Climate Innovation Center (GCIC) have consistently empowered climate entrepreneurs in various areas of climate entrepreneurship including renewable energy. Between 2017 and 2022, GCIC had incubated 119 businesses out of which, 40 are women-owned or led (GCIC, 2023). Besides, recently the NCIC empowered eight early-stage businesses in the renewable energy sector with the sum of $10,000 (NCIC, 2023). Yet, many obstacles still confront climate entrepreneurs (Oraftik, 2021; Rafi, 2021), and especially in Africa.

Renewables-based energy generation capacity in Africa increased by 7% between 2010 and 2020 with solar energy emerging as the fastest-growing renewable energy source (IRENA, 2022). An average compound annual growth rate (CAGR) of 54% in solar capacity was achieved in the continent between 2011 and 2020 (IRENA & AfDB, 2022). In Nigeria and Ghana, solar energy is gradually becoming a household name with many communities being beneficiary to government-funded solar power installations. Besides, solar energy adoption in the commercial and industrial sector remains on the increase with a year-on-year growth of 61.5% in year 2022 (Africa Solar Industry Association, Africa Solar Outlook, 2023). However, individual, home-installed solar energy consumption is still far low as many homes currently do not yet patronize solar energy. According to Madombe (2006), despite the obvious benefits solar energy provides, the majority of Africans continue to use inefficient traditional energy sources. Wood, or other biomass such as crop waste, is the dominant fuel for cooking. This scenario presents untapped climate entrepreneurial opportunities in Africa, Nigeria and Ghana in particular. More so, with the launching of the Regional Emergency Solar Power Intervention Project (RESPITE) in Africa by the World Bank, a project that seeks to rapidly increase grid-connected renewable energy capacity in the continent, motivating potential and current climate entrepreneurs to venture into the untapped areas becomes imperative.

Climate Entrepreneurship Motivations and Challenges: Empirical Review

Understanding the motivations and challenges of climate entrepreneurs is relevant in formulating policies on climate change and in the development of climate entrepreneurship theory. Martínez-Cañas et al. (2023: 4) explain entrepreneurial motivation as "the purpose or psychological cause of starting a venture, such that the greater the importance given to that motivation, the more likely the person is to form action plans aimed at starting a venture". Thus, climate entrepreneurship motivation represents the reasons and causes of establishing and starting a climate entrepreneurial venture. In line with previous research, this chapter provides empirical evidences on motivations for (climate) entrepreneurship (Agu et al., 2023a, 2023b; Shane et al., 2003).

In a study involving 16 female tourism entrepreneurs, Ratten (2023) found female tourism entrepreneurs are more likely to be motivated to take climate change action by their social network and general interest in environmental concerns. Again, a study by Prabowo et al. (2023) show that perceived behavioral control (PBC), cultural values, cognition expertise, and contextual factors have an impact on people's intentions to start green businesses. Examining what drives business owners to address climate change, Kaesehage et al. (2019) classified climate entrepreneurs into opportunists, traditional and integrative entrepreneurs. Again, Demirel et al. (2019) identified industry life cycles, knowledge spillovers, institutions, and availability of external finance as key factors in shaping decision-making and firm behaviour in green start-ups. Purwandani and Michaud (2021) found that green business practices are driven by two main factors—internal motivations and the opportunity to secure one's public image.

Furthermore, entrepreneurial challenges are problems that recur as major impediments or obstacles to a particular enterprise's success, and these issues must be solved in order to fully enable climate entrepreneurship (See Agu et al., 2023b; Todeschini et al., 2017). Previous research exploring the challenges of climate, greentech entrepreneurship reported several barriers limiting the success of climate entrepreneurs. For instance, Purwandani and Michaud (2021) reported that the main obstacle to implementing green business strategies is a lack of funding. Beyond finance, Rafi (2021) disclosed that female entrepreneurs in the greentech sector also face the challenge of structural discrimination. UNFCCC (2023) disclosed that climate entrepreneurs face challenges in undertaking successful innovation, and three key obstacles they face include limited opportunity, lack of conditions that encourage the development of innovative climate change solutions, as well as insufficient funding for such activities. In their study, Akinbami et al. (2019) explored the challenges of climate-related entrepreneurship among rural female farmers in Nigeria and found lack of fund as one of the barriers. Generally, high investment costs, lack of information and subjective awareness in the market, lack of incentive and support mechanism and poor policy support hinder successful climate entrepreneurship (Haldar, 2019; Makki et al., 2020; Oraftik, 2021; Sarvari et al., 2023).

Methodology

This study is exploratory in nature given the limited knowledge of climate entrepreneurship motivations and challenges in the context of Nigeria and Ghana, a phenomenon that is worthy of investigation (Agu et al., 2023a, 2023b; Stebbins, 2001). So far, no study has explored the entrepreneurial motivations and challenges of climate entrepreneurs operating in the renewable, solar energy sector of these African countries. By undertaking this task, this chapter provides relevant insights into climate entrepreneurship to enable theoretical development. Therefore, a qualitative research approach that is based on semi-structured online interviews was adopted. Being an iterative process, qualitative research is relevant in gaining comprehensive understanding of the behaviour of participants (Aspers & Corte, 2019; Mariampolski, 2001). In this study, informants were asked predetermined but open-ended questions derived from the research objectives and insights from literature review (Given, 2008; Madichie & Agu, 2022). The semi-structured interview guide provided by Silverman (2020) was adopted in conducting the mixed e-channels interviews. The research followed an ethical standard for qualitative, in-depth interview research. The research process and interview protocol were approved by the Research and Publications Unit of the lead author's institution with the approval number: ABSU/FEMS/MKT/2023/016.

Sample and Data Collection

The participants are registered climate entrepreneurs in Nigeria and Ghana whose information were retrieved online, especially from the NCIC and GCIC and other greentech platforms. Purposive sampling method has been adopted in recruiting the participants, and basing on the professional Judgment of the researchers in selecting best suited informants to address the research questions (Maul, 2018). Only climate entrepreneurs in the renewable, solar energy sector, and who own duly registered ventures qualified for inclusion in the study. To identify and select the right experts to participate in a survey ensures the reliability of findings (Salamzadeh et al., 2023). The interviews which were conducted via email, WhatsApp and phone lasted for 45 min on average, spanning between May to August 2023. Besides, indirect interviews in the form of testimonials of the climate entrepreneurs hosted on the NCIC, GCIC and other platforms, which are relevant in addressing the research questions, were also utilized. This is in line with extant research.

First, emails were sent to forty-one climate entrepreneurs in the renewable energy sector, explaining the purpose of the study and solicited their consent to participate voluntarily in the interview (Agu et al., 2023a, 2023b). Participants were assured that their information would be treated confidentially. Most participants, who replied via the email, also permitted a recorded phone call or WhatsApp call to clarify certain points, thus ensuring validity of the process. The process of interview was flexible, allowing informants to provide deeper insights to key research questions. In all, data was generated from twenty-four participants, and data saturation was achieved as no new or surprising insights emerged (Knott et al., 2022).

Data Analysis Procedures

The data analysis process began with the transcription of all recorded/audio data using the traditional manual process. This was follow with collective and independent review of the transcribed data by the research team. In doing this, content analysis and thematic analysis were conducted following the inductive approach. In an inductive thematic analysis, researchers start with tentative themes that eventually make their way into a developing codebook rather than with theoretical constructions or frameworks (Milhas, 2023). Therefore, the goal of this analysis is to generate firsthand data from the lived experiences of climate, renewable solar-energy entrepreneurs about their motivations and challenges. Thereby resulting in data-based framework that provides a theoretical guide to policymakers. The six iterative process of thematic analysis by Braun and Clarke (2006) guided the analysis. Thus, the researchers familiarized themselves with the data by reading the transcripts severally, and thus generated relevant codes which were manually labeled to index them (King, 2004; Madichie & Agu, 2022). Thereafter, tentative themes were generated and carefully reviewed. The process ended by defining, refining and naming the emerging themes, and using relevant participants' quotes to support their emergence. While independent research experts were engaged to evaluate the transcribed data and emerging themes to establish the reliability of the findings (Madichie & Agu, 2022), the interviewees were presented with the findings to establish validity of the results. The themes are presented in tables.

Results and Discussion

The findings of the study are presented in two parts which provide answers to the first two research questions. Therefore, part one addresses the question "What factors drive climate entrepreneurship in Nigeria and Ghana?" In part two, the second research question "What challenges do climate entrepreneurs face in Nigeria and Ghana?" was handled. The third research question "What policy measures can be deployed to help climate entrepreneurs in Africa to tackle the challenges and scale up?" is treated in Sect. 16.5 of the study. These analyses were preceded with the descriptive characteristics of participants (Table 16.1).

Table 16.1 Demographic analysis of the respondents ($N = 24$)

Participants	Country	Gender	Age bracket (yrs)	Marital status	Level of education	Enterprise years of operation
A	Nigeria	Female	Below 45	Married	First degree	11
B	Nigeria	Male	Below 45	Married	First degree	7
C	Nigeria	Male	Below 45	Single	First degree	9
D	Nigeria	Male	Below 45	Married	First degree	4
E	Nigeria	Male	Below 45	Single	Graduate degree	8
F	Nigeria	Female	Below 45	Married	First degree e	6
G	Nigeria	Female	Above 45	Married	First degree	11
H	Nigeria	Male	Below 45	Single	First degree	3
1	Nigeria	Male	Below 45	Married	First degree	6
J	Nigeria	Female	Below 45	Married	Graduate degree	4
K	Nigeria	Female	Below 45	Single	First degree	8
L	Nigeria	Male	Above 45	Single	First degree	14
M	Nigeria	Male	Below 45	Single	First degree	4
N	Nigeria	Male	Below 45	Married	Graduate degree	5
O	Nigeria	Male	Below 45	Single	Graduate degree	8
P	Nigeria	Male	Below 45	Single	First degree	4
Q	Ghana	Male	Below 45	Married	First degree	3
R	Ghana	Male	Below 45	Married	First degree	7
S	Ghana	Female	Below 45	Married	First degree	4
T	Ghana	Male	Below 45	Married	First degree	7

(continued)

Table 16.1 (continued)

Participants	Country	Gender	Age bracket (yrs)	Marital status	Level of education	Enterprise years of operation
U	Ghana	Male	Below 45	Married	First degree	2
V	Ghana	Male	Above 45	Married	Graduate degree	6
W	Ghana	Female	Below 45	Married	First degree	4
X	Ghana	Male	Below 45	Single	First degree	6

Descriptive Characteristics of Participants

Out of the 41 contacted targets, 24 responded positively, thus representing 58.54% response rate. 7 (33.33%) are females, while 17 (66.67%) are males in the study sample. All the participants have university education with 19 or 79.17% holding first degree and 5(20.83%) having postgraduate degrees. The participants are mostly young adults aging below 45 years. There were 21 (87.50%) participants in this age bracket. The rest are above 45 years. 15 or 62.50% are married, while 9 or 37.50% are single. The ventures range from below 5 years in existence (9 or 37.50%) to 5–10 years (12 or 50.00%) and above 10 years (3 or 12.50%). Majority of the online testimonials were drawn from the NCIC and GCIC platforms.

Drivers of Climate Entrepreneurship Engagement

Findings, as summarized in Table 16.2, indicate the emergence of five themes as the motivating factors for climate, renewable solar-energy entrepreneurship among the participants. These we present using the POTENT acronym. Thus, the entrepreneurs are motivated by their *passion* for the environment, the *opportunity* created by climate change which could be entrepreneurially exploited, *technology* which makes it easy to innovatively tap from the created opportunity, *empowerment* of various forms provided by different stakeholders, *network* of like-minds providing innovative insights, ideas and business connections, and *training/education* in climate-related businesses which equips the entrepreneurs with requisite skills for successful operation. Amongst all the entrepreneurs, passion for the environment leads in their list of motivations. They believe that through their entrepreneurial efforts that are rooted in green technology, they can contribute reasonably in meeting the UN's SDGs. The findings resonate with previous research results in this direction ((Demirel et al., 2019; Kaesehage et al., 2019; Prabowo et al., 2023; Purwandani & Michaud, 2021).

Table 16.2 Evidence of climate entrepreneurship motivations

Motivations	Sample quotes
Passion for the environment	"I have an inclination and passion towards sustainable environment. My company distributes mitigation solutions based on solar and renewable energy" *(Participant N, Male)*
	"I'm following my passion in renewable energy and the pursuit of SDGs. To help tackle environmental degradation caused by the use of fossil fuel, we organize trainings and teachings in renewable energy" (Participant X, Male)
	"We are passionate about clean energy environment and committed to providing renewable energy alternatives that are affordable and environmentally friendly" (Participant H, Male)
	"We provide a business model that creates a sustainable social impact and provides appropriate opportunities for a greener and cleaner environment" (Participant W, Female)
	"Our solar-powered agro solutions provide solution to the sufferings of farmers dealing on perishable item" (Participant B, Male)
Entrepreneurial opportunities	"Looking at the teaming population of communities that are off-grid, we explore the window by ensuring that these communities access clean, reliable and affordable energy that is solar-generated." (Participant R, Female)
	"There are numerous unmet clean energy needs of the society that present profitable business opportunities. So, I have chosen to engage in some of these openings" (Participant E, Male)
Technology	"We contribute to the building of sustainable economies through technology innovations that deliver solar energy solutions" (Participant C, Male)
Empowerment	"I'm so much grateful to GCIC for providing financial and enlightenment supports that have helped my enterprise to grow" (Participant W, Male)
Networking	"I built networks from the various workshops/projects which have led me to securing two contracts" (Participant U, Male)
Training/Education	"Periodic training and constant networking have really helped. We are now investor-ready and have been able to equip ourselves for the competitive market" (Participant P, Male)
	"Trainings such as participating in progarmmes organized by government and international organizations inculcate innovative ideas and give opportunity for us to link with others in the same line of business" (Participant D, Male)

Challenges Confronting Climate Entrepreneurship

Financial, environmental, and weak policy formulation/implementation (FEW) challenges, as presented in Table 16.3, confront the climate entrepreneurs. These factors, represented by the FEW acronym, limit their ability to commercialize innovative climate technological ideas that are capable of mitigating climate change threats. In

Table 16.3 Evidence of climate entrepreneurship challenges

Challenge	Sample quotes
Finance	"If I have adequate access to the needed fund, my brother, there are innovative ideas that I will launch into the market that even you will call me to say congratulations. Finance in this our business is top-most because the technological base required in most of the incubated ideas is quite expensive" (Participant M, Male)
	"Investing in solar energy is cost-involving. In some instances, contracts that require an initial huge monetary commitment to prove to the client that you can deliver the service are not won because of our limited capital base and difficulty in securing loan at reasonable terms" (Participant A, Female)
Environment	"I received a distress call in the early hours of March 2nd this year informing me that bandits had set my work station ablaze. Reaching there, Sir, I found that the fire fighters were unable to subdue the fire. I lost millions of Naira-worth of materials purchased for a federal government contract to install solar energy across some states in the north". You know insecurity thrives most in the dark. So the bandits never wanted any technology that would expose their deeds" (Participant O, Male)
	"People still see solar energy as an expensive project. There are limited households opting for solar. Some people do not still have good knowledge of the durability of solar power and what environmental gains it carries" (Participant L, Male)
	"During a business conversation with one of my prospects, the prospects clearly informed me that the reason why he will not install solar energy in his building is because he does not trust the safety of the heavy solar batteries and their health implications" (Participant J, Female)
Weak policy implementation	"The government needs to prioritize solar energy usage in order to reduce the use of fossil fuel. But there are no laws currently that enforce the transitioning to solar energy solution" (Participant K, Male)
	"Because of economic hardship caused by poor governance, solar energy appears unaffordable to many people" (Participant S, Female)

line with previous research, financial constraint emerged as the first mentioned challenge by majority of the participants. These findings provide support to previous research on the challenges of climate entrepreneurship (Akinbami et al., 2019; Haldar, 2019; Makki et al., 2020; Purwandani & Michaud, 2021; Rafi, 2021; Sarvari et al., 2023; UNFCCC, 2023).

In Table 16.4, the thematic frames are presented.

Policy Implications

Technological innovation is a key catalyst of efforts to execute national climate action UNFCCC (2023), and the role of climate entrepreneurs in the development of technological innovations for climate change mitigation and adaptation stands tall.

Table 16.4 Coded thematic frames of climate entrepreneurship motivations and challenges

Themes	Description	Emerging subthemes
Climate entrepreneurship motivations	*The reasons and causes of establishing and starting a climate entrepreneurial venture*	Passion Opportunity Technology Networking Empowerment Training/education
Climate entrepreneurship challenges	*Problems that recur as major impediments or obstacles to the success of a climate enterprise, which must be solved in order to fully enable climate entrepreneurship*	Finance Environment Weak Policy

To motivate climate entrepreneurs and address the identified challenges confronting their ability to achieve their goals and contribute to the actualization of the SDGs, there is an urgent need to deploy international, regional and local policy frameworks. Following the findings of this research, regulatory/policy frameworks are required that can accelerate potential and current climate entrepreneurs' interest in exploiting climate friendly business opportunities and also alleviate or eradicate the obstacles hindering their success. While international and regional organizations such as the World Bank, United Nations, African Development Bank etc. have consistently initiated programs and frameworks such as NCIC, GCIC and RESPITE that support, educate and create enabling environment for climate entrepreneurs, the national efforts in African countries require urgent acceleration. By adapting the support programs to the region and countries, the barriers (e.g., FEW) are addressed (Oraftik, 2021) and the POTENT (motivating factors) enhanced. To guide policymakers in drafting and implementing policies and action programs to support climate entrepreneurs, some recommendations are presented.

First, **Education for climate change mitigation and adaptation (ECCMA)**: While education for sustainable development has gained ground across educational levels and programs, the climate change component has not received attention that is commensurate with the threat caused by climate change. For instance, in Nigerian universities, there is a sustainability component to management science courses including entrepreneurship (Agu, 2021). However, the aspects relating to climate change mitigation and adaptation and the role of management education (e.g., entrepreneurship, marketing, finance, accounting, etc.) have not received the required emphasis. Evidences abound on the effectiveness of education in stimulating passion for sustainable entrepreneurship (Agu, 2021; Leal Filho, 2015), and developing entrepreneurial passion is the first step towards entrepreneurial engagement (Anjum et al., 2021). Thus, education regulatory bodies such as National Universities Commission in Nigeria and National Accreditation Board of Ghana are encouraged to take a step in this direction. Besides, periodic free or affordable trainings are required for the potential and current climate entrepreneurs to give them insights, ides and directions that will help them succeed in their chosen entrepreneurial path and equip them with skills that can convince target markets to patronize them. The trainings and workshops also create avenues for networking.

Secondly, **finance**: Governments at national, state and local government levels are encouraged to alleviate/eradicate the financial challenges/risks faced by climate entrepreneurs. This might be in terms of tax rebate for start-up climate entrepreneurs, regulation of loan terms of financial institutions to climate technological innovating firms, monitoring the deployment of international financial supports to ensure that the *real* climate entrepreneurs are beneficiaries.

Thirdly, **technological support**: Government and other stakeholders can support the development and or purchase of specific technological equipment that start-up climate entrepreneurs need to be able to carry out specific innovative projects. This will also enable them to exploit untapped climate entrepreneurial opportunities. Fourthly, **creating enabling environment**: Deepening efforts at the security of lives, businesses and investments in the African region is imperative. Climate entrepreneurial business ventures, infrastructures and properties that play relevant role in crime prevention need adequate security. Government at all levels needs to understand this important role of solar energy and make policies that promote the security of players in the sector.

Fifthly, **transparency** in the administration of international supports: By ensuring that only committed *real* climate entrepreneurs benefit from the various programs of the World Bank, United Nations, African Development Bank etc., African countries would have equipped an army of climate entrepreneurs that are ready to provide innovative solution for climate change mitigation and adaptation. Again, awarding of government-sponsored contracts for community solar projects and other climate change mitigation programs should be transparently handled and not politicized. This serves as a great empowerment to climate entrepreneurs. Following the findings and policy implications, this chapter proposes a model. Since entrepreneurial intents and motivations research remain inconclusive and context-specific (Agu et al., 2023a, 2023b), the chapter contributes to the multidimensional theory of entrepreneurial motivation (Shane et al., 2003) and challenges by proposing the ***POTENT-FEW-Policy model***. Thus, to understand the motivations and challenges of climate entrepreneurship in the context of emerging African economies and how stakeholders can deploy policies to accelerate motivations and eradicate obstacles, the **POTENT-FEW-Policy model** comes handy. The model stimulates the consciousness in policy-makers that climate entrepreneurs possess the *potential* to subdue climate change threats through their mitigation and adaptation entrepreneurial paths, but they are faced with few surmountable challenges which the government and other stakeholders can address through policymaking. Figure 16.1 depicts the model.

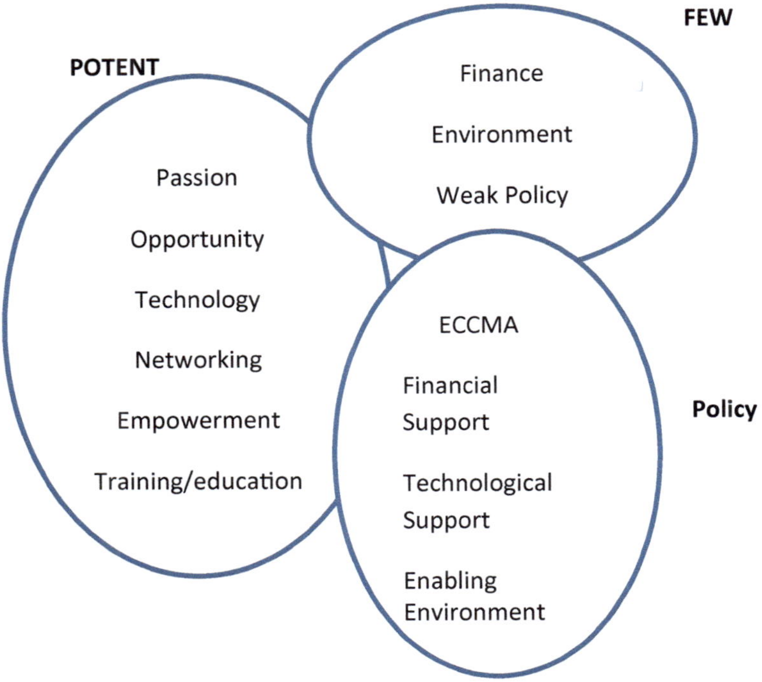

Fig. 16.1 POTENT-FEW-Policy model. *Note* **ECCMA:** Education for Climate Change Mitigation and Adaptation

Conclusions

This study aimed at exploring the motivations and challenges of climate entrepreneurship in Africa, Nigeria and Ghana in particular. The experiences of 24 climate entrepreneurs in the renewable energy sector were explored by means of online interviews. Through the process of thematic analysis, findings reveal that climate entrepreneurs in Nigeria and Ghana are motivated by entrepreneurial passion, opportunities, technological access, networking, empowerment, and training/education. They, however, face financial, environmental and weak policy implementation challenges. The chapter proposed the POTENT-FEW-Policy model which will be helpful to policymakers. Given the inevitable role of entrepreneurship in providing mitigation and adaptation solutions to climate change, the findings of the study provide insights on how to support climate entrepreneurs to scale-up. By making policies that focus on education for climate change, financing of climate entrepreneurs, providing technological support and enabling environment for climate entrepreneurship, and empowering climate entrepreneurs through transparent administration of international supports to climate entrepreneurs, climate entrepreneurs in Africa will be motivated (POTENT acceleration) and the obstacles confronting them (FEW) alleviated substantially or eradicated.

Therefore, the chapter contributes to the scarce empirical literature on climate entrepreneurship motivations and challenges in Africa. Thereby, providing directions to policy-makers and other stakeholders interested in harnessing the important role of climate entrepreneurs in responding to climate change.

The present study is limited in scope, covering only two African countries and climate entrepreneurs in the renewable solar energy sector. It also adopted the qualitative approach and utilized the inductive method. To take the study forward, future and ongoing studies are encourage to explore other countries and sectors as well as adopt the quantitative or mixed approach to test our proposed model.

Acknowledgements The authors are grateful to the editors and anonymous reviewers for the insightful suggestions.

References

Abbass, K., Qasim, M. Z., Song, H., Murshed, M., Mahmood, H., & Younis, I. (2022). A review of the global climate change impacts, adaptation, and sustainable mitigation measures. *Environmental Science and Pollution Research, 29*, 42539–42559. https://doi.org/10.1007/s11356-022-19718-6

Africa Solar Industry Association, Africa Solar Outlook. (2023). Solar outlook 2023. Retrieved from www.afsiasola.com

Agu, G. A. (2021). A survey of business and science students' intentions to engage in sustainable entrepreneurship in Nigeria. *Small Enterprise Journal, 28*(2), 206–227.

Agu, G. A., Okereafor, G. E., Omotosho, D. T., Okocha, E. R., & Uche, D. B. (2023a). Drivers of religious entrepreneurial behaviour among Pentecostal pastors: Push-pull-mooring theory and theory of planned behaviour. *Small Enterprise Research*. https://doi.org/10.1080/13215906.2023.2231408

Agu, G. A., Salamzadah, A., Omotosho, T. D., Uche, D. B., Okocha, R. E., Agu, O. A., & Eke, C. N. (2023b). Digital pentepreneurship in Nigeria: Challenges and opportunities. In F. Nezameddin (Ed.), *Digital entrepreneurship.*

Akinbami, C. A. O., Olawoye, J. E., Adesina, F. A., & Nelson, V. (2019). Exploring potential climate-related entrepreneurship opportunities and challenges for rural Nigerian women. *Journal of Global Entrepreneurship Research, 9*, 19. https://doi.org/10.1186/s40497-018-0141-3

Ani, K. J., Anyika, V. O., & Mutambara, E. (2022). The impact of climate change on food and human security in Nigeria. *International Journal of Climate Change Strategies and Management, 14*(2), 148–167. https://doi.org/10.1108/IJCCSM-11-2020-0119

Anjum, T., Heidler, P., Amoozegar, A., & Anees, R. T. (2021). The impact of entrepreneurial passion on the entrepreneurial intention; moderating impact of perception of university support. *Administrative Sciences, 11*, 45. https://doi.org/10.3390/admsci11020045

Aspers, P., & Corte, U. (2019). What is qualitative in qualitative research. *Qualitative Sociology, 42*, 139–160. https://doi.org/10.1007/s11133-019-9413-7

Braun, V., & Clarke, V. (2006). Using thematic analysis in psychology. *Qualitative Research in Psychology, 3*, 77–101.

Demirel, P., Li, Q. C., Rentocchini, F., & Tamvada, P. (2019). Born to be green: New insights into the economics and management of green entrepreneurship. *Small Business Economics, 52*, 759–771. https://doi.org/10.1007/s11187-017-9933-z

Ghana Climate Innovation Center. (2023). GCIC results. www.ghanacic.org
Given, L. M. C. (2008). Semi-structured interview. In *The Sage encyclopedia of qualitative research methods.*
Haldar, S. (2019). Green entrepreneurship in the renewable energy sector—A case study of Gujarat. *Journal of Science and Technology Policy Management, 10*(1), 234–250. https://doi.org/10.1108/JSTPM-12-2017-0070
IRENA. (2022). *Renewable energy market analysis: Africa and its regions.* www.irena.org
IRENA and AfDB. (2022). *Renewable energy market analysis: Africa and its regions—A summary for policy makers.* International Renewable Energy Agency and African Development Bank.
Kaartemo, V., & Gonzalez-Perez, M. A. (2020). Renewable energy in international business. *Critical Perspectives on International Business, 16*(4), 325–336. https://doi.org/10.1108/cpoib-08-2019-0062
Kaesehage, K., Leyshon, M., Ferns, G., & Leyshon, G. (2019). Seriously personal: The reasons that motivate entrepreneurs to address climate change. *Journal of Business Ethics, 157*, 1091–1109. https://doi.org/10.1007/s10551-017-3624-1
Karki, G., Bhatta, B., Devkota, N. R., Acharya, R. P. & Kunwar, R. (2022). Climate change adaptation (CCA) research in Nepal: implications for the advancement of adaptation planning. *Mitigation and Adaptation Strategies for Global Chang, 27*, 18. https://doi.org/10.1007/s11027-021-09991-0
King, N. (2004). Using templates in the thematic analysis of text. In C. Cassell & G. Symon (Eds.), *Essential guide to qualitative methods in organizational research* (pp. 257–270). Sage.
Knott, E., Rao, A. H., Summers, K., & Teeger, C. (2022). Interviews in the social sciences. *Nature Review Methods Primers, 2*, 73. https://doi.org/10.1038/s43586-022-00150-6
Krishnan, S. K., Kandasamy, S., & Subbiah, K. (2021). Chapter 32—Fabrication of microbial fuel cells with nanoelectrodes for enhanced bioenergy production. In R. Praveen Kumar, B. Bharathiraja (Eds.), *Nanomaterials* (pp. 677–687).
Leal Filho, W. (2015) Education for sustainable development in higher education: Reviewing needs. In *Transformative approaches to sustainable development at universities.*https://doi.org/10.1007/978-3-319-08837-2_1
Lee, S., & Ahn, Y. (2019). Climate-entrepreneurship in response to climate change: Lessons from the Korean emissions trading scheme (ETS). *International Journal of Climate Change Strategies and Management, 11*(2), 235–325.
Madichie, N., & Agu, G. A. (2022). The role of universities in scaling up informal entrepreneurship. *Industry and Higher Education.* https://doi.org/10.1177/09504222221101548
Madombe, I. (2006). Solar power: Cheap energy source for Africa. www.un.org
Makki, A. A., Alidrisi, H., Iqbal, A., & Al-Sasi, B. O. (2020). Barriers to green entrepreneurship: An ISM-based investigation. *Journal of Risk and Financial Management, 13*(11), 249. https://doi.org/10.3390/jrfm13110249
Mariampolski, H. (2001). *Qualitative market research: A comprehensive guide.* Sage.
Martínez-Cañas, R., Ruiz-Palomino, P., Jiménez-Moreno, J., & Linuesa-Langreo, J. (2023). Push versus Pull motivations in entrepreneurial intention: The mediating effect of perceived risk and opportunity recognition. *European Research on Management and Business Economics, 29*(2), 100214. https://doi.org/10.1016/j.iedeen.2023.100214
Maul, A. (2018). Judgment sampling. In B. Frey (Ed.), *The SAGE encyclopedia of educational research, measurement, and evaluation* (pp. 914–914). SAGE Publications, Inc. https://doi.org/10.4135/9781506326139.n367
Milhas, P. (2023). *Qualitative, multimethod, and mixed methods research* (4th ed.) International Encyclopedia of Education.
Nigeria Climate Innovation Centre. (2023). Green venture development. www.nigeriacic.org
Oraftik, C. (2021). Climate entrepreneurship in developing economies: Supporting entrepreneurs tackling climate change. www.andeglobal.org

Prabowo, H., Ikhsan, R. B., & Yuniarty, Y. (2023). Drivers of green entrepreneurial intention: why does sustainability awareness matter among university students? *Organizational Psychology, 12*.https://doi.org/10.3389/fpsyg.2022.873140

Purwandani, J. A., & Michaud, G. (2021). What are the drivers and barriers for green business practice adoption for SMEs? *Environment Systems and Decisions, 41*(4), 577–593. https://doi.org/10.1007/s10669-021-09821-3

Rafi, T. (2021). Why the world needs to invest in female climate entrepreneurs. www.imf.org

Ratten, V. (2023). Gender and climate change: A sustainable tourism perspective. *Thunderbird International Business Review*. https://doi.org/10.1002/tie.22351

Salamzadeh, A., Hadizadeh, M., Yazdanpanah, Y., & Agu, G. A. (2023). Marketing innovation drivers: Toward reusing and recycling. In *Fashion and environmental sustainability* (pp. 203–221). De Gruyter.

Sarvari, R., Jabarzadeh, Y., Karami, A., & Jabarnejad, M. (2023). An interpretive structural modeling—Analytic network process approach for analysing green entrepreneurship barriers. *International Entrepreneurship and Management Journal*. https://doi.org/10.1007/s11365-023-00881-2

Schaer, C., & Kuruppu, N. D. (2018). *Private-sector action in adaptation: Perspectives on the role of micro, small and medium size enterprises*. UNEP DTU Partnership Research.

Shane, S., Locke, E. A., & Collins, C. J. (2003). Entrepreneurial motivation. *Human Resource Management Review, 13*(2), 257–279.

Silverman, D. (2020). *Doing qualitative research* (5th ed.,). Sage.

Stebbins, R. (2001). Exploratory research. In *The social sciences*. Sage: https://doi.org/10.4135/9781412984249

The World Bank. (2023). Climate change overview. www.worldbank.org

Todeschini, B. V., Cortimiglia, M. N., Daniela, C., & Ghezzi, A. (2017). Innovative and sustainable business models in the fashion industry: Entrepreneurial drivers, opportunities, and challenges. *Business Horizons*. https://doi.org/10.1016/j.bushor.2017.07.00

United Nations. (2023). What is climate change? www.un.org

United Nations Framework Convention on Climate Change. (2018). Energizing entrepreneurs to tackle climate change. TEC Brief 12.

UNFCCC. (2023). *Africa Climate week 2023: Charting a fresh course for climate action*. United Nations Climate Change. www.unfccc.int

United Nations' Environmental Programme. (2023). Africa's green business opportunities are abundant. www.unep.org

World Economic Forum. (2022). Africa's businesses are its secret weapons against climate change. www.weforum.org

Agu Godswill Agu is the head of the Department of Marketing, Abia State University, Nigeria, and a facilitating scholar at the National Open University of Nigeria. Godswill holds BSc, MSc, and PhD all in Marketing, and was awarded a Post-doctoral scholarship of the University of Johannesburg in 2019. He specializes in Marketing and Entrepreneurship and currently researching on how education, marketing and entrepreneurship can contribute in accelerating sustainable development and climate change mitigation/adaptation in Africa.

Dike Onwubiko Ngozi holds BSc (Chemistry), MBA and PhD in Marketing of the University of Nigeria Nsukka. He is an Associate Professor of Marketing and the Director of Entrepreneurship Studies at Rhema University, Nigeria. His current areas of research include religious entrepreneurship and digital marketing/entrepreneurship.

Robert Ebo Hinson is a Professor of Marketing at the University of Ghana Business School and the Pro-Vice Chancellor of the Ghana Communication Technology University. He holds several academic qualifications including MBA, DPhil and PhD, and has taught undergraduate and graduate students across Africa. His current research interests cover marketing communications, information and technology management, service management, and social responsibility/sustainability management. He is also affiliated with the University of Johannesburg.

Chapter 17
Case Study of Developing Mumbai City's Disaster Management Plan Encompassing Global and National Frameworks on Disaster Risk Reduction and Climate Resilience

Chinmai Hemani

Abstract Mumbai, a bustling Indian coastal metropolis, is highly vulnerable to natural hazards like flood, cyclone; moderately vulnerable to earthquake, tsunami, drought etc. owing to its geo-climatic, geological and physical features. It is increasingly vulnerable to climate-induced disasters like heavy rainfall, sea level rise, tropical cyclones apart from various human-made hazards of building collapse, epidemics, road accidents, terrorism. Economic opportunities have led to influx of migrants thus high population densities, scarcity of land, loss of natural ecosystems like wetlands, mangroves to accommodate the increased needs of growing population only exacerbates its vulnerability. To reduce city's vulnerability and improve city's resilience, Mumbai's Disaster Management Department (DMD) undertook an initiative of revising Disaster Management Plan (DMP), making it more context specific addressing disaster at ward-level while also taking inputs from future climate impacts indicated in Mumbai Climate Action Plan (MCAP). Method adopted for this revision involved identifying existing challenges; current and projected climate impacts and what DMD would undertake to meet these future challenges by way of setting Sendai targets for the city. Sensitization meets were held to update the stakeholder and involve them for this revision. It helped each stakeholder brainstorm relevant actionable suggestions to reduce their response time, to make city safe for its most vulnerable population (women, children, elderly, physically/mentally challenged etc.), to improve resources to cater to any type and scale of disasters such that the life/property/economic loss due to impending disasters tends to be minimal. Inputs of stakeholders through several exercises and table-top exercises considering earthquake and cyclone scenario were collated in-turn leading to revision of improved processes, Standard Operating Procedures (SOP) and the DMP. Such a plan is first of its kind amongst Indian cities, an all-encompassing one with inputs from Sendai Framework, PMs 10-point agenda, MCAP and from stakeholders. With such a plan

C. Hemani (✉)
RuChiNi Environment Sustainability Planners, Rajkot, India
e-mail: chinmayhemani@gmail.com

W. Leal Filho et al. (eds.), *Climate Change Mitigation and Adaptation in Practice*, Climate Change Management, https://doi.org/10.1007/978-3-031-85217-6_17

in place, it would be a model document for all the other Indian districts and countries to follow.

Keywords Climate resilience · Disaster Risk Reduction · Sendai framework · Disaster Management Plan · Mumbai Climate Action Plan · PMs 10-point agenda

Background

Cities are economic hub offering opportunities of growth, which have attracted half the world population already (UN, 2024). For fast growing economy like India aspiring to become US$40 trillion by 2047 (NITI Aayog, 2022), it is projected that it will add another 675 million people to its urban areas by 2035 (UNHABITAT, 2022). Growing population, increased infrastructure with dwindling resources often lead to increasing disaster risk in those urban areas due to impending hazards, impacting vulnerable population with limited capacities to cope, the most (Gawler & Tiwari, 2014). Cities increased consumption needs lead to increased Greenhouse gas (GHG) emissions, a major cause of climate change, in-turn making it victims of climate change as well (Gawler & Tiwari, 2014). Most often cities regular functioning when faced with climate extremes and non-climatic disasters gets impacted due to lack of data, limited technological and institutional capacities, limited financial resources, making it difficult to take informed decisions. Thus rescue, relief, rehabilitation takes precedence over resilience building (Jain & Bazaz, 2017). However, India has been working to address such disaster risks by improving its institutional arrangements by way of enactment of Disaster Management Act (DMA) 2005, adoption of National Policy on Disaster Management 2009 and National Disaster Management Plan (NDMP) 2019 percolating down to state, districts, cities and rural areas as well (NDMA, 2019). NDMP 2019 considers inputs from three landmark international agreements signed off in 2015: (1) Sendai Framework for Disaster Risk Reduction (SF DRR); (2) Sustainable Development Goals (SDGs) (2015–30); and (3) Paris Agreement on Climate Change at the 21st Conference of Parties (COP 21). It has also considered inputs from Ten Point Agenda on DRR, enunciated by Prime Minister (PM) during Asian Ministerial Conference on DRR (AMCDRR) in 2016 (NDMA, 2019). This chapter is a case study of Mumbai's revision of its DMP considering these three international agreements and PMs Declaration to improve city's climate and disaster resilience.

Study Area

Brihanmumbai/Greater Mumbai (referred here as Mumbai)—a coastal metropolis, an economic capital of India, is reclaimed from a cluster of seven islands having mean sea level of just 10–15 m (Fig. 17.1) (MCGM, 2020).

Fig. 17.1 Map of Greater Mumbai. *Source* Google maps

With an area of about 468 km^2 and an estimated population of about 13.01 million in 2023, population density about 26,934 person/km^2 (MCGM, 2023), Mumbai is vulnerable to all major natural hazards—flood, cyclone, earthquake, tsunami, drought etc. City falls under seismic zone III (Moderate Damage Risk Zone) (BMC, 2022). Thus, city observes a high risk of liquefaction during an earthquake as large area of city is reclaimed through infilling (BMC, 2022). City also has landslide prone areas, which are foothills with loose soil and rocks or extreme slope. Informal settlements settled along unstable slopes and foothills, have limited adaptive capacity would be the most affected during occurrence of heavy rainfall, earthquake. All this together makes city vulnerable to hydro-geological hazard as well (BMC, 2022).

While it is also highly vulnerable to various human-made hazards of terrorism, epidemics, road accidents, etc. apart from other pressures that high-density population brings like sewage disposal, solid waste management, health impacts etc.

Due to human induced climate change, city is now facing extreme events more frequently and with greater intensity. Since 1973, city has observed an increased warming of 0.25 °C per decade (MCAP, 2022a, 2022b). Dense settlements, reflective building surfaces and low vegetation cover add to the risk of heat exposure. Projections of Sea Level Rise (SLR) for Mumbai show that by 2050, Arabian Sea could begin flooding most of Mumbai at least once every year, estimating the at-risk population to be three times more in the coming decades than was previously estimated (MCAP, 2022a, 2022b). It is projected that Mumbai will have to bear US$ 112–162 billion by 2050 due to SLR damages, which by 2070 could increase by a factor of 2.8–2.9 (Shaw et al., 2022).

Owing to multiple entities co-existing together, from high population densities, influx of migrants, and scarcity of land to other aspects like co-existence of natural ecosystems—Wetlands, Mangroves, invites redoubling efforts for DRR.

Objectives and Conceptual Framework

To improve city's resilience, it is extremely important to understand city's vulnerability from natural and human-made hazards, while also considering future climatic changes city is likely going to face. Thus, it is necessary to devise a dynamic Disaster Management Plan (DMP) and deploy resources to mitigate these hazards. DMP provides a framework and direction to the city governing body to respond appropriately to these risks (BMC, 2022).

Brihanmumbai Municipal Corporations (BMC) Disaster Management Department (DMD) had comprehended the lacunae about climate considerations and elements from Sendai framework in its existing DMPs and its understanding amongst its stakeholders. Thus, as a periodical exercise, DMD wanted to revise the city's DMP by bringing in elements from the Sendai Framework and PM's 10-point agenda for DRR (NDMA, 2019) while considering inputs about future climate impacts.

To reach this goal, following objectives were framed:

1. Work on process strengthening by identifying gaps in the existing processes for disaster risk management (DRM) and improving it,
2. Bring in Data-centricity based on data driven evidences through Geographical Information System (GIS) and Command and Control System (CCS)
3. Incorporate climate change perspective in the planning of project interventions
4. Incorporate elements from International and national framework on DRR.

This DMP revision would improve management of hazards and risks benefiting citizens and communities in-turn helping Mumbai meet its international obligations of C40 Cities Deadline 2020.

For a holistic approach, following conceptual framework which suits local needs has been adapted from Gawler and Tiwari (2014), NDMA (2019) consisting of three broad steps (1) understanding current and future climate and disaster risks (2) stakeholder engagement (3) Improving city's resilience strategy (Fig. 17.2).

1. To understand city's climate and disaster risk following steps were undertaken:

Literature review: Relevant inputs were taken from a range of documents:

- International frameworks—SFDRR, Intergovernmental Panel on Climate change (IPCC) AR6 report

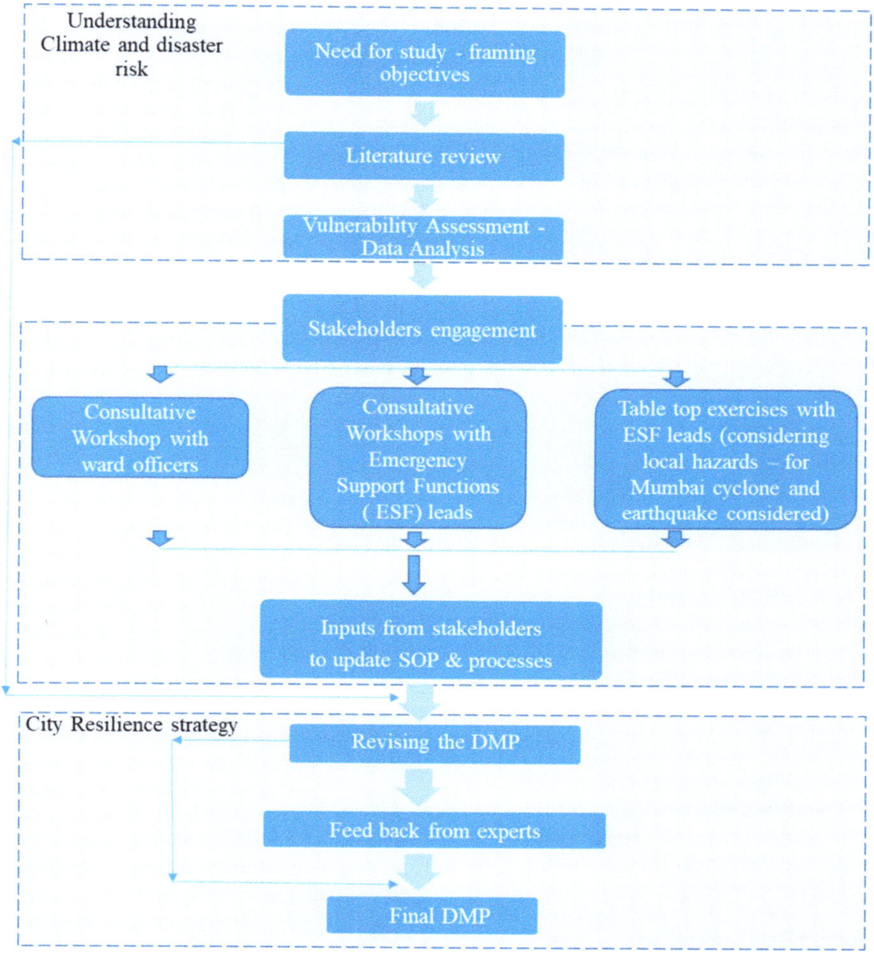

Fig. 17.2 Conceptual framework of the study

National Documents like: DMA 2005, NDMP 2019 and PM's 10-point Agenda

- Documents relating to Mumbai City and Suburbs District:
  DMP 2022, Mumbai Climate Action Plan (MCAP) 2022 etc.
- Several State, District and City DMP.

Data Analysis: to understand the city's vulnerabilities and risks, city's decadal data of hazards from the CCS and GIS along with various reports produced by DMD were analysed. DMD receives daily data through CCS, GIS, Automatic Weather Stations (AWS) and other monitoring devices. To facilitate the generation of evidence-based strategies for prevention and preparedness purposes, as well as the development of decision support systems for planning and operational purposes, tools like trend analysis, mapping of hazards on GIS are used.

Trend analysis is a foresight tool where collections of signals/events help understand patterns and likely directions of change, underlying drivers, which will be useful in making decisions (UNDP, 2022). Thus, to understand types and trends of incidences being captured in CCS, Incidence report from 2013 to 2022 was taken from the CCS. Three-year moving average was used in trend analysis which showed an increasing trend for many of noted incidences owing to various pressure city faces in-spite of efforts taken by BMC for improving city's disaster resilience. All the incidences captured in CCS gets mapped in GIS, providing valuable insights for analysing the data while helping visualize geospatial data. Data analysis from CCS and GIS would help in identifying the gaps to cater to the incidences across the city and take appropriate interventions at the grassroots.

Literature review and data analysis led to learning about gaps/areas needing interventions especially due to climate change impacts experienced currently and likely to be exacerbated in future.

2. **Engaging stakeholders** involved following steps:

Consultative workshop with ward officers: Sensitization meeting held with ward officers, consisted of few exercises to capture challenges wards face while experiencing various disasters. The findings from the exercises with ward officers further strengthened the identified gaps.

Consultative workshops with Emergency Support Function (ESF)[1] To capture the inputs from the various stakeholders responsible to address disasters, six sensitization meetings along-with various exercises including table top exercises (TTE) to streamline Standard Operating Procedure (SOP) for all Mumbai relevant disasters was undertaken. The exercises allowed stakeholders to consider climate change projections and probed them for their action plan, and interventions they would undertake to improve city's climate and disaster resilience.

Table-top exercises: To improve stakeholders understanding and facilitate them to contribute better, they were also taken through two TTEs consisting of earthquake and cyclone scenario respectively. TTE simulates a real world scenario, which depicts impact of a hazard across the city (UNDRR, 2017) to help increase readiness in case of an actual emergency. Such informal setting facilitates participants to discuss plan by brainstorming, problem solving as a group by sharing capabilities based on their organization's existing plans suiting the objectives of the exercise. In case of an Earthquake, the information about the disaster passes from citizens to the authorities as it occurs leaving no scope for preparedness, thus Earthquake is a bottom-up approach where community are the first responders. While for Cyclone,

[1] ESF: In case of any hazardous event, ESF would provide coordination mechanisms including processes relevant to planning, managing and coordinating specific response and preparedness activities amongst various responsible agencies. Depending on the function, several ESF are defined where each ESF would be headed by a lead agency and several identified support agencies which forms integral part to functioning of Emergency Operation Centres (EOC). (BMC, 2022).

it is a top-down approach as the information about tropical cyclone is well-tracked by meteorologists as it develops and there is a scope for preparedness, planned evacuation, relief and better response by the department. Both the hazards have medium and high probability of its occurrence for Mumbai respectively with probability of occurrence of multiple secondary hazards as well.

Through these two TTEs all major hazards that Mumbai faces would be discussed and thus were considered.

3. Working on **City Resilience Strategy** involved:

Expert review and revision to final DMP: After incorporating all the comments and suggestions from stakeholders, a final draft plan thus prepared was reviewed by experts from National Disaster Management Authority (NDMA). Comments from the experts were incorporated into the final plan, which after due process of approvals will be made publicly available on BMC's website.

Next section talks about the processes, tools, exercises through stakeholder engagement and tables generated for updating the DMP keeping objectives in mind for improving climate and disaster resilience. Existing governmental policies, techno-legal regime and provisions for financial resources were also tabulated and incorporated in the DMP but are out of scope for this paper. All the tables depicted here are for academic/reference purposes only to highlight the processes used. The wards have been marked numerically here for representation sake. The final tables considering all hazards and actual impacted wards would appear in the published DMP. It would be a useful aid for local governments on how to revise the DMP while adopting Sendai targets, Local nationalised DRR agenda like PMs 10-point agenda and adhering to localised Climate Action Plans for making city climate and disaster resilient. Such pro-active steps help move away from relief centric approach to an integrated approach inculcating a culture of preparedness, promoting innovative strategies of mitigation and emergency response.

Building Climate and Disaster Resilience

Building resilience of people, economies and natural resources against impacts of extreme weather and climate is the common aim for DRR and CCA programs, thus it is prudent to converge both the programs with the onslaught of increasing climate disasters (UNDRR, 2019). Even a small relative increase of + 0.5 °C in global warming can lead to statistically significant changes in intensity and frequency of climate extremes globally (IPCC, 2019). Thus, for city to be resilient, it needs to create enabling environment for integrated risk governance at local level. Implementation of strategies would require not only commitment but also sufficient resources, capacity and effective engagement of relevant stakeholders. With data available from secondary literature, CCS, GIS etc. effort made to build case for increasing linkage for Climate change adaptation (CCA) and disaster resilience along with need for stakeholder engagement is elaborated below.

Increasing Linkage of Climate Change and Disaster Risk Reduction

Compounded impacts of climate change that Mumbai faces makes it one of the top five most vulnerable districts in Maharashtra state (MCAP, 2022a, 2022b). IPCC AR6 report states that under the worst emission scenario—RCP 8.5 (projects a 3–4 °C temperature rise) Mumbai coast will likely experience SLR of 60 cm by 2100 considering benchmark from 1995 to 2014. Under the best possible emission scenario—RCP 2.6 (projects a 1.8 °C temp rise) likely SLR of 40 cm is projected (Basu, 2021). W.r.t total rainfall, under the worst possible emission scenario a probable increase of about 45% by 2081–2100 (benchmark considered is 1995–2014) will be observed. Moreover, w.r.t maximum one-day rainfall (short duration high intensity rainfall), city is likely going to experience 53% increase compared to pre-industrial period where average about 102 mm rainfall is expected on such days under the worst emission scenario (Basu, 2021). W.r.t heat under worst emission scenario, there would be about 117 days (i.e. about one out of every three days.) annually where the temperature will exceed 35 °C (Basu, 2021). Considering increasing climate hazards and need for convergence of CCA and DRR interventions for Mumbai, Table 17.1 was tabulated. This table at one-glance highlights: multi-hazard risks, corresponding wards facing current vulnerabilities, existing technologies used to cater to these risks along with wards likely to be impacted with future projections of climate risks as indicated in IPCC AR6, MCAP simultaneously would allow relevant authorities to plan such interventions to address multiple future vulnerabilities.

Approaching Multi stakeholders for Strategic Engagement and Insights

Disaster risk manifests when hazard patterns, exposure, coping capacity and vulnerabilities of people alter. Thus, understanding complex risks, adopting systems thinking, using interdisciplinary insights and knowledge to effectively address the uncertainty becomes a crucial step to adhere to while developing DRR strategies (UNDRR, 2019).

Interdisciplinary insights emerge when tools like theory of change are used where several initiatives amongst stakeholders are undertaken which provide platform for dialogue and reflection on ideas of change and their underlying assumptions. Such tools will help create better understanding and plan for long-term change (UNDRR, 2019). DRR should be everyone's responsibility and not department specific. Such collective decision-making involving all sections of society having relevant knowledge and experience to decide upon issues related to DRR while finding solutions

Table 17.1 Hazard faced by Mumbai city and Suburbs districts considering current and future probability

Hazard category	Hazard	Wards impacted based on history, current vulnerability	Early warning technology	Frequency of occurrence	Probability of occurrence in Future	Assumption-Future probability	Probable Ward impacted (in future)
Hydrological	Flood	6, 11, 12, 13,14, 15, 18, 19, 21, 22	a) 2 Doppler Radar (S band and C band), 100 Flow level sensors are used in identified flood prone areas	High	High	Large part of Mumbai is at risk of being submerged by 2050 and that by 2080, the likelihood of urban floods, such as the 2005 event, is more than double (Source: MCAP, 2022a, 2022b)	
			b) 120 AWS are installed				
			c) Hydrological model Hydraulic and hydro-dynamic model developed for flood forecasting				
Hydro -geological	Landslide	4, 9, 14, 15, 18, 23		High	High		
Meteoro-logical / Climato-logical	Cyclone / storm surge		2 Doppler Radar (S band and C band)	Low	Moderate		All

(continued)

Table 17.1 (continued)

Hazard category	Hazard	Wards impacted based on history, current vulnerability	Early warning technology	Frequency of occurrence	Probability of occurrence in Future	Assumption-Future probability	Probable Ward impacted (in future)
	Heat waves	All	120 AWS	High	High		All
Climato-logical	SLR / coastal inundation	1, 3,4, 8, 9, 11, 13, 18, 21	Hydraulic, Hydrological and, Hydro-dynamic model (I-FLOWS)	Low	Moderate	Projections of SLR for Mumbai show that by 2050, the Arabian Sea could begin flooding most of Mumbai at least once every year, estimating at-risk population to be thrice more in coming decades than was previously estimated (Source: MCAP, 2022a, 2022b)	

Source DMD

accepted by all the concerned stakeholders proves useful process of drafting or developing DRR strategies and plans (NIDM, 2016). Thus, with this forethought, stakeholders (lead and supporting ESF agencies) including municipal sectoral departmental officials, private sector, NGOs, Academic institutions etc. were invited for consultative workshops with a purpose to gain perspectives on:

- Building understanding on city's current and future vulnerabilities
- Strategies to improve city's resilience, policy and planning
- Identifying gaps in the existing procedures,
- Identifying gaps in coordination,
- Identifying their actionable suggestions to reduce response time, to make city safe for its most vulnerable population (women, children, elderly, physically/mentally challenged etc.), to improve resources to cater to any type and scale of disasters such that the life loss/property/economic loss due to impending disasters tends to be minimal.

Exercises with stakeholders led to diagnosis, inference and actions needed in local context in order to anticipate, respond and adapt to potential risks. While workshop exercises with ward officials captured voices from the ground, workshops with ESF lead and supporting agencies led to identification of some lacunae in operations, management and policy, which hamper city's resilience.

Exercises with ward officials:

Exercise 1: Hazard risk matrix is ward representative's perception about the hazards faced by respective wards and ranking hazards based on severity and likelihood of hazard impacts. This hazard risk matrix and scale depicted in Figs. 17.3 and 17.4 was adapted from UNISDR (2017), GDPC (n.d.) to suit the local needs.

A	B	C	D
Hazard	**Likelihood**	**Severity**	**Multiply**
	1 - 5 rank	**1 - 5 rank**	**(C X D)**

Fig. 17.3 Hazard risk matrix

Likelihood scale:

Likelihood of the hazards happening	
	Almost certain
	5
	Will probably occur
	4
	Possible occur
	3
	Remote possibility
	2
	Extremely unlikely
	1

Severity scale:

insignificant damage to property, equipment or minor injury	Non- reportable injury, minor loss of process or slight damage to property	Reportable injury moderate loss of process or limited damage to Property	Major injury, Single Fatality critical loss of process/damage to property	Multiple fatalities catastrophic loss of services and infrastructure
1	2	3	4	5

Fig. 17.4 Likelihood, severity scale

A table for hazard assessment with four columns labeled A to D. Column A is "Hazard," column B is "Likelihood" with a rank from 1 to 5, column C is "Severity" with a rank from 1 to 5, and column D is "Multiply" showing the product of columns C and D.

Risk assessment matrix (Fig. 17.3) would give a micro-level perspective of challenges and help determine the possibility and the magnitude of disaster and its impact at ward level.

For likelihood column: representatives were asked to rank considering the worst-case scenario for all hazards (e.g. intense rainfall leading to floods, landslide etc.)

For Severity column: They were asked to consider the hazards impact on business, people especially the most vulnerable (women, children, physically challenged etc.), to property, livelihoods etc.

By multiplying two scores of likelihoods and impact, Total on the scale of 1–25 is obtained. Higher score implies higher risk w.r.t that particular hazard.

On putting colour code pertaining to score number as depicted in Fig. 17.5, a multi-graded scale will start emerging for each respective ward which when collated as a table for the entire city would indicate disaster risks based on ward representative's perception as depicted in Fig. 17.6.

0-5 = Low risk
6-10 = Moderate risk
11-15 = High risk
16-25 = Extremely high unacceptable risk

Fig. 17.5 Colour code scale

	Meterological				Hydrological		Geological		Biological		Human made disasters															
Hazards -> Wards ↓	Cloud brust	Cyclone	Lightining	Air pollution	Flooding	Landslide	Earthquake	Tsunami	Tree Fallen	Human Animal conflict	Dilapilated Bldg	Building Collapse	Fire	Electricity related issue	Gas leakage	chemical Hazards	Bomb Blast	Water Logging	Drawing	Crowd	High Tention Line	Pedestrain Bridge collapse	High Power Transmission zone	Heavy pedistrin Congnation Road	slum area	Sewer Manholes Caves
4		12				16		6	10			9	8					6								
9		4			20	2	1					8	20					12	6							
11		2			15		1		12			20	10	12							25					
13		4				6			4		8	4	16								1					
17		4	2		3	8	5		6			9	15													
19		1			2	4			4		4	1	4								1					
20		6				2			6			9	10			1		2				2	12	2		4
21	1	9			12	12	1	2				9	4													
23		1				20			3			6	6					3	4							
24		9		3	3	9	9		6	4		12	6			4	10									

Fig. 17.6 Overall ward level perception analysis of hazard (worst 10 wards)

Table 17.2 Identification of issues and suggestive actions

A	B	C	D
Issues	**Scale**	**Vulnerable Spots**	**Suggestive Actions**
	1 – 10		

Scale: 1 = low severity and 10 = high severity

Overall City scaling based on ward representatives perception:

Exercise 2: Hazard specific ward wise issues and suggestive actions

This exercise helped understand grassroots issues pertaining to socio-economic vulnerability (water, health, shelter, livelihoods, property damage etc.) while seeking suggestions from ward officials about the probable solution to the identified problem and vulnerable spots.

A guiding note (Table 17.3) based on the local situation was created and provided to the participants which guided them in scaling the impacts or concerns faced by their wards due to impending hazards in terms of service continuity, environmental, urban heat, flood related concerns, methods used for information gathering and sharing about good practices adopted.

Such bottom-up perspective about issues each ward is facing would need to be further ground-proofed with the analysis by EOC team from events captured in GIS and CCS system. Subsequent relevant identified gaps would help in taking appropriate interventions.

Exercise 3: Exercises with the ESF leads and Supporting agencies

Similar to the exercises with ward officials, DMD conducted workshops sensitising all ESF stakeholders about data gaps identified in the existing DMP, conducted two TTE seeking data gaps and inputs on improving existing SOPs to tackle hazards and revising it.

Conducting table-top exercises (earthquake and cyclone scenario): Stakeholders were invited for TTE where the scenario and exercise sheets were sent to the stakeholders over email before the actual TTE for them to think through, brainstorm amongst their own team members and come prepared. Duration of TTE was about 5 h. To save on time, scenarios with all injects were described at one go to the group prior to starting the exercise. Stakeholders were asked to first individually respond about their understanding of their roles and responsibilities as an individual agency based on the scenario described and later as a group exercise, where 14 groups were formed based on 14 Lead ESFs. Each corresponding supporting agency was asked to sit with their respective lead agency to brainstorm on the same scenario. They were asked to think through:

- If their roles and responsibilities w.r.t each hazard is clear for Pre-disaster (Prevention|Mitigation|Preparedness), During Disaster and Post-Disaster (Recovery|Rehabilitation|Reconstruction) phases

Table 17.3 Key guiding points w.r.t mapping hazard specific vulnerabilities at ward level

Service continuity	1. Do you have adequate water supply for your ward population?
	3. What are the major incidences faced in your wards?
	5. How do you cater to incidences in your surrounding?
	7. Due to various incidences what is extreme stress on livelihoods
	9. What is the response time by various ESFs in times of disaster
	11. Have you identified rehab shelters in the ward?
	13. Is the identified rehab centre in low risk area of the ward?
	15. What is its capacity? Is there enough space to accommodate people during times of disaster?
	17. Are there enough resources available to meet the people's basic needs?
Environmental concerns	18. What are the environmental concerns in your ward?
	19. Do you face human-animal conflict in your ward?
Urban heat	20. Are there urban heat island hotspots?
	21. Do you have suggestion to address the increased urban heat island effect?
Floods	22. What are resources to handle flooding in ward?
	23. How do you update the EOC?
	24. What is the planning you do during the peacetime? (normal/non-disaster times)
	25. Are you aware of the increased frequency of intense rainfalls in future due to climatic changes?
	26. Are you ready to handle increased frequency of rainfall events like July 26th, 2005 (with more than 900 mm of rainfall in a day)
	27. What are the recorded water borne diseases post each flooding event
	28. Have you identified landslide prone zones in your ward if they exist?
	29. Have you identified the vulnerable community living near the landslides?
Information gathering and sharing	30. How do you pass on the information to any citizen about disaster (Pre|during|Post disaster)
	31. What are the bottle necks/constraints in passing the information to the citizen (last mile people)
	32. What are your suggestions for the same?
W.r.t good practices	33. Do you share your good practices with other wards? If so, how?
	34. Do you share your good practices to EOC? If so, how?

- If there is any gap (in terms of additional information needed for action, coordination, know-how, resources, finances etc.),
- Any addition/deletion to their roles and responsibilities as mentioned in the existing SOP.

Well-conducted TTEs provides an opportunity for participating stakeholders to rehearse their roles and responsibilities to cater to an incident. It not only helps build capacity of an organization but also helps them identify their strengths, scope for improvement while helping them evaluate their business continuity plans (NYC Emergency Management, n.d). These elaborate TTEs considering the case of Mumbai is out of scope for this paper.

Outcome of such exercises are better identification of roles and responsibilities, process strengthening like streamlining of stakeholders, (removing unnecessary, adding new necessary ones, reshuffling of roles as leads, creating revised contact list of key officials for communication), inclusion of data analysis, tabulating relevant text in DMP for easier inference etc. It also led to improved SOPs and ideating for better response to all the disaster cycle phases apart from finding gaps in the existing procedure. Moreover, it also helped each agency to look into their SOPs, evaluate them, rephrase it after participating and learning from such exercises with the given context.

Project Findings

Critical analysis of the various reports, past DMP and several consultative workshops and exercises with stakeholders led to identification of some areas for the scope of improvement w.r.t to climate and disaster resilience. Lacunae was also observed in stakeholders understanding to plan and execute medium- and long-term time frame interventions to address medium and long-term impacts of climate change as indicated in MCAP and achieve adherence to Sendai framework.

Identification of Gaps

DMD has been working to improve the city's resiliency however, it is a work in progress. Acknowledging gaps can help create effective action plans and take subsequent action and interventions to bridge the gap. Revised DMP grouped key concern areas around those requiring planning provisions, operational responses, improving communication strategy for improved gathering and sharing of information and policy changes as mentioned below:

A. Planning provisions: are those where relevant sectoral departments would need to plan considering about current and future risks exacerbated due to climate change and impacts due to developmental interventions.

1. Water security: Water supplied to Mumbai comes from lakes outside Mumbai. In case of water scarcity or drought in the catchment areas of these lakes, alternative arrangement to cater to growing consumption needs to be planned.
2. Alternate electricity arrangement: In case of blackout situation, alternate electricity arrangement needs to be planned.
3. E-waste management: Increasing promotion and use of Electric Vehicles and other electronic devices would need e-waste management.
4. Landfill concerns: Treatment of landfill waste and working on vulnerabilities of people living in its vicinity needs to be planned.
5. Green cover: Decreasing green cover to accommodate increasing infrastructure needs reconsideration.
6. DRM of livestock and wild animals during disaster: Alternative animal shelter for domestic/wild animals on higher ground needs considerations.
7. Seismic micro-zonation.

B. Operational responses: These interventions would require relevant sectoral departments to work on addressing operational issues which city is facing due to high population, high influx of floating population and high population density.

1. About floating population in the Central Business District (CBD): large population from neighbouring districts travel to work during daytime leading to high population density in CBD to cater to in case of a disaster.
2. Temporary shelters, rehabilitation shelters and their carrying capacity: Existing identified temporary shelters need to be marked with its carrying capacity and available resources w.r.t corresponding probable ward population needing such shelters in case of a hazard.
3. About impacts of SLR and sanitation challenge: SLR will impact most of the city low-lying wards, understanding need for relocation of community living in low-lying areas as well as sanitation challenges in case of SLR, needs to be understood.
4. Human-animal conflict: interventions to cater to human-animal conflict for people living in wards around Sanjay Gandhi National Park as well as for attacks from stray animals.
5. Need for dedicated corridor for emergency services: in order to cater to high population densities a dedicated corridor to carry essential emergency services needs to be operationalised.
6. Protecting critical infrastructure: making strategies to protect water, transportation, electricity, schools, hospitals etc. from disasters.

C. Information Gathering and sharing: To increase resilience to long-term risks and sudden emergencies it would be prudent to focus on increased connection between relevant sectors (UNDRR, 2019). Lack of communication amongst relevant sectoral departments and with the community increases community-level vulnerability. Though, updates about impending hazards, ongoing disasters are put regularly on BMCs websites, all medias/social media handles, these exercises revealed need for creating improved Communication strategy.

D. Policy level decisions: Through discussions, gaps in existing policy related to compensation emerged. Need to provide compensation, insurance to non-governmental agencies staff engaged for volunteering during mass gathering events or whenever their support is required was identified. Accommodating the need would lead to policy changes, which would be taken up by BMC in due course. Such exercises often lead to policy level interventions as well as.

These gaps when addressed will help frame and realize Sendai targets for the city. Next section suggests processes adopted to realize Sendai targets.

Realizing Sendai Target Indicators and Leveraging Existing Policies

Sendai Framework embodies a risk-informed approach where achieving Sendai targets would need reducing the existing risk, working to prevent creation of new risk by building the resiliency in all sectors, working on recovery while building back better (UNDRR, 2019). In the given context, stakeholders were asked to note interventions their department will undertake based on City's current and future vulnerability, also try to cater to identified gaps. These interventions were categorised as Prevention & Mitigation|Preparedness|Response for short term (2023–2029), Medium term (2030–2039), Long-term (2040–2049) time frames respectively.

For the project timeframe, it was possible to obtain short-term measures only from the stakeholders. An example of short-term measure is presented in Table 17.4. Each measure mentioned was correspondingly matched to see which indicator from Sendai framework (UNDRR, 2015), or which point from PMs 10 point agenda (NDMA, 2016) will be achieved through that measure and whether that measure is structural (those needing physical construction) or non-structural and who would be the responsible authority to take action.

For obtaining Medium and long-term measures from all stakeholders considering climate impacts based on future projections and identified gaps would require further handholding, thus this exercise is a work in progress, achieved by using several tools as mentioned in way forward section. Moreover, efforts were also made to tabulate: Financial resources available at the National, State, Local level and what are the clauses if any to avail it; Relevant applicable laws/sections—would help in evoking appropriate law in times of crises in order to manage disasters better. Such efforts will provide evidence-based guidance on DRM with renewed commitments requiring behaviour change and practice w.r.t data analysis, policy creation and implementation; planning protocols and collaboration mechanisms. Such methods would help in effective decision-making and improve technical and functional implementation capacities needed in order to achieve Sendai targets (UNDRR, 2019).

Table 17.4 Short-term measure indicated by DMD and compliance to Sendai framework and PMs 10-point agenda

Hazard	Short term measures (2023–2029)	Category: structural/ non-structural	Stakeholder	Target (Sendai framework)	Compliance to PMs 10-point agenda
Multi-hazard	In order to investigate city level multi-hazard vulnerability through GIS based tools Multi-Hazard Risk Vulnerability & Capacity Assessment (MHRVCA) project is undertaken (Nov'23–Nov'25)	Non-structural	DMD	A-1, B-1, E-2, G-3	Pt. 4, Pt. 5

Conclusion

This revision to DMP played a crucial role in simplifying the processes of coordination between the sectoral departments, NGOs and the private sector and built capacities of stakeholders to design and implement these plans while incorporating elements to address city's climate change concerns and Sendai targets, which did not exist in previous DMPs. Such exercises when conducted periodically with stakeholders would not only help in process strengthening but also in building City's resilience by integrating DRR and CCA and build back better. As an outcome of this project, identified gaps were duly acknowledged by DMD and effective strategies to address it were noted as indicated in way forward.

Way Forward

DMD will coordinate with the concerned department in order to address the gaps and corresponding identified indicative aspirational initiatives and make city truly climate and disaster resilient. These are compiled and categorised in Table 17.5 as: No-regret|Low-regret solutions. No-regret solutions are those, which even in absence of climate change adaptation provide net benefits for society (GCCA, 2022). These were further sub-categorised into those, which need enhanced efforts for citizen awareness, communication, those needing identification of new land or other resources, and those around capacity building. While Low-regret solutions are those expected to have an acceptable cost for society providing climate change adaptation (GCCA, 2022). These were further sub-categorised into improved coordination and implementation of existing/creating new byelaw, policy and creating new infrastructure. If action initiative corresponding to each gap mentioned in rows is same as column heading, a 'yes' is marked otherwise specific corresponding initiative is mentioned under each relevant row.

To achieve Sendai framework targets by 2030, and climate resilience matching MCAP timeline of 2050, DMD would use a combination of futuristic tools like Visioning, Back casting and Road mapping to frame medium and long-term measures. While visioning tool through collaborative process with its stakeholders would help co-create future vision based on the trends for desired reduction in city's vulnerability, Back-casting would be used to envision a future where city's goals and strategic objectives are achieved in a desired time frame and then work backwards to incorporate steps to bring back to the current starting point. Resulting output would be a set of steps and actions required by all the relevant stakeholders to actualise the vision of desired future (UNDP, 2022). Tools like Road mapping would help achieve a preferred future by mapping practical actions, strategies, regulations, policies, programmes, interventions and resources highlighting need for better understanding of the interconnections and interdependencies existing amongst policies, programmes, and its various components. (UNDP, 2022). BMC does use road mapping and other tools for short-term measures however, it will be strengthened while also working on improving capacity of stakeholders and handholding them to help frame Sendai targets as a way forward. Using such tools will help DMD from foresight into anticipatory governance. Such efforts of revising DMP seems futuristic moreover it will also create a base for its upcoming project of Multi-Hazard Vulnerability Risk and Capacity Assessment (MHVRCA) for Mumbai. Such an exercise is first of its kind for any Indian city and it will set as an example for other nations as well.

Table 17.5 Mapping identified gaps with suggestive initiatives

Suggested initiatives → Identified gap↓	No-regret solutions			Low regret solutions		
	Creating enhanced efforts for citizen awareness, communication	Identification of new land/other resources/agencies	Capacity building/ enhancing knowledge about climate change impacts and future climate projections to various concerned sectoral departments	Improving coordination and implementation of existing bye law and policy	Creating new policy/bye law	Creation of new infrastructure
Water security	On water conservation		Yes	About ground water recharge, and better implementation of rain water harvesting bye-law	New policy to curb land use changes and stopping deforestation in the catchment areas of water sources supplying water to Mumbai	
Alternate electricity arrangement	About renewable energy, installing solar at individual residences and offices and government premises as well		Yes	Yes		Yes

(continued)

Table 17.5 (continued)

Suggested initiatives → Identified gap↓	No-regret solutions			Low regret solutions		
	Creating enhanced efforts for citizen awareness, communication	Identification of new land/other resources/agencies	Capacity building/ enhancing knowledge about climate change impacts and future climate projections to various concerned sectoral departments	Improving coordination and implementation of existing bye law and policy	Creating new policy/bye law	Creation of new infrastructure
Catering to e-waste treatment and management	Yes				Collecting and treating of e-waste	
Catering to solid waste treatment and management	Segregation and treatment of wet waste at source			Implementing solid waste segregation collection at treatment centres		

(continued)

Table 17.5 (continued)

Suggested initiatives → Identified gap↓	No-regret solutions			Low regret solutions		
	Creating enhanced efforts for citizen awareness, communication	Identification of new land/other resources/agencies	Capacity building/ enhancing knowledge about climate change impacts and future climate projections to various concerned sectoral departments	Improving coordination and implementation of existing bye law and policy	Creating new policy/bye law	Creation of new infrastructure
Improving Green cover	About local species, tree care etc.), training about scientific pruning of trees such that trees remains stable, withstand strong winds	Identifying spots in Mumbai which create urban heat islands, pollution hotspots and potential nearby places for tree plantation to address urban heat island and air pollution	Yes		Creating new policy/bye law to protect existing trees and plantation of new trees at the identified spots	

(continued)

Table 17.5 (continued)

Suggested initiatives → Identified gap↓	No-regret solutions			Low regret solutions		
	Creating enhanced efforts for citizen awareness, communication	Identification of new land/other resources/agencies	Capacity building/ enhancing knowledge about climate change impacts and future climate projections to various concerned sectoral departments	Improving coordination and implementation of existing bye law and policy	Creating new policy/bye law	Creation of new infrastructure
Disaster management for livestock/wildlife	Challenges animals face due to impending disaster	Identifying places for animal rescue in case of any disaster	Yes			Creating animal rescue and temporary shelters on high ground
Seismic micro-zonation	Yes	Create seismic micro-zonation maps				

(continued)

Table 17.5 (continued)

Suggested initiatives → Identified gap↓	No-regret solutions			Low regret solutions		
	Creating enhanced efforts for citizen awareness, communication	Identification of new land/other resources/agencies	Capacity building/ enhancing knowledge about climate change impacts and future climate projections to various concerned sectoral departments	Improving coordination and implementation of existing bye law and policy	Creating new policy/bye law	Creation of new infrastructure
Carrying capacity of temporary shelters	Making citizens aware about the available temporary shelter available in a ward and its usage as the impending disaster needs through various media platforms	Identifying of new shelters on high ground, better structural stability and basic amenities				Identifying the carrying capacity of the existing shelters and creation of as new needed depending on the vulnerable ward population, also making rules of admission to the shelters, its usage etc.

(continued)

Table 17.5 (continued)

Suggested initiatives → Identified gap↓	No-regret solutions			Low regret solutions		
	Creating enhanced efforts for citizen awareness, communication	Identification of new land/other resources/agencies	Capacity building/ enhancing knowledge about climate change impacts and future climate projections to various concerned sectoral departments	Improving coordination and implementation of existing bye law and policy	Creating new policy/bye law	Creation of new infrastructure
Catering to Sea level rise (SLR)	Yes	Identifying places to shift the vulnerable population living in wards which would likely be inundated due to SLR	Yes	With relevant sectoral departments, identifying vulnerable wards and specific spots due to SLR	W.r.t SLR and the sanitation challenge it will bring. Also making policies about relocation of vulnerable population, compensation etc.	Creation of green infrastructure by way of planting trees, mangroves in the identified vulnerable coastal wards

(continued)

Table 17.5 (continued)

Suggested initiatives → Identified gap↓	No-regret solutions			Low regret solutions		
	Creating enhanced efforts for citizen awareness, communication	Identification of new land/other resources/agencies	Capacity building/ enhancing knowledge about climate change impacts and future climate projections to various concerned sectoral departments	Improving coordination and implementation of existing bye law and policy	Creating new policy/bye law	Creation of new infrastructure
Catering to human-animal conflict	About reason for human animal conflict (people living near SGNP)	Yes	Yes	Working to improve buffer zone surrounding areas prone to human-animal conflict especially near SGNP boundary		

(continued)

Table 17.5 (continued)

Suggested initiatives → Identified gap↓	No-regret solutions			Low regret solutions		
	Creating enhanced efforts for citizen awareness, communication	Identification of new land/other resources/agencies	Capacity building/ enhancing knowledge about climate change impacts and future climate projections to various concerned sectoral departments	Improving coordination and implementation of existing bye law and policy	Creating new policy/bye law	Creation of new infrastructure
Creation of dedicated corridor for emergency services	Yes	Identification of spaces near existing roads leading to major hospitals/critical infrastructure to create a green corridor to be used for emergency services		On identifying—identifying all relevant stakeholders for the implementation of the green corridor	Creation and implementation of rules/penalties for citizens breaking into green corridor	On identifying—creation of green corridor

(continued)

Table 17.5 (continued)

Suggested initiatives → Identified gap↓	No-regret solutions			Low regret solutions		
	Creating enhanced efforts for citizen awareness, communication	Identification of new land/other resources/agencies	Capacity building/ enhancing knowledge about climate change impacts and future climate projections to various concerned sectoral departments	Improving coordination and implementation of existing bye law and policy	Creating new policy/bye law	Creation of new infrastructure
Creating communication strategy	Engaging with various media agencies, influences on social media to pass important messages through various kinds of media		Yes			
Creating compensation/medical assistance policy		Identifying support agency staff which provide services during disaster who would need medical assistance/ compensation			Creation of policy for compensation/ medical assistance to such agencies	

Acknowledgements This work is an extension of the work that author had done as a technical consultant for BMC DMD, supported by UNICEF. Exercises 1 & 2 were created by brainstorming with Praveen kumar Pawar from RedR India. Author would like to sincerely thank them for this opportunity and support respectively.

References

Basu, J. (2021, Sept 3). *IPCC report warning: Mumbai will be worst hit among Indian metros.* Retrieved from Down to Earth : https://www.downtoearth.org.in/news/climate-change/ipcc-report-warning-mumbai-will-be-worst-hit-among-indian-metros-78823

BMC. (2016). *Report on draft development plan 2034.* BMC.

BMC. (2022). *District disaster management authority Mumbai suburban district.* Retrieved from MCGM: https://dm.mcgm.gov.in/disaster-management-plan-suburbs

EPA. (2022, March 23). *Reduce urban heat island effect.* Retrieved from USEPA: https://www.epa.gov/green-infrastructure/reduce-urban-heat-island-effect

Gawler, S., & Tiwari, S. (2014). *ICLEI ACCCRN process building urban climate change resilience: A toolkit for local governments.* Retrieved from ICLEI: https://southasia.iclei.org/wp-content/uploads/2021/08/ICLEI-ACCCRN-Process-IAP-A-toolkit-for-Local-Governments.pdf

GCCA. (2022). *MODULE 4 understanding and planning under uncertainty.* Retrieved from UNFCCC: https://www4.unfccc.int/sites/NAPC/Documents%20NAP/General/Module4_Handout_EN_2011-10-22.pdf

GDPC. (n.d.). *Global disaster prepradness center.* Retrieved from https://preparecenter.org/topic/risk-assessment/

IPCC. (2019, June). *Global warming of 1.5 C.* Retrieved from IPCC: https://www.ipcc.ch/site/assets/uploads/sites/2/2019/06/SR15_Headline-statements.pdf

Jain, G., & Bazaz, A. (2017). *Urban risks and resilience in India.* Retrieved from IIHS: https://iihs.co.in/knowledge-gateway/wp-content/uploads/2017/10/Urban-Risk-and-Resilience-in-India.pdf

MCAP. (2022a, March). *Climate & air pollution risks and vulnerability assessment.* Retrieved from https://drive.google.com/file/d/1fItAQZ3qPsNEFe4bsq24orbWqRHPp6d3/view

MCAP. (2022b). *Mumbai climate action plan.* Retrieved from Mumbai Climate Action Plan: https://drive.google.com/file/d/1gU3Bnhk3UJ_wCFaMC1ognZBdsdDkQBY1/view

MCGM. (2023). *Environment status report 2022–2023.* Retrieved from MCGM: https://portal.mcgm.gov.in/irj/go/km/docs/documents/MCGM%20Department%20List/Environment/Docs/English%20ESR%20for%20the%20year%202022-23.pdf

NDMA. (2016). *Prime Minister's ten point agenda on DRR.* Retrieved from NDMA: https://ndma.gov.in/Governance/PM-10-Agenda

NDMA. (2019). *National disaster management plan.* Retrieved from NDMA: https://ndma.gov.in/sites/default/files/PDF/ndmp-2019.pdf

NIDM. (2016, Dec). *Disaster risk reduction—A handbook for urban managers* . Retrieved from NIDM: https://ndmindia.mha.gov.in/images/pdf/DRRHandbookUrbanManagers.pdf

NITI Aayog. (2022, May). *Cities as engines of growth* . Retrieved from NITI Aayog: https://www.niti.gov.in/sites/default/files/2022-05/Mod_CEOG_Executive_Summary_18052022.pdf

NYC Emergency Management. (n.d). *NYC emergency management table top exercise tool kit facilitator guide.* Retrieved from NYC: https://www.nyc.gov/assets/em/downloads/pdf/business/spe_facilitator_guide.pdf

Shaw, R., Luo, Y., Cheong, T. S., Abdul Halim, S., Chaturved, S., et al. (2022). Asia. In *Climate Change 2022: Impacts, Adaptation and Vulnerability. Contribution of Working Group II to the Sixth Assessment Report of the Intergovernmental Panel on Climate*

Change. Retrieved from IPCC: https://www.ipcc.ch/report/ar6/wg2/downloads/report/IPCC_AR6_WGII_Chapter10.pdf

TERI. (2015, March). *Environment status report of Mumbai Metropolitan Region March 2015.* Retrieved from TERI: https://www.mmreis.org.in/images/research/Environmental%20Status%20Report%20for%20Mumbai%20Metropolitan%20Region-ilovepdf-compressed.pdf

UN. (2024). *Urbanization.* Retrieved from United Nations Population division : https://www.un.org/development/desa/pd/content/urbanization-0#:~:text=The%20world%20is%20becoming%20increasingly,around%20two%2Dthirds%20in%202050.

UNDP. (2022, July). *UNDP RBAP foresight playbook.* Retrieved from UNDP: https://www.undp.org/asia-pacific/publications/undp-rbap-foresight-playbook

UNDRR. (2015). *Sendai framework indicators.* Retrieved from Prevention Web: https://www.preventionweb.net/sendai-framework/sendai-framework-indicators

UNDRR. (2017, May). *Disaster resilience scorecard for cities.* Retrieved from https://www.unisdr.org/campaign/resilientcities/assets/toolkit/Scorecard/UNDRR_Disaster%20resilience%20%20scorecard%20for%20cities_Preliminary_English.pdf

UNDRR. (2019). *Global assessment report on disaster risk reduction.* Retrieved from UNDRR: https://gar.undrr.org/report-2019.html

UNHABITAT. (2022). *World cities report 2022 envisaging the future of cities.* Retrieved from UNHABITAT: https://unhabitat.org/sites/default/files/2022/06/wcr_2022.pdf

UNISDR. (2017). *Words into action guidelines national disaster risk assessment.* Retrieved from UNISDR: https://www.unisdr.org/files/52828_nationaldisasterriskassessmentpart1.pdf

Ms. Chinmai Hemani is an MTech in Climate change and Sustainable Development and is a climate change consultant from India. She has rich experience of over twelve years in desk and field research on Climate change Vulnerability assessment, adaptation planning for semi-arid and coastal regions, Disaster Risk Reduction, Policy analysis w.r.t Climate Change and Sustainable Development and is engaged with many multilateral NGOs. She has motivated many citizens to adopt pro-environment lifestyle changes by conducting awareness and action research projects about climate change and environment conservation through her Ecology club of Rajkot and Eco-lab initiative.

Chapter 18
Modelling the Role of Urbanization in Environmental Degradation in North Africa: Implications for Climate Change Adaptation

Mohsin Raza and Assem Abu Hatab

Abstract In recent years, North African countries have undergone rapid urbanization processes, which concentrated close to 70% of their population in metropolitan centers. While urbanization has been recognized as a catalyst for economic growth and development in these countries, unplanned urban growth presents profound environmental challenges that threaten to jeopardize the existing development challenges and exacerbate the impacts of climate change. At the same time, studies on the environmental consequences of urban growth in the region are noticeably limited. This chapter utilizes a panel dataset covering the period 1990–2019 and employs a cross-sectional augmented autoregressive distributed lag (CS-ARDL) modeling to examine the relationship among urbanization and carbon emissions (CO_2) in four North African countries, namely Egypt, Tunisia, Sudan, and Morocco. The empirical results reveal that rapid urbanization processes have been associated with significant increases in CO_2 emissions in these countries. Furthermore, the results indicated that foreign direct investment flows, economic growth, and international trade, partly brought about by urbanization, have also contributed to rising CO_2 emissions. Conversely, the findings point out that expanding renewable energy use holds significant potential to reduce environmental degradation in the region. Overall, the findings emphasize the need for effective measures and policy interventions that prioritize sustainable urban planning and promote low-carbon development strategies to mitigate the negative impacts of urban growth on environmental degradation.

M. Raza · A. Abu Hatab
Department of Economics, Swedish University of Agricultural Sciences, Uppsala, Sweden

M. Raza
Institute for Food and Resource Economics (ILR), University of Bonn, Bonn, Germany

A. Abu Hatab
Nordic Africa Institute, Uppsala, Sweden

M. Raza (✉)
Business Economics Group, Wageningen University & Research, Wageningen, Netherlands
e-mail: mohsin.raza@wur.nl

W. Leal Filho et al. (eds.), *Climate Change Mitigation and Adaptation in Practice*, Climate Change Management, https://doi.org/10.1007/978-3-031-85217-6_18

Keywords Urbanization · Environmental quality · North Africa · CS-ARDL

Introduction

With over half of the world's population already residing in urban areas, a proportion projected to surpass 70% by 2050, urbanization emerges as a paramount global and regional concern, embodying one of the defining challenges of the twenty-first century (IPCC, 2022; United Nations, 2023). As urbanization continues its rapid expansion, particularly in regions like Africa and Asia, the demand for energy and resources from industries and other economic sectors grows significantly and places immense strain on existing infrastructure, leading to congestion, inadequate transportation systems, and overwhelmed utilities such as water and sanitation (Liu & Bae, 2018; Abu Hatab et al., 2019; Zhou et al., 2021). Moreover, the uncontrolled growth of urban areas contributes to environmental degradation through land use changes and encroachment on agricultural land, deforestation, and increased pollution from industrial activities and transportation (Badreldin et al., 2019). Social inequality is exacerbated as rapid urbanization often leads to the formation of informal settlements and slums, where marginalized populations struggle with inadequate access to basic services and face heightened vulnerabilities. In addition, the concentration of population and economic activities in urban centers intensifies the emission of greenhouse gases (GHG), further contributing to climate change (Hanif, 2018; Liu et al., 2023; Sun et al., 2022). Consequently, Abu Hatab et al. (2022a) point out that the United Nations Sustainable Development Goals (SDGs) and their targets explicitly recognize the interconnectedness between urbanization and the social, economic, and environmental dimensions of sustainable development, and thus achieving the SDGs in developing countries will greatly depend on how urbanization processes are planned and managed.

Similar to many developing regions, North Africa has experienced rapid urbanization since the second half of the twentieth century, which has resulted in profound changes in demographics and spatial distribution and led to the concentration of close to 70% of the population in metropolitan centers (World Bank, 2023). As shown in Fig. 18.1, urban expansion in North African countries has seen an annual average increase of 2.3% over the last two decades, nearly twice the rate observed in other African regions. Such rapid urbanization can be attributed to several interlined development trends, including accelerating economic growth, a surge in population, a significant influx of individuals migrating to urban areas, and a pronounced rise in the youth demographic (OECD, 2021). In this context, the Neoclassical economic theory posits that urbanization reshapes both the population distribution and resource allocation within a country. This makes migration to urban areas a rational choice to maximize the economic well-being and overall welfare of migrants and their households (Yang et al., 2020). For instance, the GDP per capita in North African cities tends to surpass the respective national averages, GDP per capita in Tunis is twice that of Tunisia, in Casablanca exceeds that of Morocco by more than twice, and the same holds true for Cairo concerning Egypt, and Omdurman, Sudan (Mezran & Sanguini, 2021). These cities, in contrast to the rural areas within the region, provide many livelihood prospects, including employment and income-generating options, food security, access to financial services, education, social connections, and social safety nets. Therefore, Abu Hatab et al. (2022a, 2022b) note that urbanization processes in North Africa led to regional development inequalities and wage variations. This motivated people from rural areas to migrate to cities, resulting in the expansion of urban areas. Essentially, these imbalances induced by urbanization lead to surplus labor being drawn to more developed regions, primarily urban ones, due to the availability of higher wages and better living conditions (Wu et al., 2020).

Nevertheless, North Africa's rapid urbanization has brought about a range of negative environmental impacts. As displayed in Fig. 18.2, CO_2 emissions have shown a significant upward trend over the last few decades in North African countries, with Egypt exhibiting a particularly pronounced increase. This escalation in emissions can be ascribed to interlinked factors related to urbanization, industrialization, increased energy consumption, and changes in land use. For instance, rapid urbanization has led to increased air and water pollution levels in North African cities (Nathaniel et al., 2020). As urban areas expand, so do industrialization and vehicular traffic, both significant contributors to air pollution. Similarly, inadequate wastewater treatment infrastructure and improper waste disposal practices contribute to water pollution, contaminating rivers and coastal areas, and posing risks to aquatic ecosystems and human health (Singh et al., 2022). Furthermore, urbanization processes in the region have resulted in the loss of natural habitats and green spaces (Bai et al.,

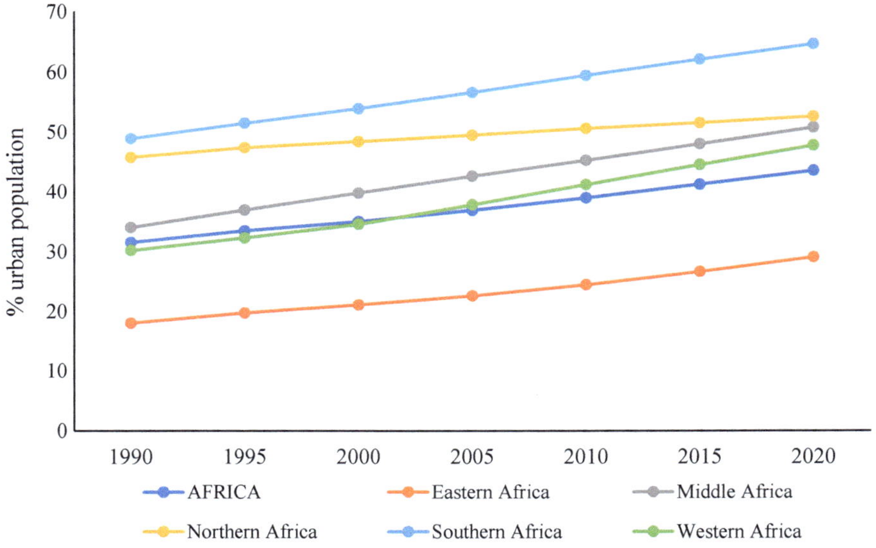

Fig. 18.1 Urban population growth (%) across African regions. *Source* Plotted by Author(s) based on UN World Urbanization Prospects 2023 data

2019). As North African cities expand, they encroach upon previously undeveloped areas, leading to habitat destruction, and loss of biodiversity (Ayeni et al., 2023). This disrupts ecosystems and threatens the survival of plant and animal species native to the region. Moreover, the rapid urbanization of North Africa has led to the depletion of natural resources by increasing demand for land, water, and energy in urban areas and placing additional pressure on finite resources, leading to their overexploitation and degradation (Badreldin et al., 2019). For instance, agricultural land is often converted into urban infrastructure to accommodate population growth, resulting in soil degradation, and reduced agricultural productivity. Similarly, the growing urban population requires more water for domestic, industrial, and agricultural purposes, leading to the unsustainable extraction of water from rivers, aquifers, and other sources. On the top of these negative effects, urbanization processes in North Africa exacerbates the effects of climate change on the region and amplifies its contribution to global warming and climate change (Abdullah et al., 2021; Haouraji et al., 2021). In particular, urban expansion leads to a heat island effect in North African cities, where they experience higher temperatures compared to surrounding rural areas due to the absorption and retention of heat by buildings and infrastructure. This phenomenon exacerbates heat-related health problems and energy consumption for cooling, further contributing to greenhouse gas (GHG) emissions and leading to increased carbon emissions. Hence, urbanization trends in the region have been characterized as socioeconomically and environmentally unsustainable that is posing a significant risk to urban sustainability, despite the evidence demonstrating that it has played an important role in improving development and economic growth in North African countries (Chen et al., 2022; Musah et al., 2021).

From a literature standpoint, a substantial body of theoretical and empirical literature has sought to clarify the environmental implications of urbanization. For instance, Poumanyvong and Kaneko (2010) describe the connection between urbanization and its environmental consequences in context to ecological modernization theory. They highlight that as countries progress, they often prioritize economic growth over environmental sustainability, resulting in increased environmental degradation. However, as further modernization unfolds, these issues tend to diminish as countries increasingly emphasize environmental sustainability. Conversely, the work of Liu et al. (2021) and Marcotullio and Lee (2003) align well with the urban environmental transition theory's perspective. This theory suggests that as countries transition toward a manufacturing-based economy, the environmental impact escalates with the expansion of cities. Nevertheless, as cities become more prosperous, these effects generally diminish due to enhanced environmental legislation, technological advancements, and changes in the economic landscape. Moreover, it is essential to acknowledge that affluent cities may also increase their need for urban infrastructure and energy-intensive goods, which can further burden the environment. Conversely, Borck and Pflüger (2019) and De Roo (2000) embrace the compact city theory, contending that urbanization's advantages lie in the concept that increasing urban density leads to economies of scale in public infrastructure, resulting in reduced environmental strain. However, insufficient urban infrastructure provision in areas with increased urban density can result in environmental damage.

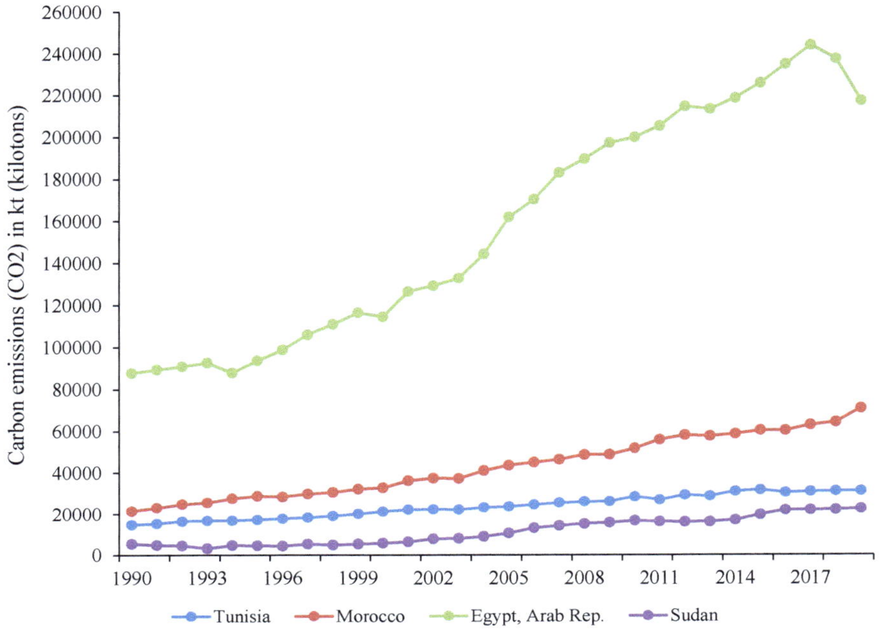

Fig. 18.2 Carbon emissions (CO_2) in kilotons across sample countries. *Source* Plotted by Author(s) based on World Development Indicators (WDI) 2023 data

Empirically, several recent studies have focused on the associations between urbanization and its environmental consequences. Multiple studies have demonstrated a direct relationship between urbanization and increased CO_2 emissions, indicating environmental quality and degradation (Poumanyvong & Kaneko, 2010; Liddle & Lung, 2010; Al-Mulali & Ozturk, 2015). By and large, these studies examined the impacts of urbanization on a range of environmental degradation indicators, including GHG emissions (Liu et al., 2023; Wu et al., 2021), and examined the interlinkages among urbanization, renewable energy, economic growth, carbon emissions, and environmental sustainability (e.g., Raihan et al., 2022; Sun et al., 2022). Nevertheless, the results of these studies offer equivocal evidence about the effect of urbanization on CO_2 emissions. Specifically, some studies indicate a positive correlation between urbanization and CO_2 emissions (e.g., Hanif, 2018; Poumanyvong & Kaneko, 2010); other studies point out a negative association between them (e.g., Sadorsky, 2014; Sharma, 2011).

Both theoretical propositions and empirical evidence suggest that urbanization has diverse environmental effects, encompassing both detrimental and beneficial features, depending on a suite of sociodemographic, economic, environmental, and institutional factors (Azam & Khan, 2016; Long et al., 2017). This implies the necessity for regional and more disaggregated analyses that account for the unique characteristics and contexts of specific socioeconomically and environmental regions.

However, until now, the investigation of the link among urbanization and environmental degradation in North African nations remains an emerging field of study (Fatima et al., 2021; Mehmood & Mansoor, 2021). That is, the policy discussion on urbanization in North Africa has been without solid evidence despite the presence of both obstacles and opportunities associated with urbanization (ECA/AfDB, 2022). The lack of reliable evidence and empirical studies on the association between urbanization and environmental degradation in North Africa creates critical knowledge gaps that hinder the development of effective policies and informed decision-making concerning the sustainable management of urbanization, which hampers the ability of governments to tackle the difficulties and maximize the advantages brought about by urbanization (Effiong, 2016).

As a contribution to the existing body of research and the ongoing discourse on the environmental consequences of urbanization, this chapter employs a panel dataset comprising four North African nations, namely Egypt, Tunisia, Morocco, and Sudan, spanning the years 1990–2019, to investigate the connections between urbanization and CO_2 emissions. The empirical analysis is based on estimating a cross-sectional augmented autoregressive distributed lag model (CS-ARDL). The empirical findings offer valuable insights that could guide policymakers and stakeholders in the region about the implications of urbanization on climate change adaptability and environmental sustainability. Furthermore, the results emphasize the need for proactive measures and policy interventions that prioritize sustainable urban planning and promote low-carbon development strategies in these countries to mitigate the adverse effects of urbanization on the environment.

Materials and Methods

In this section, the selection of variables and data sources utilized to investigate the impact of urbanization on environmental degradation in North Africa are described. The model specification and deliberate on the estimation techniques employed are also presented.

Data and Variables

The selection of the variables was based on previous studies on the nexus between urbanization and environmental degradation (e.g. Raihan et al., 2022). Secondary-sourced panel dataset covering the period 1990–2019 for four North African countries (Egypt, Tunisia, Morocco, and Sudan) was compiled from the World Development Indicators (WDI). CO_2 emissions kt (kilotons) served as a proxy variable for environmental quality (Musah et al., 2021), designating the dependent variable, whereas urbanization (urban population in millions), GDP (millions of US$), net inflows of foreign direct investment (FDI net inflows as % of GDP), trade-to-GDP ratio

Table 18.1 Descriptive statistics

Variables	Symbol	Measurement unit	Mean	Std. Dev	Min	Max
CO_2 emissions	CO_2	kt (kilotons)	58,975	65,105	3210	244,540
Urban population	UP	Millions	16.94	11.26	4.89	45.13
GDP	GDP	Millions (US$)	74,783	72,977	7032	332,442
FDI net inflows	FDINI	(% of GDP)	2.36	1.77	– 0.21	9.42
Renewable energy	REC	(% of total final energy consumption	26.87	26.38	4.93	85.63
Trade	Trade	% of GDP	56.32	24.48	11.09	114.34

Note Number of observations = 120

(Trade as a % of GDP), and renewable energy consumption (% of total final energy consumption), comprised the independent variables (Haouraji et al., 2021; Raihan et al., 2022). However, study used the logarithmic form of the CO2 emissions (ln_ CO_2) and GDP variable (ln_GDP) to linearize the relationships and stabilize the variance. The descriptive statistics of the variables utilized in the investigation are displayed in Table 18.1.

Model Specifications

The following econometric equation was estimated to measure the magnitude and direction of the relationship between environmental quality and urbanization in the presence of the control variables. The functional form contains the selected variables as follows:

$$ln\_CO_{2i,t} = f\left(UP_{i,t}, ln\_GDP_{i,t}, FDINI_{i,t}, REC_{i,t}, Trade_{i,t}\right)$$

where ln_CO_2 represents the log-transformed CO_2 emissions, UP is the urban population, ln_GDP is the log of GDP, FDINI accounts for foreign direct investment net inflows, RES for renewable energy, and Trade is the ratio of Trade to GDP. The econometric specification of the model is as below:

$$ln\_CO_{2i,t} = \alpha_i + \beta_{i,1}UP_{i,t} + \beta_{i,2}ln\_GDP_{i,t} + \beta_{i,3}FDINI_{i,t} + \beta_{i,4}REC_{i,t} + \beta_{i,5}Trade_{i,t} + \varepsilon_{it}$$

Here, i represents the number of cross sections/countries, whereas t highlights the time span and intercept term in the model is represented by α_i with ε_{it} as the residual term.

Estimation Methods

To test for the existence of heterogeneity across samples, authors used the slope homogeneity test by Pesaran and Yamagata (2008). The study detects cross-sectional dependence (CD) based on Pesaran (2015, 2021), and Xie and Pesaran (2022) methods. CD test is applied due to countries' associations by several aspects such as cultural, social, political, and economic, and consequently, CD may be present. Moreover, cross-sectional dependence, or the correlation of residuals across countries, can arise due to shared shocks, common policies, or spillover effects between countries (Hussain, 2022). The presence of CD might result in biased and inconsistent estimators while conducting panel data analysis. In addition, as first-generation unit root tests (Im et al., 2003; Levin et al., 2002) do not consider CD, authors employed second-generation unit root tests, namely Cross-Sectional Augmented Im, Pesaran and Shin (CIPS) and Cross-sectional Augmented Dickey-Fuller (CADF), to assess the stationarity in the presence of CD.

To address the challenges of slope heterogeneity, CD, and mixed order of integration among relevant variables, authors employ the Cross-Sectionally Augmented Autoregressive Distributed Lag (CS-ARDL) model, as proposed by Pesaran (2015, 2021), to examine both the long-run and short-run relationships among ln_CO_2, UP, ln_GDP, $FDINI$, REC, $Trade$. Nevertheless, the CS-ARDL model has faced criticism for enforcing a homogeneity constraint in the long term, disregarding the diverse economic and social systems observed among countries. Therefore, as a robustness check of the estimates obtained from CS-ARDL, two more models, namely the common correlated effects mean group (CCEMG) and augmented mean group (AMG) by Pesaran (2006) are employed. These models account for cross-sectional dependence by allowing parameters to be heterogeneous in the long run.

While the CS-ARDL, CCEMG, and AMG models yield reliable results, they do not assess the causality between the variables under investigation, which is crucial for policy considerations. Hence, to identify the causal link among urbanization and CO_2 emissions, study utilize the improved version of the panel Granger Causality by Juodis et al. (2021) that cater to CD, heterogeneity, and Nickell bias, overlooked by Dumitrescu and Hurlin (2012) Granger Causality test.

Results and Discussion

The results and discussion section is organized first to present the critical empirical findings derived from the estimation methods and data analysis, followed by an in-depth interpretation that contextualizes these results within the existing body of literature and theoretical framework.

Table 18.2 Slope Homogeneity test

$\tilde{\Delta}$	$\sqrt{N}\left(\frac{N^{-1}\overline{S}-k}{\sqrt{2k}}\right) \sim X_k^2$	11.11*
$\tilde{\Delta}_{adj}$	$\sqrt{N}\left(\frac{N^{-1}\overline{S}-k}{v(T,k)}\right) \sim N(0,1)$	12.69*

*Denotes statistical significance at the 1% level. The terms $\tilde{\Delta}$ and $\tilde{\Delta}_{adj}$ denote the slope homogeneity tests for both the "simple" and "mean–variance bias adjusted" cases

Slope Homogeneity

The slope homogeneity test assesses whether the relationship (or slope) between the independent and dependent variables is consistent across panel units, in this case, countries. If this relationship varies significantly between countries, the panel regression results might not be reliable if homogeneity is assumed. Table 18.2 displays the findings of the slope homogeneity tests, while "N" represents the total number of cross sections, "S" indicates the Swamy test statistic, and "k" represents the independent variables. If the p-value exceeds 5%, the null hypothesis at a 5% significance level is accepted, treating the cointegrating coefficients as homogeneous. $\tilde{\Delta}$ and $\tilde{\Delta}_{adj}$ are tailored for analyzing small and large samples, respectively, where $\tilde{\Delta}_{adj}$ serves as the 'mean–variance bias adjusted' version of $\tilde{\Delta}$, offering refined accuracy in smaller sample contexts by adjusting for potential bias in mean and variance estimates.

As shown in Table 18.2, study find a significant p-value for the slope homogeneity test, indicating heterogeneity in the association among CO_2 emissions and urbanization across the countries studied, which is theoretically mentioned in Herbert (2019). This suggests that urbanization's effect on environmental quality varies by country, as measured by CO_2 emissions. Such variations could be influenced by different urbanization processes, policies, and other country-specific factors not captured by control variables (Mehmood et al., 2022). Hence, based on heterogeneity, authors employ heterogeneous panel techniques.

Cross-Sectional Dependence (CD)

In order to identify the cross-sectional dependence (CD) in the data, authors used the Pesaran (2015, 2021), and Xie and Pesaran (2022) methods. The results for the CD for relevant variables are presented in Table 18.3. The study found strong evidence of CD in the panel based on the test results. This implies that there are unobserved common factors influencing CO_2 emissions across countries. Such factors could include global economic trends, international environmental policies, or worldwide technological advancements (Hussain, 2022). Moreover, the process of demographic transition is marked by national and regional variations due to social, political, and economic differences (OECD, 2021). Therefore, those estimation tools are selected that includes CD in the estimation process.

Table 18.3 Cross-sectional dependence test

CD	$ln\_CO_{2i,t}$	$UP_{i,t}$	$ln\_GDP_{i,t}$	$FDINI_{i,t}$	$REC_{i,t}$	$Trade_{i,t}$
	13.09*	13.33*	11.63*	4.23*	9.47*	4.76*

*Represent statistical significance at 1%

Stationarity Characteristics

In the presence of cross-sectional dependence, as evidenced by the findings in Table 18.3, it becomes necessary to check the stationarity of variables. Thus, study used the second-generation unit of root tests (Cross-Sectional Augmented Im, Pesaran and Shin (CIPS) and Cross-sectional Augmented Dickey-Fuller (CADF)) to identify the order of integration. The CADF and CIPS tests results are displayed in Table 18.4, explicitly evaluating unit root tests for individual variables. These tests are essential for ascertaining the existence of unit roots and estimating the level of integration for the variables in the panel dataset.

The CADF test findings show the mixed order of integration, i.e., $\overbrace{\left[ln\_CO_{2i,t}, UP_{i,t}, ln\_GDP_{i,t}, FDINI_{i,t}, REC_{i,t}, Trade_{i,t}\right]}^{\left[I(0),\, I(1),\, I(0),\, I(1),\, I(1),\, I(1)\right]}$. Similarly, the CIPS test results support this finding, also indicating the mixed order of integration among

Table 18.4 CADF and CIPS test estimates

Variables	CADF	Cross-sectional IPS (CIPS)
$ln\_CO_{2i,t}$	− 2.39*	− 2.91***
$\Delta ln\_CO_{2i,t}$	–	–
$UP_{i,t}$	− 1.59	− 0.58
$\Delta UP_{i,t}$	− 3.2***	− 4.41***
$ln\_GDP_{i,t}$	− 2.39*	− 1.68
$\Delta ln\_GDP_{i,t}$	–	− 4.11***
$FDINI_{i,t}$	− 2.16	− 4.03***
$\Delta FDINI_{i,t}$	− 4.36***	–
$REC_{i,t}$	− 1.85	− 2.04
$\Delta REC_{i,t}$	− 3.76***	− 5.28***
$Trade_{i,t}$	− 1.67	− 1.56
$\Delta Trade_{i,t}$	− 3.71***	− 5.96***

Note By definition: $CIPS = \frac{\sum_{i=1}^{N} t_i(N,T)}{N} = \frac{\sum_{i=1}^{N} CADF_i}{N}$

***, *Denotes statistical significance at the 1%, and 10%, respectively the variables, i.e., $\overbrace{\left[ln\_CO_{2i,t},\ UP_{i,t},\ ln\_GDP_{i,t},\ FDINI_{i,t},\ REC_{i,t},\ Trade_{i,t}\right]}^{\left[I(0),\ I(1),\ I(1),\ I(0),\ I(1),\ I(0)\right]}$. The presence of mixed orders of integration can sometimes complicate econometric analysis. However, the subsequent use of the CS-ARDL in this chapter proves advantageous, as the ARDL approach is adept at handling both first and mixed orders of integration (Mehmood et al., 2023). This ensures robustness in the econometric analysis despite the mixed integration nature of the data.

CS-ARDL Estimates for the Relationship Among CO_2 Emissions and Urbanization

The results of the estimated CS-ARDL model, as shown in Table 18.5, indicate that urbanization has a statistically significant positive impact on CO_2 emissions in both the short and long run. This finding is aligned with the prior investigations carried out in other regions of Africa (e.g., Mignamissi & Djeufack, 2022; Musah et al., 2021; Salahuddin et al., 2019), which found a statistically significant correlation between urban expansion and GHG emissions. This also supports the results of Zhou et al. (2021) and Sun et al. (2022), implying that rapid urban growth can add to significant environmental degradation without sufficient environmental protection policies. In this regard, Mezran and Sanguini (2021) and Salahuddin et al. (2019) argue that the rapid urbanization in North Africa has resulted in increased CO_2 emissions. This is primarily due to the growth of industries, which has led to higher energy consumption. Additionally, expanding the private housing sector and providing essential public services like transportation, water and sanitation, and healthcare have further contributed to the rise in emissions. Furthermore, the consumption patterns of urban dwellers in North Africa are more carbon-intensive than those in rural areas.

Furthermore, GDP has a statistically significant positive relationship with CO_2 emissions in the long and short run. This indicates that as a country's economy grows, there is a corresponding rise in emissions. Economic growth is often accompanied by increased industrial activities, consumption, and energy needs. If green technologies and sustainable practices do not complement this growth, it can lead to higher emissions (Akorede, 2020; Anwar et al., 2020). Similarly, the findings suggest that increased foreign investment leads to substantial increases in CO_2 emissions in both the long and short run, which indicates that foreign investments in North African countries have been directed towards sectors with a high carbon footprint (Shahbaz et al., 2019). Additionally, the results show that countries with higher trade activities as a proportion of their GDP tend to have higher CO_2 emissions. This is attributed to the argument that Trade involves transportation, production, and often the consumption of goods. Countries with high trade activities might have bustling ports, increased vehicular activities, and more industrial outputs, leading to higher emissions (Asongu & Odhiambo, 2021). Conversely, using renewable energy

Table 18.5 CS-ARDL estimates

Variables	Long run estimates	Short run estimates
$UP_{i,t}$	0.0723* (0.0406)	0.0643* (0.035)
$ln\_GDP_{i,t}$	0.0589* (0.0305)	0.052** (0.026)
$FDINI_{i,t}$	0.0026* (0.0014)	0.0022** (0.001)
$REC_{i,t}$	− 0.0356** (0.0179)	− 0.0316** (0.0149)
$Trade_{i,t}$	0.0027*** (0.0002)	0.0025*** (0.0002)
$ECM\,(-1)$	− 0.9033*** (0.0686)	
Observations	120	

Dependent variable: $\boldsymbol{ln\_CO_{2i,t}}$
***, **, *Represent statistical significance at 1%, 5% and 10%, respectively. Parentheses contain standard errors

is significantly associated with a decrease in CO_2 emissions, with a statistical significance threshold of 5%. This implies that a rise in the utilization of renewable energy is directly linked to reduced emissions. This finding aligns with the research conducted by Dauda et al. (2021), which demonstrates that when countries increase their renewable energy consumption, their dependence on coal, oil, and gas decreases, resulting in reduced emissions.

The disequilibrium correction {ECM (-1)} result suggests that around 90% of any disequilibrium in the relationship between the variables is corrected every year. This highlights the dynamic adjustment process in the relationship between urbanization and environmental quality. Therefore, CS-ARDL results provide a strong case for the impact of urbanization, economic activities, and energy choices on North African countries' environmental quality, which underscores the importance of sustainable urban development, green economic policies, and the adoption of renewable energy to mitigate environmental degradation.

Robustness Check Using CCEMG and AMG

Table 18.6 displays the findings of the CCEMG and AMG estimators. Upon examining the results of CCEMG and AMG, authors realize that findings from these estimators generally align with those obtained from CS-ARDL. The consistent sign and significance of the urbanization coefficient across all models underscore the substantive impact of urban growth on CO_2 emissions. The analysis supports the long-established relationship between economic growth and environmental stress, as depicted by the positive coefficient of GDP, further echoed in the robustness checks. This trend is indicative of the environmental Kuznets curve hypothesis, where economic development initially leads to environmental degradation before improving at higher income levels (He & Richard, 2010). Moreover, the negative coefficient for REC in both the CS-ARDL and CCEMG and AMG suggests that renewable energy consumption is a critical mitigating factor for CO_2 emissions. Furthermore, the findings substantiate the intricate role of trade in environmental dynamics. The positive sign of the trade coefficient across all models highlights that, while trade can promote economic growth, it may also entail environmental costs, likely due to the scale of production and associated emissions.

This stability of the results indicates the reliability and consistency of CS-ARDL estimations and supports the robustness of the findings. This also strengthens the validity of CS-ARDL estimations and reinforces confidence in the relationships observed between explanatory variables and CO_2 emissions. Moreover, under AMG estimates, the common Dynamic process (CDP) is statistically significant at 1%, $\hat{\mu}_t^{va\cdot} = 0.625$, which can be attributed to regional and international agreements, technological diffusion, standard policies of countries, and innovation across countries that can motivate to attain low carbon emissions (Sun et al., 2022).

Table 18.6 CCEMG and AMG estimation results

Variables	CCEMG	AMG
$UP_{i,t}$	0.0975*	0.0364*
	(0.0617)	(0.0192)
$ln\_GDP_{i,t}$	0.0523**	0.0709***
	(0.0256)	(0.0237)
$FDINI_{i,t}$	0.00247	0.00394*
	(0.0016)	(0.0024)
$REC_{i,t}$	− 0.0342**	− 0.0356**
	(0.0158)	(0.0170)
$Trade_{i,t}$	0.0022***	0.0006
	(0.0002)	(0.0006)
Constant	2.608***	8.835***
	(0.855)	(0.838)
CDP		0.625***
		(0.161)
Observations	120	120

The dependent variable is ln_CO_2. Standard errors in parentheses
***, **, *Denotes statistical significance at the 1%, 5% and 10%, respectively while CDP is common dynamic process

Table 18.7 Panel Granger non-causality test results

Causality direction	$\hat{W}_{HPJ}$	*Coef*	Results	Remarks
$UP_{i,t} \rightarrow ln\_CO_{2i,t}$	11.16*	−.0397*** (.0118)	Causality from *UP* to *ln*_CO_2	Uni-causality
$CO_{2i,t} \rightarrow ln\_UP_{i,t}$	3.82	−.0368 (.0188)	Causality from *ln*_CO_2 to *UP*	

*Show statistical significance at 1%. Parentheses contain standard errors. $\hat{W}_{HPJ}$ acronym for Wald statistic Half Panel Jack-knife estimator

Granger Non-causality

The findings displayed in Table 18.7 demonstrate the Granger Non-Causality test conducted on the association between CO_2 emissions and urbanization. The estimations factored into account cross-sectional heteroskedasticity-robust standard errors. The results manifest a clear uni-causality from urbanization to CO_2 emissions and indicate that the expansion of urban centers in North African countries corresponds to a substantial rise in CO_2 emissions. These results align with the research conducted by Liu et al. (2023), which identified a direct causal relationship between urbanization and environmental quality. The robustness of these results is ensured by accounting for heteroskedasticity across sections, providing a more accurate and reliable inference as the statistical significance level is 1%. Interestingly, the absence of causality from CO_2 emissions to UP implies that the level of CO_2 emissions is not a significant factor in determining the pace or scale of urbanization. This may indicate that urban expansion is largely driven by economic and demographic factors, rather than environmental constraints or considerations. In essence, the findings underscore the environmental ramifications of unchecked urban expansion and the imperative for sustainable urban development strategies. Policies should aim to curtail emissions through the promotion of green spaces, green infrastructure, sustainable waste management practices, efficient public transport, and sustainable energy sources.

Conclusion and Policy Implications

The North African region, marked by its rapid urbanization over recent decades, stands at the crossroads of environmental challenges and socio-economic growth. Although, urban expansion is widely acknowledged as a driving force for economic development and growth within these nations, unstructured and uncontrolled urban development poses significant environmental dilemmas, potentially undermining ongoing development efforts and intensifying the effects of climate change. While the environmental ramifications of urban expansion in North Africa have not been extensively researched, this chapter utilizes a panel data spanning from 1990 to

2019 to examine the relationship between urbanization and environmental degradation across four North African nations: Egypt, Tunisia, Sudan, and Morocco. The empirical findings underscore a positive relationship among rapid urban expansion and escalating CO_2 emissions in North African countries. Furthermore, the results point out that changes in macroeconomic variables that often go hand in hand with urbanization, such as GDP, FDI net inflow, and Trade, are linked with increased CO_2 emissions. Conversely, authors discover that the utilization of renewable energy has a detrimental and statistically noteworthy impact on releasing CO_2 emissions, aiding in alleviating environmental distress.

The findings of this chapter offer several crucial policy suggestions for urban planners and policymakers in North Africa in relation to addressing the environmental outcomes of urbanization processes. First, it is essential for these countries to pivot towards sustainable urban planning by encouraging local governments to implement policies that consider the environment to ensure that urban expansion has naturalized or minimized the negative impacts on the environment. This requires the adoption of holistic approaches to urban planning to address issues that hinder progress, including insufficient regulations, limited resources, stakeholder resistance, existing urban sprawl, and uncontrolled urban migration. To overcome these challenges, raise awareness, implement supportive policies, increase investments in sustainable infrastructure, incentivize businesses and developers, retrofit existing urban areas, and promote balanced regional development to mitigate the negative impacts and promote sustainable urban development in North African nations. Second, given the positive association between economic growth and trade openness with environmental degradation, green trade policies could incentivize the exchange of eco-friendly technologies and penalize high-emission imports, encouraging sustainable practices. Coupled with economic strategies that promote low-carbon infrastructure and circular economy principles, such measures can decouple growth from environmental harm. Rigorous environmental regulations for businesses and strategic investments in green technology can further ensure that development enhances rather than exhausts environmental quality, ultimately fostering a sustainable and resilient economic future. Third, the positive effect of renewable energy on environmental quality implies that North African countries should enhance sustainable energy efficiency by investing in renewable energy sectors, potentially partnering with international bodies for technological and financial support. This includes engaging with green climate funds, global environmental facilities, and technology-sharing platforms, which can facilitate the transfer of cutting-edge renewable technologies and offer financial aid for large-scale renewable projects. Moreover, creating favorable regulatory frameworks to attract private investments in the renewable sector can accelerate this transition. Fourth, while foreign direct investment is a significant source of financing and crucial for economic growth in North Africa, attracting such investment needs to be aligned with environmental goals. This entails formulating and enforcing stringent environmental regulations that foreign investors must adhere to, thereby ensuring that their contributions do not come at the expense of environmental integrity. Countries in the region could consider the establishment of 'green FDI' standards, mandating that incoming investments support sustainable projects or incorporate sustainable practices. For example, offering incentives for investments in low-carbon technologies and industries that pursue energy efficiency can direct FDI flows into areas that reinforce both economic growth and environmental sustainability. Finally, regular environmental audits and feedback mechanisms should be institutionalized, which will help track progress and tweak policies based on real-time data and outcomes.

Although the study provides valuable insights, it is not without its limitations. The scope is geographically restricted to just four North African countries, which may not capture the full spectrum of urbanization's environmental impacts across more diverse African contexts, potentially affecting the wider applicability of the findings. Moreover, the environmental measures employed are limited to CO_2 emissions, overlooking other critical indicators such as water pollution, biodiversity loss, and land degradation, which are equally important in assessing the state of environmental degradation. Urbanization is treated as a singular concept based on urban population growth, failing to account for its multifaceted nature, including migration patterns and the expansion of city boundaries. Additionally, the timeframe of the study, spanning from 1990 to 2019, does not account for recent global shifts and the implications of the COVID-19 pandemic, that could significantly influence current and future trends in urbanization and environmental dynamics. Furthermore, the study does not account for the variability in policy and institutional frameworks that can significantly influence the observed relationship. Such factors might include environmental regulations, urban planning policies, and the enforcement of sustainability initiatives, all of which can affect how urbanization translates into environmental outcomes. The absence of these considerations means that study may not fully reflect the complexities of governance that are crucial to understanding and addressing the multifaceted challenges posed by urbanization. These limitations suggest a need for a more nuanced approach in future research to comprehensively understand the interplay between urbanization and environmental quality.

References

Abdullah, H., Elgendy, K., & Knaepen, H. (2021). *Climate resilience in cities of the EU's Southern Neighbourhood*. http://citesunies.s3.amazonaws.com/pages/619cfacd12a2c.pdf

Abu Hatab, A., Ravula, P., Nedumaran, S., & Lagerkvist, C.-J. (2022a). Perceptions of the impacts of urban sprawl among urban and peri-urban dwellers of Hyderabad, India: A Latent class clustering analysis. *Environment, Development and Sustainability, 24*, 12787–12812.

Abu Hatab, A., Amuakwa-Mensah, F., & Lagerkvist, C.-J. (2022b). Who moves and who gains from internal migration in Egypt? Evidence from two waves of a labor market panel survey. *Habitat International, 124*, 102573. https://doi.org/10.1016/j.habitatint.2022.102573

Akorede, Y. F. (2020). The relationship between urbanization, CO2 emissions, economic growth and energy consumption in Nigeria. *International Journal of Energy Economics and Policy*.

Al-Mulali, U., & Ozturk, I. (2015). The effect of energy consumption, urbanization, trade openness, industrial output, and the political stability on the environmental degradation in the MENA (Middle East and North African) region. *Energy, 84*, 382–389.

Anwar, A., Younis, M., & Ullah, I. (2020). Impact of urbanization and economic growth on CO2 emission: A case of far east Asian countries. *International Journal of Environmental Research and Public Health, 17*(7), 2531.

Asongu, S., & Odhiambo, N. M. (2021). Trade and FDI thresholds of CO2 emissions for a Green economy in sub-Saharan Africa. *International Journal of Energy Sector Management, 15*(1), 227–245.

Ayeni, A. O., Aborisade, A. G., Onuminya, T. O., Soneye, A. S. O., & Ogundipe, O. T. (2023). Urban development in Africa and impact on biodiversity. *Current Landscape Ecology Reports, 8*(2), 73–89.

Azam, M., & Khan, A. Q. (2016). Urbanization and environmental degradation: Evidence from four SAARC Countries—Bangladesh, India, Pakistan, and Sri Lanka. *Environmental Progress and Sustainable Energy, 35*(3), 823–832. https://doi.org/10.1002/ep.12282

Badreldin, N., Abu Hatab, A., & Lagerkvist, C. J. (2019). Spatiotemporal dynamics of urbanization and cropland in the Nile Delta of Egypt using machine learning and satellite big data: Implications for sustainable development. *Environmental Monitoring and Assessment, 191*, 1–23.

Bai, L., Xiu, C., Feng, X., & Liu, D. (2019). Influence of urbanization on regional habitat quality: A case study of Changchun City. *Habitat International, 93*, 102042.

Borck, R., & Pflüger, M. (2019). Green cities? Urbanization, Trade, and the environment. *Journal of Regional Science, 59*(4), 743–766. https://doi.org/10.1111/jors.12423

Chen, M., Chen, L., Cheng, J., & Yu, J. (2022). Identifying interlinkages between urbanization and sustainable development goals. *Geography and Sustainability, 3*(4), 339–346. https://doi.org/10.1016/j.geosus.2022.10.001

Dauda, L., Long, X., Mensah, C. N., Salman, M., Boamah, K. B., Ampon-Wireko, S., & Dogbe, C. S. K. (2021). Innovation, trade openness and CO2 emissions in selected countries in Africa. *Journal of Cleaner Production, 281*, 125143.

De Roo, G. (2000). Environmental conflicts in compact cities: Complexity, decisionmaking, and policy approaches. *Environment and Planning b: Planning and Design, 27*(1), 151–162. https://doi.org/10.1068/b2614

Dumitrescu, E. I., & Hurlin, C. (2012). Testing for Granger non-causality in heterogeneous panels. *Economic Modelling, 29*(4), 1450–1460.

ECA/AfDB, O. (2022). *Africa's urbanisation dynamics 2022: The economic power of Africa's cities, West African studies.* OECD Publishing Paris.

Effiong, E. (2016). *Urbanization and environmental quality in Africa.* https://mpra.ub.uni-muenchen.de/id/eprint/73224

Fatima, T., Karim, M. Z. A., & Meo, M. S. (2021). Sectoral CO 2 emissions in China: Asymmetric and time-varying analysis. *Journal of Environmental Planning and Management, 64*(4), 581–610. https://doi.org/10.1080/09640568.2020.1776691

Hanif, I. (2018). Impact of economic growth, non-renewable and renewable energy consumption, and urbanization on carbon emissions in Sub-Saharan Africa. *Environmental Science and Pollution Research, 25*(15), 15057–15067.

Haouraji, C., Mounir, B., Mounir, I., & Farchi, A. (2021). Exploring the relationship between residential CO2 emissions, urbanization, economic growth, and residential energy consumption: Evidence from the North Africa region. *Energies, 14*(18), 5849.

Hatab, A. A., Cavinato, M. E. R., Lindemer, A., & Lagerkvist, C. J. (2019). Urban sprawl, food security and agricultural systems in developing countries: A systematic review of the literature. *Cities, 94*, 129–142.

He, J., & Richard, P. (2010). Environmental Kuznets curve for CO2 in Canada. *Ecological Economics, 69*(5), 1083–1093.

Herbert, S. (2019). *Development characteristics of Small Island Developing States.*

Hussain, Z. (2022). Environmental and economic-oriented transport efficiency: The role of climate change mitigation technology. *Environmental Science and Pollution Research, 29*(19), 29165–29182.

Im, K. S., Pesaran, M. H., & Shin, Y. (2003). Testing for unit roots in heterogeneous panels. *Journal of Econometrics, 115*(1), 53–74.

IPCC. (2022). *Climate change 2022: Impacts, adaptation and vulnerability*. Retrieved from https://report.ipcc.ch/ar6/wg2/IPCC_AR6_WGII_FullReport.pdf

Juodis, A., Karavias, Y., & Sarafidis, V. (2021). A homogeneous approach to testing for Granger non-causality in heterogeneous panels. *Empirical Economics, 60*(1), 93–112.

Levin, A., Lin, C. F., & Chu, C. S. J. (2002). Unit root tests in panel data: Asymptotic and finite-sample properties. *Journal of Econometrics, 108*(1), 1–24.

Liddle, B., & Lung, S. (2010). Age-structure, urbanization, and climate change in developed countries: Revisiting STIRPAT for disaggregated population and consumption-related environmental impacts. *Population and Environment, 31*, 317–343.

Liu, F., Khan, Y., & Hassan, T. (2023). Does excessive energy utilization and expansion of urbanization increase carbon dioxide emission in Belt and Road economies? *Environmental Science and Pollution Research*, 1–26.

Liu, X., & Bae, J. (2018). Urbanization and industrialization impact of CO2 emissions in China. *Journal of Cleaner Production, 172*, 178–186.

Liu, X., Guo, P., Yue, X., Zhong, S., & Cao, X. (2021). Urban transition in China: Examining the coordination between urbanization and the eco-environment using a multi-model evaluation method. *Ecological Indicators, 130*, 108056.

Long, X., Ji, X., & Ulgiati, S. (2017). Is urbanization eco-friendly? An energy and land use cross-country analysis. *Energy Policy, 100*, 387–396.

Marcotullio, P. J., & Lee, Y.-S.F. (2003). Urban environmental transitions and urban transportation systems: A comparison of the North American and Asian experiences. *International Development Planning Review, 25*(4), 325–354.

Mehmood, B., Khan, S. A., & Raza, M. (2023). Econometric evidence of catalytic effect of seaport activity in OECD countries: Getting it right. *Maritime Transport Research, 4*, 100090.

Mehmood, B., Raza, M., & Ramzan, A. (2022). Forestry and governance as two pillars of sustainable development: Empirical verification. *Sustainable Business and Society in Emerging Economies, 4*(2), 573–580.

Mehmood, U., & Mansoor, A. (2021). CO2 emissions and the role of urbanization in East Asian and Pacific countries. *Environmental Science and Pollution Research, 28*(41), 58549–58557. https://doi.org/10.1007/s11356-021-14838-x

Mezran, K., & Sanguini, A. (2021). North Africa 2030: What the future holds for the region. *North Africa, 2030*, 1–158.

Mignamissi, D., & Djeufack, A. (2022). Urbanization and CO2 emissions intensity in Africa. *Journal of Environmental Planning and Management, 65*(9), 1660–1684.

Musah, M., Kong, Y., Mensah, I. A., Antwi, S. K., & Donkor, M. (2021). The connection between urbanization and carbon emissions: A panel evidence from West Africa. *Environment, Development and Sustainability, 23*, 11525–11552.

Nathaniel, S., Anyanwu, O., & Shah, M. (2020). Renewable energy, urbanization, and ecological footprint in the Middle East and North Africa region. *Environmental Science and Pollution Research, 27*(13), 14601–14613.

OECD. (2021). *Urbanisation and demography in North and West Africa, 1950–2020*. OECD.

Pesaran, M. H. (2006). Estimation and inference in large heterogeneous panels with a multifactor error structure. *Econometrica, 74*(4), 967–1012.

Pesaran, M. H. (2015). Testing weak cross-sectional dependence in large panels. *Econometric Reviews, 34*(6–10), 1089–1117.

Pesaran, M. H. (2021). General diagnostic tests for cross-sectional dependence in panels. *Empirical Economics, 60*(1), 13–50.

Pesaran, M. H., & Yamagata, T. (2008). Testing slope homogeneity in large panels. *Journal of Econometrics, 142*(1), 50–93.

Poumanyvong, P., & Kaneko, S. (2010). Does urbanization lead to less energy use and lower CO2 emissions? A cross-country analysis. *Ecological Economics, 70*(2), 434–444.

Raihan, A., Muhtasim, D. A., Farhana, S., Pavel, M. I., Faruk, O., Rahman, M., & Mahmood, A. (2022). Nexus between carbon emissions, economic growth, renewable energy use, urbanization, industrialization, technological innovation, and forest area towards achieving environmental sustainability in Bangladesh. *Energy and Climate Change, 3*, 100080.

Sadorsky, P. (2014). The effect of urbanization on CO2 emissions in emerging economies. *Energy Economics, 41*, 147–153.

Salahuddin, M., Gow, J., Ali, M. I., Hossain, M. R., Al-Azami, K. S., Akbar, D., & Gedikli, A. (2019). Urbanization-globalization-CO2 emissions nexus revisited: Empirical evidence from South Africa. *Heliyon, 5*(6).

Shahbaz, M., Balsalobre-Lorente, D., & Sinha, A. (2019). Foreign direct Investment–CO2 emissions nexus in Middle East and North African countries: Importance of biomass energy consumption. *Journal of Cleaner Production, 217*, 603–614.

Sharma, S. S. (2011). Determinants of carbon dioxide emissions: Empirical evidence from 69 countries. *Applied Energy, 88*(1), 376–382.

Singh, N., Poonia, T., Siwal, S. S., Srivastav, A. L., Sharma, H. K., & Mittal, S. K. (2022). Challenges of water contamination in urban areas. In *Current directions in water scarcity research* (vol. 6, pp. 173–202). Elsevier.

Sun, Y., Li, H., Andlib, Z., & Genie, M. G. (2022). How do renewable energy and urbanization cause carbon emissions? Evidence from advanced panel estimation techniques. *Renewable Energy, 185*, 996–1005.

United Nations. (2023a). *The sustainable development goals report: Special edition. Towards a rescue plan for people and planet*. United Nations. https://unstats.un.org/sdgs/report/2023/The-Sustainable-Development-Goals-Report-2023.pdf

World Bank. (2023). *Urban population (% of total population)*. World Bank.

Wu, J., Abban, O. J., Boadi, A. D., Haris, M., Ocran, P., & Addo, A. A. (2021). Exploring the relationships among CO2 emissions, urbanization, economic growth, economic structure, energy consumption, and trade along the BRI based on income classification. *Energy, Ecology and Environment, 6*, 213–231.

Wu, Y., Zhou, Y., & Liu, Y. (2020). Exploring the outflow of population from poor areas and its main influencing factors. *Habitat International, 99*, 102161.

Xie, Y., & Pesaran, M. H. (2022). *A bias-corrected Cd test for error cross-sectional dependence in panel data models with latent factors*. Available at SSRN 4198155.

Yang, R., Zhang, J., Xu, Q., & Luo, X. (2020). Urban-rural spatial transformation process and influences from the perspective of land use: A case study of the Pearl River Delta Region. *Habitat International, 104*, 102234.

Zhou, Y., Chen, M., Tang, Z., & Mei, Z. (2021). Urbanization, land use change, and carbon emissions: Quantitative assessments for city-level carbon emissions in Beijing-Tianjin-Hebei region. *Sustainable Cities and Society, 66*, 102701.

Mohsin Raza is a Ph.D. candidate in Business Economics at the Business Economics Group, Wageningen University & Research, Netherlands. His research focuses on the economic sustainability and resilience of food systems, with a broader interest in sustainable agricultural practices and resource economics. He holds an Erasmus Mundus Double Degree Master's in Agricultural, Food, and Environmental Policy Analysis (AFEPA) with affiliations to the Swedish University of Agricultural Sciences (SLU), Sweden, and the University of Bonn, Germany. Mohsin has diverse research experience, having worked at the Center for Development Research (ZEF), University of Bonn, where he contributed to projects on environmental sustainability and economic development. Additionally, he collaborated with the Chair Group of Agricultural Production and Resource Economics at the Technical University of Munich (TUM), focusing on pest management strategies and control measures in European agriculture. He has also been involved in research collaborations with the Nordic Africa Institute (NAI), Uppsala, Sweden, on projects related to sustainable management of natural resources, farmer's vulnerability and resilience to natural hazards, and urbanization & environmental degradation in low- and middle-income countries. His research interests include sustainable agricultural practices and food systems, bio-economic modelling, economic resilience, and environmental sustainability. Mohsin has presented his research at international conferences and has (co)-authored several peer-reviewed publications.

Assem Abu Hatab is a professor in development economics with broad empirical interests in economics and sustainable management of food systems in low- and middle-income countries. He is affiliated with the Nordic Africa Institute (NAI) and the Swedish University of Agricultural Sciences (SLU) in Uppsala, Sweden. His current research revolves around examining the resilience of social-ecological systems in the face of nature-induced changes and external shocks. Assem has authored over 90 articles in international peer-reviewed journals, covering topics related to agricultural and food economics, risk perception and management in agrifood value chains, and food system sustainability and food security. From 2018 to 2022, Assem served as a lead author for the 6th IPCC Assessment Report. In 2023, Assem was selected as a member of the drafting team of the report of the High-Level Panel of Experts on Food Security and Nutrition (HLPE-FSN) on "Strengthening urban and peri-urban food systems to achieve food security and nutrition in the context of urbanization and rural transformation." He was also selected as a Coordinating Lead Author for the special report of the Mediterranean Experts on Climate and Environmental Change (MedECC) on the climate-water-energy-food-ecosystems nexus. In 2022, he was appointed by the Government of Sweden as a member of the Scientific Advisory Board of the Swedish International Development Cooperation Agency.

Chapter 19
Vulnerability and Adaptation to Climate Change: Understanding and Awareness Among Coastal Communities in the Gambia

Bintou Dibba, Sidat Yaffa, and Mamma Sawaneh

Abstract The coastal zone of The Gambia provides important ecosystem services, including habitats and breeding grounds for aquatic animals, plants, and migratory birds. Therefore, its vulnerability to climate change hazards can have major implications on the coastal inhabitants. This study analysed the level of vulnerability and adaptation to climate change among the various coastal cells in The Gambia. The findings of the research demonstrated that while majority of households were aware (80%) and believed (88) that the climate is changing, the level of understanding (30%) of the climate change concept is low. Furthermore, the coastal zone of The Gambia on average is vulnerable to climate change with a vulnerability index score of 0.58 and an exposure index of 0.57. However, the differences in adaptive capacity and socioeconomic characteristics of households within the coastal communities resulted in differences in sensitivity, exposure and vulnerability observed among the study cells. Cell 8 scoring the lowest adaptive capacity index of 0.32 ranks highest in terms of its vulnerability index score of 0.73 while Cell 1, the least vulnerable scores the lowest exposure index of 0.31. Therefore, there is an urgent need to mobilise an appropriate level of capacity, funding, diverse skills and knowledge systems within the coastal communities and relevant stakeholders in the country to enhance adaptive capacity and minimise the socioeconomic and environmental hazards affecting the coastal inhabitants.

Keywords Adaptation · Awareness · Coastal zone · Climate change · Knowledge

B. Dibba (✉)
School of Arts and Sciences, University of The Gambia, Serekunda, The Gambia
e-mail: bdibba@utg.edu.gm

S. Yaffa · M. Sawaneh
School of Agriculture and Environmental Sciences, University of The Gambia, Serekunda, The Gambia

W. Leal Filho et al. (eds.), *Climate Change Mitigation and Adaptation in Practice*, Climate Change Management, https://doi.org/10.1007/978-3-031-85217-6_19

Introduction

The coastal zone of The Gambia is a region of socioeconomic importance where various activities including tourism, artisanal fishing, cultural and aesthetic and religious activities are prevalent. It also provides important ecosystem services, including habitats and breeding grounds for aquatic animals, plants and migratory birds. Socioeconomic activities including fisheries and hotel industries that contributed 36% of the country's GDP were negatively impacted due to accretion and erosion of coastline (Bojang et al., 2023) thereby increasing the vulnerability of households within the area given their low adaptive capacity (Gomez et al., 2020). Consequently, in 2021, these socio-economic activities contribute 34% of the country's GDP (GBOS 2022, UNDP, n.d.). In addition, understanding the effects of increased urbanisation and industrial activities on the livelihood of the population living in this environment is limited. As a result, the coastal inhabitants face several climate change challenges. The demand for marine and freshwater resources and their habitats has resulted in an increase in coastal population which is complemented by high infrastructural development, mining, and exploitation of coastal resources ultimately posing significant challenges to the coastal population.

As a result, various studies have highlighted that The Gambia coastal environment is vulnerable to climate change effects (Amuzu et al., 2018a; Belford et al., 2020; Gomez et al., 2020; Komma, 2019). These studies further highlighted significant levels of exposure and sensitivity of the households as well as the limited adaptive capacity of the majority of the households to address climate change challenges. Besides, the inability of households to adapt to these climate change effects may seriously affect the livelihood and health of individuals living around the coast (Saleem Burden et al., 2020; Khan et al., 2020).

With the limited assessment of the vulnerability and the level of climate change understanding and awareness of coastal households in literature in The Gambia, there is a need for further research that specifically aims to assess the vulnerability of households within the coastal region of The Gambia. This will help to guide the government and relevant stakeholders in providing more reliable and effective programs that would protect the coastal environment and inhabitants from climate change hazards and help in allocating priorities for funding and development projects within communities along the coastline.

Climate change vulnerability is the tendency to be negatively impacted by climate hazards and it is measured through the exposure, sensitivity and adaptive capacity of a system (Füssel, 2010). As the African settlements increase around the coast, they will be subjected to various climate hazards associated with increased health risks therefore increasing their vulnerability (Williams et al., 2019). As socioeconomic development and urbanisation continue to increase, climate change is expected to threaten Sub-Saharan Africa's progress in human well-being (Dickerson et al., 2022). This is because physical conditions especially of households with lower educational levels and insufficient income to meet household demands can lead to situations of vulnerability by increasing their exposure to flood and storm surges (Bera et al.,

2020; Fatemi et al., 2020). Similarly, increased socioeconomic development and urbanisation are expected to negatively impact sub-Saharan Africa as it continues to experience the largest exposed population to climate hazards due to its inadequate capacity to address climate change challenges (Ofori et al., 2021).

It is evidenced that most of the climate-induced pressures experienced by coastal inhabitants were facilitated by the exploitation of coastal resources by humans for socio-economic development, urbanisation, food and other essential services obtained along the coasts (Hossain et al., 2020). Coastal ecosystems have been significantly disturbed by various human activities including settlements, transport systems, tourism, aquaculture, fisheries and mining (Lu et al., 2018; Zerebecki et al., 2022). In addition, human activities have been the determinant factors in ocean surface temperature rise around the world since the 1970s and atmospheric carbon dioxide concentrations were higher in 2019 than any period in about two million years, leading to the increased negative effect of temperature rise and ocean-atmospheric temperature changes on species distribution and abundance in coastal environments (Pörtner et al., 2022). Therefore, there is a need for improved socioeconomic, technical skills, climate change awareness and understanding and living conditions of individuals at community levels especially around low-lying coastlines to address the impact of climate change (Riede et al., 2016; Zhou et al., 2021).

Incidentally, proper environmental policy developments and implementation can help to increase public and political awareness of climate change impacts and improve the well-being of individuals and communities (Pörtner et al. 2022). This will also help to address the conflicting interest between human socioeconomic activities that negatively affect the coastal environment and human well-being (Herrmann et al., 2020). Moreover, the evaluation of socioeconomic vulnerability to climate change must be based on the understanding of human populations which determines their ability to minimise the effects of climate extremes (Mafi-Gholami et al., 2020). The coastline that represents the interface between land and ocean has experienced significant transformation that modifies greatly their original landscape leading to the vulnerability of these areas to climate extremes in the last two centuries (De Serio et al., 2018).

Moreover, the assessment of social vulnerability to climate change can be based on an understanding of urbanisation and increased human populations which determines their ability to cope with or adapt to stress (Mesta et al., 2022). In this regard, as the human-induced climate change reality has increased over the past decades, effective and efficient evaluations of climate change in coastal areas are necessary to enhance the capacities of all sectors in developing a vibrant coastal management policy by incorporating substantial vulnerabilities, adaptation, and resilience building issues (Dale et al., 2019; Schipper et al., 2021). Thus, this study adopts the IPCC concept of vulnerability which represents the extent to which a system is exposed or sensitive to or unable to adapt to the effects of climate change hazards (IPCC, 2022).

The review was conducted with reference to an integrated assessment approach based on diverse knowledge and practices from a broader discipline. This model according to UNFCCC (2022) provides an understanding of the global climate change issues. This approach explores the computation of important processes of human-earth interaction by utilising information from different scientific domains. The assumption from these disciplines helps in identifying the appropriate assessment methods for this study by integrating available information from various studies. In this regard, this study investigates the awareness and understanding and the level of vulnerability and adaptation to livelihood stress among coastal communities in The Gambia.

Methodology

Area of Study

The Gambia has a total land area of about 11,000 km^2 and is located within the tropical sub-humid eco-climatic zone with about one-fifth of its surface area water. The country is considered to have a Sudano Sahelian climate usually identified by an irregular dry Harmattan Northern Winds from the Sahara Desert and South-Westerly Monsoon winds from the Atlantic Ocean (NEA, 2010). Its 80 km coastline extends from Buniadou Point and Kerenti Bolong in the north to the mouth of Allanhein in the south. The river Gambia runs 680 km from the Futa Jallon in the Republic of Guinea to the Atlantic Ocean dividing the country into North and South Banks. The coastal region is divided into nine subunits called cells based on the geomorphic characteristics and conditions of the sandy beaches and the characteristic of each of the coastal cells described by Jallow et al. (1999) and Amuzu et al. (2018b) is presented in Table 19.1.

Sampling Technique and Sample Size

Guided by household and population data of The Gambia by Local Government Area, District and Settlement, 2013 compiled by the Gambia Bureau of Statistics (GBOS, 2013), the major communities near the coastline within the nine cells were outlined. The population of the study comprised the households within the communities of the nine coastal cells with household heads as the target respondents using a multistage sampling method. In the first stage, 5 cells of 8 coastal cells were selected using a simple random technique. Cell 3 was exempted during cell sampling because this area is between Banjul and Bakau Old Cape without significant settlement. Two communities with the highest population were selected from each sampled cell. The sample size (Table 19.2) was determined using Raosoft software. Using the margin of error of 5%, confidence interval of 95% and population size of 16,990 households. The sample size n and the margin of error E were calculated using the formula:

$$n = Nx/\left((N - 1)E^2 + x\right)$$

Table 19.1 Characteristic of the coastal zone of The Gambia

Cell name	Characteristics	Length in km along the coastline
Cell 1:	This area is identified by sandy beaches with extensive mangrove vegetation and wetlands which harbour different fauna and flora and provide a spawning environment for aquatic organisms. This area extend from Kerenti bolong to Barra	12.0
Cell 2:	This cell starts from Buniadou point to about 4 km wide into Banjul. The area is characterised by mangrove vegetation, mudflats and marshes. The area provides an important spawning and habitat for various marine and freshwater organisms	11.0
Cell 3:	This area extends from Banjul to Bakau Old Cape with no significant settlement and is threatened by sea level rise and beach erosion	13.5
Cell 4:	This cell extends from Bakau Old Cape to Faraja and is highly vulnerable to coastal erosion and sea level rise	3.5
Cell 5:	This cell is between Fajara and the bald point of the coastal area and is characterised by sandy cliffs. With most of the hotels and other tourism establishments, the area is believed to be highly vulnerable to sea level rise and coastal erosion	2.5
Cell 6:	This area extends from Kerr Serigne to Tanjeh and is identified as the most significant environment with sandy pits which attract sand miners	11
Cell 7:	Sand beaches and flat plain strands of sand are observed in this cell and it is believed to be the least affected by industrialisation extending from Tanjeh to Tintinto	4.5
Cell 8:	This cell extends from Sanyang to Gunjur and is observed to be exposed to coastal erosion	7.5
Cell 9:	This cell is the southern end of The Gambia coastline extending from Gunjur to Allenhain and is characterised by a series of sand dunes and beach ridges that run parallel to the shoreline from Kartong	24

Sources Jallow et al. (1999), Amuzu et al. (2018b)

where

N the population size
r the fraction of responses interested in, and
$z(c/100) =$ Critical value for the confidence level c.

$$x = z(c/100)^2 r(100 - r)$$

$$E\text{ (Margin of Error)} = \mathrm{Sqrt}[(N - n)x/n(N - 1)]$$

n sample size

Table 19.2 Population and sample of the study area

Coastal cells	Settlement	N.O. HHs	Target HHs	Number of HHs interviewed	Sampled HHs per cell	Total HHs interviewed per cell
Cell 1	Essau	1409	31	33	50	52
	Barra	867	19	19		
Cell 4	Bakau Old Cape	2011	45	35	98	95
	Bakau New Town	2405	53	60		
Cell 5	Kotu	2008	44	60	72	88
	Kololi	1240	27	28		
Cell 6	Kerr Seringe	1512	33	31	92	85
	Brufut	2654	59	54		
Cell 8	Sanyang	1368	30	26	64	60
	Gunjur	1516	34	34		
Total		16,990	376	380		380

Using the proportion to population size (PPS) method, where the probability of selecting a unit is proportional to its sample size, the number of households selected in each community is proportional to the number of households as follows:

$$\text{Number of Target Household} = \frac{\text{Number of Household per Community}}{\text{Total Household}} \times \text{sample size}$$

Data Collection and Data Source

The household data collected during the survey includes socio-demographic information on household members and data on exposure, sensitivity and adaptive capacity of households within the selected coastal cells. In addition, climate variability data (rainfall (mm), Maximum and minimum temperatures (°C), wind speed (knots) and relative humidity (%)) from 1990 to 2020 were also obtained to determine changes in climate variability within the 30 years. In addition, community outreach (Focus group and key informant discussions) was held in 10 coastal communities where community members including males and females discussed their perceptions, and opinions on climate change issues affecting their environment and livelihood (Fig. 19.1).

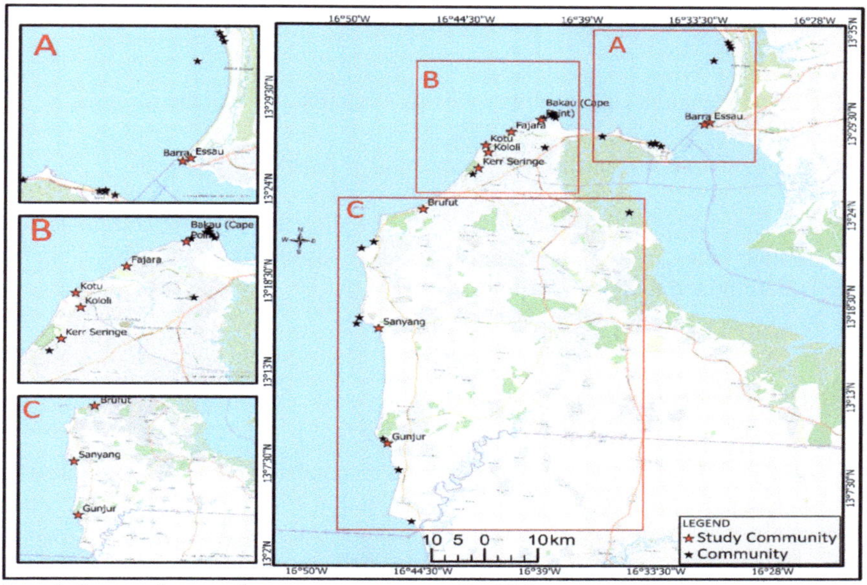

A and C= rural coastal **B= Urban coastal**

Fig. 19.1 Map of the study communities. *Source* Author's Illustration, 2020

Method of Data Analysis

To measure the vulnerability index, a numerical scale which lies between 0 and 1 was calculated from a set of selected variables and was used to compare regions. Taking into account the functional relationship between vulnerability and the indicators (where vulnerability increases with an increase or decrease in the values of the indicator), the vulnerability indexes were calculated using the formula:

1. When vulnerability has a positive relationship with an indicator value.

$$\text{Positive relationship } x_{ij} = \frac{X_{ij} - \min\limits_{i}\{X_{ij}\}}{\max\limits_{i}\{X_{ij}\} - \min\limits_{i}\{X_{ij}\}} \tag{19.1}$$

2. When vulnerability has a negative relationship with indicator value.

$$\text{Negative relationship } y_{ij} = \frac{\max\limits_{i}\{X_{ij}\} - X_{ij}}{\max\limits_{i}\{X_{ij}\} - \min\limits_{i}\{X_{ij}\}} \tag{19.2}$$

where,

X_{ij} is the value of the indicator,

j is the corresponding value to the region,
i is the region.

The climate change vulnerability index was calculated using the formula;

$$\text{CCVI} = (\text{Sensitivity} + \text{Exposure}) - \text{Adaptive Capacity}$$

The vulnerability indicators were then used to rank the different coastal cells in terms of their index score in the order of highest to lowest. The cell with the highest index is determined to be the most vulnerable.

Result and Discussions

The Socio-demographic characteristics of the respondents presented in Table 19.3 show that 38 and 25% of the respondents were within 46–65 years and 36–45 years, respectively. This is because the target respondents for this study were individuals who were mature enough to understand and respond appropriately to the issues being discussed. There were more male (66%) headed households than female (34%). This can be because of the culture and religion of the country where unless there is no adult male living within the household, women-headed households are usually frowned on by society and thus the dominance of the male gender as household heads. This is corroborated by Badjie et al. (2019) in their study on climate change and availability of food among households in The Gambia. The result highlighted that 20% of the respondents had no formal education, 28% had attended Madarasa/Arabic school, and only 2% had university-level education. Climate change education is important in determining people's level of understanding of the concept of climate change in The Gambia. This was observed during the focus group discussion and key informants' interview where individuals who had beyond secondary level education were able to explain better the concept of climate change than those with Madarasa/Arabic, secondary level and below. Furthermore, salary jobs (28%) and trading (24%) were the primary occupations of most of the respondents as a result of increased industrial development and urbanisation from 1990 to 2020. Almost half of the respondents had 6–10 household members and the monthly household income level for 48% of households was D5001-D10000, which majority (83%) of respondents revealed that their household income was inadequate to cover basic household expenses. Low-income levels, large household sizes, and lack of quality amenities contribute to the increased vulnerability of households to climate hazards (Bera et al., 2020; Fatemi et al., 2020).

Table 19.3 Proportion of respondents' socio-demographic characteristics in the studied cells within the coastal zone of The Gambia

Characteristics		Cell 1	Cell 4	Cell 5	Cell 6	Cell 8	AV.
Age	Below 25 yrs	12	6	8	2	2	6
	26–35 yrs	15	16	25	14	10	16
	36–45 yrs	25	23	22	28	25	25
	46–65 yrs	33	38	32	42	47	38
	above 65 yrs	15	17	14	13	17	15
Gender	Male	58	58	58	74	82	66
	Female	42	42	42	26	18	34
Educational level	No formal education	25	17	28	12	17	20
	Arabic or Madrassa	17	19	23	41	42	28
	Primary education	15	8	9	8	8	10
	Secondary education	25	43	32	30	23	31
	Tertiary education	13	11	7	7	8	9
	University education	4	2	2	2	2	2
Primary occupation	Salary work	31	43	30	20	18	28
	Fishing	1	4	3	7	5	4
	Farming	6	5	6	4	17	8
	Trader	35	16	15	25	28	24
	Carpentry	10	9	17	28	18	16
	Student	0	5	8	2	3	4
	Mining	19	18	23	14	8	16
Household size	1–5 members	12	25	25	26	10	20
	6–10 members	48	57	56	49	50	52
	11–15 members	31	16	13	19	27	21
	16–20 members	7	2	7	6	10	6
	21–25 members	2	0	0	0	3	1
Household income level (month)	D1000–D5000	19	11	23	14	12	16
	D5001–D10000	35	54	59	44	48	48
	D10001–D15000	29	20	15	32	25	24
	D15001–D20000	6	8	2	6	7	6
	Above D20000	6	8	2	6	7	6
Adequacy of household income	Inadequate to cover basic household expenses	81	80	92	82	80	83
	Just enough to cover basic household expenses	19	17	8	15	17	15
	More than enough for basic household expenses	0	3	0	2	3	2

Source Fieldwork (2021)

Awareness, Belief and Understanding of Climate Change

The results indicated that 80% and 88% of respondents were respectively aware and believed that the climate is changing (Table 19.4). Unfortunately, the level of understanding of the climate change concept is low considering that only 30% of the respondents on average understood climate change and 20% understood the concept to some extent. The low level of understanding of climate change among the respondents might be due to the complex nature of the concept of climate change which requires a sufficient level of education to comprehend. This finding was supported by Kellstedt et al. (2008), Masson-Delmotte et al. (2021) and Lee et al. (2022) that one of the reasons why the public has an inadequate understanding of climate change might be because the scientific nature of the concept is constantly evolving making emerging concepts of climate change complex for public understanding. Deforestation (77%) and burning fossil fuels (41%) were identified as the main causes of climate change among the respondents (Table 19.5). Less privileged household members in The Gambia rely on plants as firewood and charcoal as resources for household functions contributing to the overexploitation of woody plants. Therefore poverty reduction strategies including the promotion of renewable energy resources will be effective in reducing the demand for plants as fuel for household functions, and thus reduce deforestation (Ferraro & Simorangkir, 2020; Miyamoto, 2020). Furthermore, the belief among coastal households in The Gambia that deforestation and burning of fossil fuels are the main causes of climate change may enhance their motivation and behavioural changes towards reducing greenhouse gas emissions and enhance their support and discussions about climate change and its effects on coastal inhabitants.

A climate change vulnerability index assessment was conducted among coastal cells in The Gambia to evaluate the exposure, sensitivity, and adaptive capacity and vulnerability levels of coastal residents to extreme climate events from 1990 to 2020. Using the vulnerability assessment indicators presented in Table 19.6, the exposure, sensitivity, adaptive capacity and vulnerability indexes of households were calculated

Table 19.4 Households' awareness, belief and understanding of climate change

Characteristic (%)	Responses	Cell 1	Cell 4	Cell 5	Cell 6	Cell 8	AV.
Awareness	Yes	79	78	63	86	92	80
	No	21	20	32	13	8	19
	No response	0	2	6	1	0	1
Believe	Yes	87	91	81	88	93	88
	No	0	1	3	2	0	1
	No response	13	8	16	9	7	11
Understanding	Yes	33	37	19	35	28	30
	To some extend	13	12	20	24	30	20
	No	54	51	61	41	42	50

Source Fieldwork, 2021

Table 19.5 Households' understanding of the causes of climate change in the coastal zone of The Gambia

Causes (%)	Cell 1	Cell 4	Cell 5	Cell 6	Cell 8	AV.
Burning fossil fuels	38	44	35	41	43	40
Deforestation	78	72	77	63	75	73
Natural events	31	14	1	1	4	10
Agriculture	9	23	27	32	29	24
God	31	21	18	27	18	23
I don't know	9	14	10	12	18	13
Others (e.g., sand mining)	0	0	0	0	2	0

Source Fieldwork, 2021

for the 5 coastal cells under consideration and the results were presented in Table 19.7 and Fig. 19.2.

Exposure, Sensitivity and Adaptive Capacity of Households

The result revealed that the coastal zone on average is exposed to the impacts of climate change recording an exposure index of 0.57 on average. High exposure indexes of 0.59, 0.69, 0.64, and 0.63 were recorded for cell 4, cell 5, cell 6 and cell 8 respectively (Table 19.7). The level of exposure of these households was based on exposure to natural hazards such as maximum temperature (°C) and tropical windstorms (knots) (Table 19.5). In line with these findings, Gomez et al. (2020) highlighted that households within the coastal zone of The Gambia were exposed to the impact of natural hazards such as high temperatures and tropical windstorms. Furthermore, the physical conditions of houses where 78% of houses were made of mud (Table 19.6) and insufficient income to meet household demand (80%) (Table 19.3) were part of the main factors contributing to the exposure level of households in Cell 8 (Table 19.6). These factors according to Fatemi et al. (2020) and Bera et al. (2020) contribute to increased exposure of households to temperature extremes, floods and Tropical windstorms.

Furthermore, the sensitivity index computed indicates that households within cell 5 and cell 6 are the most sensitive to climate change scoring sensitivity indexes of 0.55 and 0.58 respectively (Table 19.7 and Fig. 19.2). This area has been also classified by Komma (2019) as a high-risk zone for climate change effects. The sensitivity level of these households was determined by their socio-economic status where 92% and 82% of households in these cells do not have adequate income to cover important household expenses (Table 19.6). Regarding livelihood activities, 70% and 71% of the household members in Cell 5 and Cell 6 respectively do not engage in any economic activity (Table 19.6). Households with inadequate livelihood support and income are more sensitive to climate extremes as their socioeconomic situation limits

Table 19.6 Vulnerability assessment indicators for the Coastal Zone of The Gambia from 1990 to 2020

Indicator	Components	Coastal cells				
Description		Cell 1	Cell 4	Cell 5	Cell 6	Cell 8
Change in annual precipitation for the last 30 years	E1	68%	26%	30%	35%	46%
Change in annual MAX TEMP for the last 30 years	E2	3%	6%	6%	6%	6%
Change in annual MIN TEMP for the last 30 years	E3	− 8%	4%	4%	4%	4%
Change in annual RH for the last 30 years	E4	− 3%	− 3%	− 3%	− 3%	− 3%
Change in annual WS for the last 30 years	E5	34%	59%	59%	59%	59%
Proportion of households exposed to coastal flood events	E6	13%	14%	26%	18%	10%
Proportion of households exposed to coastal erosion events	E7	13%	4%	10%	1%	3%
Proportion of households exposed to maximum temperature	E8	81%	87%	80%	88%	83%
Proportion of households exposed to tropical windstorms	E9	90%	75%	72%	76%	82%
Proportion of households exposed to colder temp	E10	12%	23%	26%	32%	32%
Proportion of households exposed to increased land erosion	E11	13%	17%	30%	11%	10%
Proportion of households exposed to less rain	E12	42%	43%	33%	24%	32%
Proportion of households exposed to more rain	E13	6%	16%	36%	42%	35%
Proportion of households exposed to changes in rainy and dry seasons	E14	6%	21%	26%	32%	33%
Average number of people per household	S1	10.00	8.00	8.00	8.00	9.00
Proportion of household members who are female	S2	59%	55%	58%	52%	40%
Proportion of household members who are 7 years and below	S3	18%	20%	19%	19%	12%
Proportion of household members 65 years and above	S4	4%	4%	4%	4%	5%

(continued)

Table 19.6 (continued)

Indicator	Components	Coastal cells				
Description		Cell 1	Cell 4	Cell 5	Cell 6	Cell 8
Proportion of households living in mud/corrugated house	S5	27%	35%	61%	41%	78%
Proportion of households whose income is usually not enough to cover the important household expense	S6	81%	80%	92%	82%	80%
Proportion of households who migrated to the communities they are residing	S7	52%	55%	78%	88%	32%
Proportion of household members who depend on skilled labour (fishing, farming, trading, carpentry, etc.) for their livelihood	S8	20%	17%	17%	22%	22%
Proportion of household members who don't engage in any economic activities (students, housewives…)	S9	69%	68%	70%	71%	69%
Proportion household members who a literate (at least a primary level)	A1	75%	81%	80%	73%	84%
Proportion of household members who are working	A2	31%	32%	30%	29%	31%
Proportion workers who earn a salary	A3	11%	15%	13%	7%	8%
Proportion of households who own the house they are residing in	A4	75%	57%	51%	53%	83%
Proportion of households who have livestock, poultry and other as household assets	A5	42%	19%	17%	26%	48%
Proportion of households with cement or concrete as the main construction materials of the walls of the houses they were residing in	A6	73%	63%	39%	59%	22%
Proportion of households with corrugated iron sheets as the main construction materials of the roof of the houses they were residing in	A7	75%	72%	76%	100%	100%

(continued)

Table 19.6 (continued)

Indicator	Components	Coastal cells				
Description		Cell 1	Cell 4	Cell 5	Cell 6	Cell 8
Proportion of households who have access to health care centres	A8	100%	100%	91%	85%	98%
Proportion of households with flush toilets in their home	A9	21%	33%	22%	27%	7%
Proportion of households with electricity	A10	94%	96%	82%	96%	82%
Proportion of households with safe drinking water	A11	85%	93%	78%	96%	97%
Proportion of households with improved tools for production	A12	0%	0%	0%	13%	16%
Proportion of households with radio, television or internet	A13	94%	99%	99%	100%	98%
Proportion of households with vehicles for transportation	A14	23%	24%	12%	16%	13%

Table 19.7 Households' vulnerability to climate change in the studied cells within the coastal zone of The Gambia

Components	CELL 1	CELL 4	CELL 5	CELL 6	CELL 8	Total AV.
Exposure	0.31	0.59	0.69	0.64	0.63	0.57
Sensitivity	0.48	0.29	0.55	0.58	0.43	0.47
Adaptive capacity	0.44	0.33	0.71	0.50	0.32	0.46
Vulnerability index	0.35	0.55	0.52	0.72	0.73	0.58
Vulnerability index ranking	**5**	**3**	**4**	**2**	**1**	

Source Fieldwork, 2021

their participatory adaptive capacity (Yimam & Holvoet, 2023). In addition, a high proportion of households that migrated to the coastal area was observed in Cell 5 (78%) and Cell 6 (88%) (Table 19.6). Declining growth in the agriculture sector and dependence on traditional farm tools (Loum and Fogarassy, 2015), coupled with low household income and large household size were among the reasons that enhanced the movement of individuals toward the more developed and industrialised coastal region in search of paid employment (Lucas, 2015; Swe et al., 2015). Migration has been also observed to reduce the capacity of households to adapt to climate hazards and thereby increasing their sensitivity levels (Singh Jatav, 2020).

In terms of adaptive capacity, cell 5 scored the highest adaptive capacity index of 0.71 and the lowest (0.32) recorded in cell 8 (Table 19.7 and Fig. 19.2). The majority of households within cell 5 had access to basic facilities such as electricity, safe drinking water and climate information through radio, television and internet,

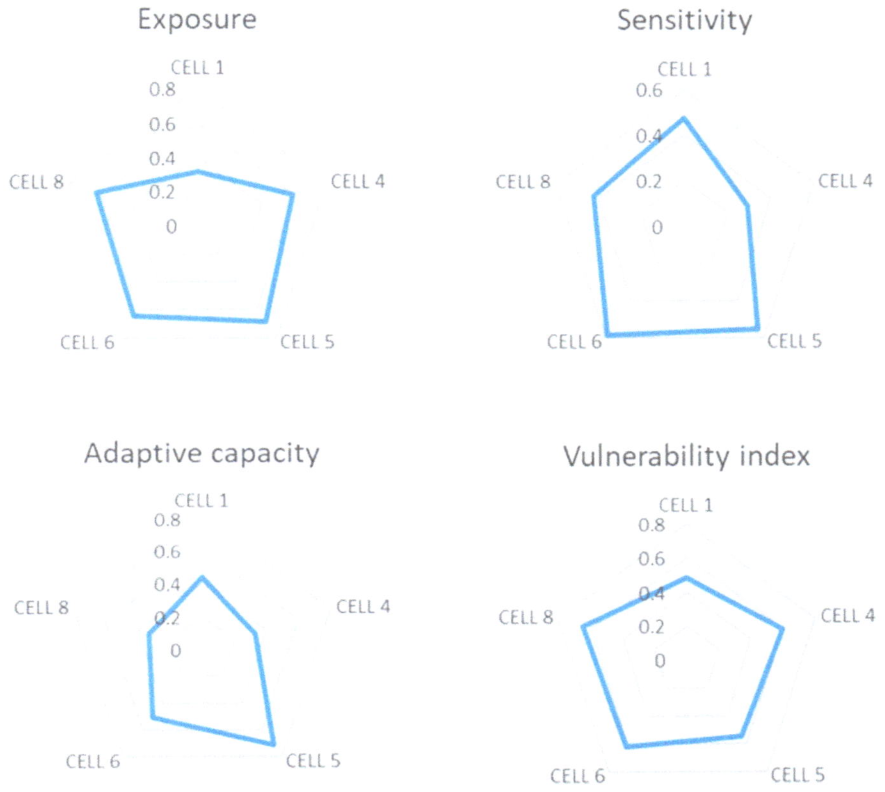

Fig. 19.2 Exposure, sensitivity and adaptive capacity within the coastal zone of The Gambia. *Source* Fieldwork, 2021

access to health care centres and literacy where most household members had at least a primary or secondary education (Table 19.6). This area was also observed to have the least proportion of households exposed to tropical windstorms (72%) and maximum temperature (80%). Typically, many low-income people were the most who built mud houses in The Gambia which is common among residential houses in cell 8 (Table 19.6). During the survey, these were mostly the type of structures that were observed to have collapsed in the country when hazards such as floods from heavy downpours and tropical windstorms occur. Therefore, advocating for climate-resilient and sustainable human settlements should be encouraged by relevant authorities in The Gambia to enhance the adaptive capacity of individuals within the coastal cells in addressing climate change (Urquhart, 2016). In addition, Nunes and Giglio (2022) highlighted that building designs that include natural ventilation, and thermal transmitters, especially for low-income households in tropical environments are crucial for reducing the impact of temperature on households.

Vulnerability Index Score of Households

The vulnerability index assessment conducted to compare the level of climate change vulnerability among the coastal cells is presented in Table 19.7 and Fig. 19.2. The result highlighted that the coastal zone on average is vulnerable to climate change with a vulnerability index score of 0.58. This corroborates with the findings of various researchers in The Gambia (Belford et al., 2020; Gomez et al., 2020; Komma, 2019). The result showed that households within Cell 8 and Cell 6 ranked highest in terms of their vulnerability index scores of 0.73 and 0.72 respectively, while cell 1 ranked lowest with a vulnerability index score of 0.35.

Cell 8 showed high exposure (0.63) with the lowest adaptive capacity index of 0.32 and thus a high vulnerability level of households. Cell 6 on the other hand showed sensitivity, exposure and adaptive capacity indexes of 0.64, 0.58 and 0.50 respectively and thus, a high vulnerability of households. Cell 1 scoring the lowest vulnerability index score of 0.35 also registered the lowest exposure index of 0.31. Furthermore, cell 5 scoring the highest exposure index of 0.69 and ranking 4th in terms of the vulnerability index of 0.52 has the highest adaptive capacity index of 0.71. This recognises that, while high levels of exposure increase the vulnerability level of households, increased adaptive capacity significantly minimises the effects of climate change and increases the potential of households to mitigate future climate hazards which is also corroborated by Choden et al. (2020).

Additionally, the findings of this study further reveal that each of the coastal cells of The Gambia has different socio-economic, natural hazards and demographic characteristics resulting in differences in sensitivity, exposure, adaptive capacity and level of vulnerability. The identification of the areas with the highest exposure (Cell 5), highest sensitivity (cell 6) and lowest adaptive capacity (cell 8) is essential in identifying the areas for intervention. Therefore, enhancing the socioeconomic, technical skills, living conditions and climate change understanding of individuals in these cells will reduce their vulnerability level and the climate change impact as highlighted by various researchers around the globe (Riede et al., 2016; Zhou et al., 2021).

Limitations of the Study

The tools for the household data collection were constructed in English Language and administered using local languages. Despite training the enumerators, there might still exist interpretation bias as some climate change terminologies and concepts are difficult to translate. The respondents were also asked to provide experience of past events which resulted in recall bias. However, the results were put into better perspective to give insights into the study.

Conclusion

As a region of socioeconomic importance, the coastal zone also serves important ecosystem services including habitats and breeding grounds for aquatic animals, plants and migratory birds. This study utilised an integrated assessment model to characterise the main factors that affect the vulnerability of coastal communities. The findings of the study highlighted that although coastal inhabitants believed and were aware that the climate had changed within the last thirty years, the majority of them lacked sufficient knowledge and understanding required to relate these changes to the various harmful activities, and environmental degradation observed within their environment. This limits their knowledge of the main causes of climate change to deforestation. The low-income levels, low literacy levels, high dependents and poor infrastructure, among others, have also resulted in numerous impacts of climate hazards given their limited adaptive capacity. The differences in socio-economic characteristics have resulted in different degrees of exposure, sensitivity, adaptive capacity and vulnerability of coastal communities to climate change effects.

Recommendation

This study recommends that;

- To adequately address the socioeconomic and environmental hazards affecting their livelihood, there is a need for community members to reorganise and re-structure to better serve the needs of its inhabitants and work together to merge different ideas on climate change adaptation and mitigation practices.
- The situation of vulnerability in terms of exposure and sensitivity to climate hazards within the coastal zone of The Gambia has called for an urgent need to mobilise appropriate levels of capacity, funding, diverse skills and knowledge systems in adaptation, mitigation and building resilience to climate hazards within communities, state and non-state organisations.
- There is a need for research on the development of environmental policies, centred on increased public and political awareness of climate change impacts, exposure and sensitivity and adaptive capacity of coastal communities and relevant stakeholders to improve the well-being of individuals and communities.
- Building low-cost housing which is climate resilient and sustainable should be encouraged by relevant authorities to minimise the impact of climate change, especially among poorer members of the community.

References

Amuzu, J., Jallow, B. P., Kabo-Bah, A. T., & Yaffa, S. (2018a). The climate change vulnerability and risk management matrix for the coastal zone of the Gambia. *Hydrology, 5*(1), 14.

Amuzu, J., Jallow, B. P., Kabo-Bah, A. T., & Yaffa, S. (2018b). *The socio-economic impact of climate change on the coastal zone of the Gambia.*

Badjie, M., Yaffa, S., Sawaneh, M., & Bah, A. (2019). Effects of climate variability on household food availability among rural farmers in Central River Region-South of The Gambia. *Climate Change, 5*(17), 1–9.

Belford, C., Delin, H., Ceesay, E., Ahmed, Y. N., Sanyang, L., & Jonga, R. H. (2020). The impact of climate change on economic growth based on time series evidence. *International Journal of Human Capital Urban Management, 5*(4), 305–318.

Bera, S., Guru, B., & Oommen, T. (2020). Indicator based approached for assigning physical vulnerability of the houses to landslide hazards in the Himalayan region of India. *International Journal of Disaster Risk Reduction, 50*, 101891.

Bojang, A., Oyedotun, T. D. T., Sawa, B. A., & Isma'il, M. (2023). Spatio-temporal coastline dynamics of the Gambia littoral zone from 1989 to 2019. *Geosystems and Geoenvironment, 2*(4), 100194.

Burden, A., Smeaton, C., Angus, S., Garbutt, A., Jones, L., Lewis, H., & Rees, S. (2020). Impacts of climate change on coastal habitats, relevant to the coastal and marine environment around the UK. *MCCIP Science Review 2020.*

Choden, K., Keenan, R. J., & Nitschke, C. R. (2020). An approach for assessing adaptive capacity to climate change in resource dependent communities in the Nikachu watershed, Bhutan. *Ecological Indicators, 114*, 106293.

Dale, P., Sporne, I., Knight, J., Sheaves, M., Eslami-Andergoli, L., & Dwyer, P. (2019). A conceptual model to improve links between science, policy and practice in coastal management. *Marine Policy, 103*, 42–49.

De Serio, F., Armenio, E., Mossa, M., & Petrillo, A. F. (2018). How to define priorities in coastal vulnerability assessment. *Geosciences, 8*(11), 415.

Dickerson, S., Cannon, M., & O'Neill, B. (2022). Climate change risks to human development in sub-Saharan Africa: A review of the literature. *Climate and Development, 14*(6), 571–589.

Fatemi, M. N., Okyere, S. A., Diko, S. K., Kita, M., Shimoda, M., & Matsubara, S. (2020). Physical vulnerability and local response to flood damage in peri-urban areas of Dhaka, Bangladesh. *Sustainability, 12*(10), 3957.

Ferraro, P. J., & Simorangkir, R. (2020). Conditional cash transfers to alleviate poverty also reduced deforestation in Indonesia. *Science Advances, 6*(24), eaaz1298.

Füssel, H. M. (2010). *Review and quantitative analysis of indices of climate change exposure, adaptive capacity, sensitivity, and impacts.*

GBOS. (2013). *The Gambia Population and Housing Census data.* The Gambia Bureau of Statistics 2013. Accessed March 10, 2021 from https://www.gbosdata.org/downloads/8-census-2013

GBOS. (2022). The Gambia Bureau of Statistics 2022, National Accounts Data; Kanifing institutional layout, P. Accessed February 13, 2024 https://www.gbosdata.org/data/13-national-accounts/177-fishing-and-aquaculture-current

Gomez, M. L. A., Adelegan, O. J., Ntajal, J., & Trawally, D. (2020). Vulnerability to coastal erosion in The Gambia: Empirical experience from Gunjur. *International Journal of Disaster Risk Reduction, 45*, 101439. Accessed February 2, 2024.

Herrmann, S. M., Brandt, M., Rasmussen, K., & Fensholt, R. (2020). Accelerating land cover change in West Africa over four decades as population pressure increased. *Communications Earth and Environment, 1*(1), 1–10.

Hossain, M. S., Gain, A. K., & Rogers, K. G. (2020). Sustainable coastal social-ecological systems: how do we define "coastal"? *International Journal of Sustainable Development and World Ecology, 27*(7), 577–582

IPCC. (2022). Climate change 2022: Impacts, adaptation and vulnerability. In H.- O. Pörtner, D. C. Roberts, M. Tignor, E. S. Poloczanska, K. Mintenbeck, A. Alegría, M. Craig, S. Langsdorf, S. Löschke, V. Möller, A. Okem, & B. Rama (Eds.), *Contribution of working group II to the sixth assessment report of the intergovernmental panel on climate change* (p. 3056). Cambridge University Press. https://doi.org/10.1017/9781009325844

Jallow, B. P., Toure, S., Barrow, M. M., & Mathieu, A. A. (1999). Coastal zone of the Gambia and the Abidjan region in Côte d\'Ivoire: Sea level rise vulnerability, response strategies, and adaptation options. *Climate Research, 12*(2–3), 129–136.

Kellstedt, P. M., Zahran, S., & Vedlitz, A. (2008). Personal efficacy, the information environment, and attitudes toward global warming and climate change in the United States. *Risk Analysis: An International Journal, 28*(1), 113–126.

Komma, L. (2019). Towards implementing integrated coastal zone management in The Gambia: Coastal adaptation to climate change and human impacts in the high-risk zone (cell 6). World Maritime University Dissertations, 119.

Lee, S., Paavola, J., & Dessai, S. (2022). Towards a deeper understanding of barriers to national climate change adaptation policy: A systematic review. *Climate Risk Management,* 100414.

Loum, A., & Foragassy, C. (2015). The effects of climate change on cereal yield of production and food security in Gambia. *Applied Studies in Agribusiness and Commerce, 9*(4), 83–92.

Lu, Y., Yuan, J., Lu, X., Su, C., Zhang, Y., Wang, C., Cao, X., Li, Q., Su, J., Ittekkot, V., Garbutt, R. A., & Sweijd, N. (2018). Major threats of pollution and climate change to global coastal ecosystems and enhanced management for sustainability. *Environmental Pollution, 239*, 670–680.

Lucas, R. E. (2015). Internal migration in developing economies: An overview. *KNOMAD's Working Paper 6*.

Mafi-Gholami, D., Jaafari, A., Zenner, E. K., Kamari, A. N., & Bui, D. T. (2020). Vulnerability of coastal communities to climate change: Thirty-year trend analysis and prospective prediction for the coastal regions of the Persian Gulf and Gulf of Oman. *Science of the Total Environment, 741*, 140305.

Masson-Delmotte, V., Zhai, P., Pirani, A., Connors, S. L., Péan, C., Berger, S., Caud, N., Chen, Y., Goldfarb, L., Gomis, M. I., Huang, M., & Zhou, B. (2021). Climate change 2021: The physical science basis. In *Contribution of working group I to the sixth assessment report of the intergovernmental panel on climate change* (p. 2).

Mesta, C., Cremen, G., & Galasso, C. (2022). Urban growth modelling and social vulnerability assessment for a hazardous Kathmandu Valley. *Scientific Reports, 12*(1), 6152.

Miyamoto, M. (2020). Poverty reduction saves forests sustainably: Lessons for deforestation policies. *World Development, 127*, 104746.

NEA. (2010). *State of the environment report The Gambia* (2nd ed.). National Environment Agency.

Nunes, G., & Giglio, T. (2022). Effects of climate change in the thermal and energy performance of low-income housing in Brazil—assessing design variable sensitivity over the 21st century. *Renewable and Sustainable Energy Reviews, 168*, 112885.

Ofori, S. A., Cobbina, S. J., & Obiri, S. (2021). Climate change, land, water, and food security: Perspectives from Sub-Saharan Africa. *Frontiers in Sustainable Food Systems, 5*, 680924.

Pörtner, H. O., Roberts, D. C., Adams, H., Adler, C., Aldunce, P., Ali, E., Begum, R. A., Betts, R., Kerr, R. B., Biesbroek, R., Birkmann, J., & Ibrahim, Z. Z. (2022). Climate change 2022: Impacts, adaptation and vulnerability (p. 3056). IPCC.

Riede, J. O., Posada, R., Fink, A. H., & Kaspar, F. (2016). What's on the 5th IPCC Report for West Africa? In *Adaptation to climate change and variability in rural West Africa* (pp. 7–23). Springer, Cham.

Saleem Khan, A., Sabuj Kumar, M., Sudhir Chella, R., & Devdyuti, B. (2020). Chennai city and coastal hazards: Addressing community-based adaptation through the lens of climate change and sea-level rise (CBACCS). *Climate Change, Hazards and Adaptation Options: Handling the Impacts of a Changing Climate,* 777–798.

Schipper, C. A., Dekker, G. G., de Visser, B., Bolman, B., & Lodder, Q. (2021). Characterisation of SDGs towards coastal management: Sustainability performance and cross-linking consequences. *Sustainability, 13*(3), 1560.

Singh Jatav, S. (2020). Bridging the gap between biophysical and social vulnerability in rural India: A community livelihood vulnerability approach. *Area Development and Policy, 5*(4), 390–411.

Swe, L. M. M., Shrestha, R. P., Ebbers, T., & Jourdain, D. (2015). Farmers' perception of and adaptation to climate-change impacts in the Dry Zone of Myanmar. *Climate and Development, 7*(5), 437–453.

UNDP. (n.d). The Gambia 2019–2022 report. Accessed February 13, 2024 https://www.undp.org/gambia/publications/undp-gambia-2019-2021-report

UNFCCC. (2022). *Integrated Assessment Models (IAMs) and Energy Environment Economy e3 Models*

Urquhart, P. (2016). National climate change policy of the Gambia. *Consultancy report for Ministry of Environment, Climate Change, Water, Forestry and Wildlife.*

Williams, D. S., Máñez Costa, M., Sutherland, C., Celliers, L., & Scheffran, J. (2019). Vulnerability of informal settlements in the context of rapid urbanisation and climate change. *Environment and Urbanisation, 31*(1), 157–176.

Yimam, D. A., & Holvoet, N. (2023). Assessing climate change vulnerability of smallholder farmers in northwest Ethiopia: Application of a household intrinsic vulnerability index. *International Journal of Climate Change Strategies and Management, 15*(4), 537–560.

Zerebecki, R. A., Heck, K. L., Jr., & Valentine, J. F. (2022). Biodiversity influences the effects of oil disturbance on coastal ecosystems. *Ecology and Evolution, 12*(1), e8532.

Zhou, Y., Chen, M., Tang, Z., & Mei, Z. (2021). Urbanisation, land use change, and carbon emissions: Quantitative assessments for city-level carbon emissions in Beijing-Tianjin-Hebei region. *Sustainable Cities and Society, 66*, 102701.

Chapter 20
Enhancing the Farm Household Resilience to Climate Change Through the Intwasa/Pfumbudza Programme. Insights from Umguza District, Zimbabwe

Douglas Nyathi, Mbongeni Nhliziyo, Joram Ndlovu, and Manase K. Chiweshe

Abstract Climate change and variability are global issues that significantly impact rural populations, particularly smallholder farmers, who rely on rain-fed agriculture. In response, the Government of Zimbabwe launched the *Intwasa/Pfumvudza* programme, a nationwide programme aimed at enhancing the climate resilience of smallholder farmers. This initiative was based on the concept of conservation agriculture, which intensifies crop production by ensuring efficient resource use on small plots of land. This chapter evaluates the viability of the programme in building resilience for smallholder farmers in Umguza District, Zimbabwe, and assesses the extent to which the programme strengthens farmers' capacity to withstand climate impact. Using a qualitative case study approach, we conducted 40 in-depth interviews, six key informant interviews, and document analysis to collect the data. The findings indicate that the *Intwasa/Pfumvudza* initiative significantly enhances farmers' resilience to climate change in Umguza District. Beneficiaries of the initiative reported increased resilience against drought impacts and improved harvest yields, reversing the previous trends of declining agricultural productivity. However, challenges such as labour shortages, lack of mulch supplies, and unclear government agrarian policies were identified as obstacles to the programme's success. This study

D. Nyathi (✉)
School of Public Management, Governance and Public Policy, University of Johannesburg, Johannesburg, South Africa
e-mail: dnyathi@uj.ac.za

M. Nhliziyo
Department of Development Studies, Lupane State University, Lupane, Zimbabwe

J. Ndlovu
School of Social Sciences, University of KwaZulu Natal, Howard College, Durban, Republic of South Africa
e-mail: Ndlovuj1@ukzn.ac.za

M. K. Chiweshe
Department of Social and Community Development, University of Zimbabwe, Harare, Zimbabwe

W. Leal Filho et al. (eds.), *Climate Change Mitigation and Adaptation in Practice*, Climate Change Management, https://doi.org/10.1007/978-3-031-85217-6_20

underscores the importance of a supportive policy framework and active participation from the government, business community, and civil society to ensure successful implementation and accountability in resource management. It also highlights the need for farmers to fully embrace the *Intwasa/Pfumvudza* initiative to enhance their resilience to climate change, improve harvests, and achieve food security. These findings have important implications for the design and execution of climate-resilience programmes in similar contexts.

Keywords Climate change · Intwasa/Pfumvudza · Vulnerability · Resilience · Smallholder farmers · Umguza

Introduction

The introduction of climate-smart agricultural technologies presents an opportunity to achieve the goals outlined in the 2015 Paris Agreement, which aims to build resilience against climate change, limit global temperature rise to below 2 °C, boost agricultural productivity, and enhance progress towards national food security objectives (UN, 2016). Additionally, these technologies can significantly contribute to the realisation of Africa's Agenda 2063, which serves as a model for the UN's Agenda 2030, particularly in addressing climate change threats to development in the Global South, including Zimbabwe (UN, 2016). The threat of climate change to agriculture, compromising livelihoods and food security in climate-sensitive areas, is no longer an emerging issue, but a confirmed reality (IPCC, 2014, 2019). If left unaddressed, smallholder farmers may face significant challenges owing to the detrimental impacts of climate change on agriculture, exacerbating poverty, inequalities, food insecurity, and health complications, potentially reversing development progress many countries make (FANRPAN, 2017). Tackling agriculture within food security and sustainable development necessitates incorporating climate change considerations into agricultural operations, implementing climate-resilient practices, adopting sustainable farming methods that mitigate environmental impact, and fostering social inclusivity to ensure long-term economic sustainability and societal well-being.

In pursuit of climate-smart agriculture objectives, the Government of Zimbabwe (GoZ) has identified the *Intwasa/Pfumvudza* programme as an emerging strategy to effectively address the intertwined challenges of achieving food security amid severely constrained natural resources (GoZ, 2020). The programme is based on vital agricultural principles for conservation developed by the Foundations for Farming (Oldreive, 2006; FAO, 2020) and focuses on intensifying crop production on a small plot of land to concentrate resources and achieve higher yields with low investment, resulting in high profitability. The programme's input pack allows smallholder farmers to feed their families for up to one year with minimal investment. Climate-smart agriculture aims to adapt and protect the agricultural sector against climate change, reconfiguring the risks associated with food insecurity by aligning agricultural practices with new realities (Mujere, 2021).

The *Intwasa/Pfumvudza* initiative, led by the government, combines financial, technological, and policy alternatives to enhance resilience and mitigate the impact of climate change on agricultural development. By explicitly incorporating resilience and mitigation measures into planning and implementation processes, this approach seeks to address the challenges faced by the agricultural sector due to climate change (GoZ, 2020). Despite the challenges posed by rising food costs and climate change, integrating climate change adaptation and mitigation techniques into agricultural development planning and implementation strives to achieve broader developmental goals. Resilience is crucial in addressing extreme weather events and is essential for comprehensive climate action programmes, as climate change is both a global and hyper-local issue (CIAT & World Bank, 2017). According to C2ES (2019), resilience is the capacity of a system and its components to anticipate, mitigate, adapt to, or recover quickly and effectively from the impacts of hazardous events. Climate resilience involves anticipating, recovering from, and adapting to climate-induced impacts to manage change, minimise disruptions, and capitalise on opportunities (C2ES, 2019; IPCC, 2014).

Zimbabwe's agricultural sector, which is heavily reliant on rain-fed agriculture, is already suffering from the effects of climate change, including temperature increases, decreased rainfall, and periodic droughts (CIAT & World Bank, 2017; Rusinamhodzi, 2015). The country's low average annual rainfall and many rivers in drier regions further exacerbate the challenges faced by the agricultural sector (CIAT & World Bank, 2017). Perennial droughts affect the water supply, agriculture, food access, and essential health- and agriculture-based livelihoods (Phiri et al., 2021; Unganai & Murwira, 2010). Climate change impacts have significantly affected agriculture in Zimbabwe, leading to various challenges, including inadequate financing, high operating costs, pricing distortions, difficulties in managing variable inputs, insufficient investment, inadequate irrigation infrastructure, and significant postharvest losses. These challenges highlight the urgent need for sustainable approaches to address the vulnerabilities of Zimbabwe's agricultural sector, such as incorporating climate resilience and mitigation measures into agricultural development planning and execution (GoZ, 2020).

Background

Climate Change Adaptation in Sub-Saharan Africa

Like other global regions, development in sub-Saharan Africa (SSA) has occurred in response to climate change. The region is expected to experience further climate change, with increased temperatures likely to affect the temporal and spatial distribution of rainfall. Therefore, careful planning and adaptation to flood and drought severity are imperative (England et al., 2018; IPCC, 2019). Climate change effects occur concurrently with sociocultural, geographical, governance, biophysical, institutional, economic, ecological, and technological processes that interconnect in complex ways, potentially decreasing farmers' ability to withstand climate change. Smallholder farmers, particularly female farmers with limited assets and adaptive capacities, are more vulnerable to climate change (Nyasimi et al., 2014; Nyathi et al., 2024). Like the rest of Africa, SSA is impacted by changes in health, the environment, water, agriculture, and other relevant areas as well as the ecological system (Kassie et al., 2015; Alemaw and Matondo, 2020).

Changes in rainfall frequency and intensity, water bodies drying up, increased cyclones, and unprecedented dry periods are signs of how climate change affects the water resources of SSA. These climatic changes have broader impacts on national development, agriculture, energy, food production, ecosystems, and other social and economic factors of vulnerable communities in different parts of the region (Alemaw and Matondo, 2020). Owing to the higher pace of global climate change, the SSA region is expected to become drier in future decades (Gbegbelegbe et al., 2018; IPCC, 2019). Socioeconomically, the region is one of the world's poorest and most vulnerable regions. As the land and water resources of Southern Africa are already under stress from climate change, developing strategies to reduce vulnerability to current and future weather-related events and environmental concerns is critical.

Adaptation involves altering natural or human systems in response to current or projected climate changes or impacts to minimise harm or take advantage of favourable circumstances. Several adaptation strategies can be identified, including anticipatory, planned, and autonomous. Anticipatory or proactive adaptation occurs before the evident effects of climate change are observed (Dube et al., 2018). Autonomous adaptation, often called spontaneous adaptation, occurs naturally without a deliberate response to climate change, and is triggered by changes in the environment in natural systems or changes in the market and welfare in human systems. Planned adaptation involves a conscious policy choice that necessitates action to maintain, achieve, or return to the intended condition, considering evolving circumstances (IPCC, 2007). According to the UNDP (2018), adaptation refers to making changes in response to actual or anticipated climate conditions and their impacts. In human societies, adaptation aims to mitigate or prevent harm caused by climate change or to take advantage of potential opportunities. In certain natural ecosystems, human intervention may be necessary to facilitate adjustments to expected changes in climate and its effects on these systems.

Climate-resilient strategies are expected to support the sustainability of planning for the effects of climate change on livelihoods and standards of living (ACT, 2018). Adaptation strategies encompass various measures, such as developing the response of the built environment to rising sea levels, expanding planning and land use to integrate climate warning systems, preserving vital ecosystems and biogeochemical zones at risk from climate change, establishing landscape corridors to facilitate migration, minimizing groundwater withdrawal from shallow aquifers, controlling water use, fortifying dikes, and strengthening seawalls. These strategies aim to mitigate the adverse effects of climate change on ecosystems and populations by increasing resilience (Zolnikov, 2019). The availability, usability, and accessibility of water are vital to millions of people in sub-Saharan Africa (SSA), and these factors are directly influenced by the region's vulnerability to climate change. Future initiatives must prioritise livelihood-centred integrated resource and water management while addressing climate change responses (Alemaw, 2020: 71). This strategy involves better risk management through crop insurance and early warning systems, wastewater reuse for agriculture, improved land use and natural resource management, and increased utilisation of sustainable energy sources (Matondo et al., 2020).

Ndlovu et al. (2020) have highlighted the benefits of field rainwater harvesting systems, including reducing surface runoff, promoting water conservation, improving soil fertility, and increasing agricultural yields. Additionally, communities in Zimbabwe adapt to climate change through water-harvesting technologies, such as contour ridges, runoff rainfall collection, zero tillage farming, irrigation, stream bank crop cultivation, and animal transhumance (Dube et al., 2018). They also used heat- and drought-tolerant crops. A study conducted in multiple countries (Malaysia, Mozambique, South Africa, Zambia, and Zimbabwe) by Vincent et al. (2013) demonstrated the application of multiple strategies for mitigating climate variability and changes in the agricultural sector. The crisis response strategies included diversification, modification of farming practices, modification of crop types and varieties, and resource management (Nyathi, 2024).

Unpacking the Intwasa/Pfumvudza Concept

The *Intwasa/Pfumvudza* programme is research-based and has the qualities of Climate Smart Agriculture (CSA). *Intwasa/Pfumvudza*, originating from the indigenous languages of Zimbabwe (isiNdebele/Shona), translates to the burgeoning of fresh leaves in spring, symbolising the beginning of a new agricultural season. According to a scenario presented by the Foundation of Farming, a local non-governmental organisation in Zimbabwe, a simple plot of land could supply an average household of six people with all the grain they would need for a year. Mujere (2021) claims that *Intwasa/Pfumvudza* denotes a season of increasing production, in which more is produced with fewer resources and on less land, a time when the implementation of conservation agriculture methods is the focus of agriculture ready for climate change.

FAO (2020) argues that this model's uniqueness lies in the plot size, which measures just 16 m × 39 m. This small plot size makes it easy to prepare and maintain with mulch, weeds, and hand-applied rainwater harvesting in the case of a drought or mid-season dry spell. The compact size enables efficient resource management and utilisation, contributing to the overall success and sustainability of the agricultural model. The model has effectively assisted farmers in harvesting more grains, including millet, sorghum, and maize, while encouraging the alternation of legumes, such as cowpeas, groundnuts, and beans. The government of Zimbabwe chose the *Intwasa/Pfumvudza* approach to address problems including low farming productivity and profitability that hurt the nation's food security. These challenges are attributed to factors such as poor agronomic practices, inadequate rainfall, prolonged mid-season dry spells influenced by climate change, warmth during the growing season, and an early end to the rainy season. The failure of agricultural extension workers and smallholder farmers to view agriculture as a business enterprise contributes to these challenges. By embracing the *Intwasa/Pfumvudza* concept grounded in conservation agriculture principles, agricultural production can be climate-proofed and improved (Mujere, 2021). Kassim and Nkomoki (2023) summarise the purposes of *Intwasa/Pfumvudza*, arguing that it is a smallholder farmers' programme that promotes planting basins, often incorporated with improved seed varieties and mulching. This initiative encourages novel sustainable agriculture, crop diversification, and conservation practices, and its goal is to enhance food security and agricultural productivity.

What is Climate Change Resilience?

Broadly, resilience is the capacity to overcome adversity. Collaboration among individuals, corporations, communities, and industries is necessary to manage climate change effects effectively. Furthermore, transformability, the capacity of a social-ecological system to follow a desired path to change from one state to another, can be understood as a subset of resilience (Folke et al., 2010). Likewise, resilience is an ongoing approach to navigating environmental changes using adaptable, creative, and inclusive methods (Wilbanks et al., 2014). Resilience emphasises the critical connection and overlap among survival, adaptation, and transformation as essential elements in preserving the positive traits of a system (Berbés-Blázquez et al., 2017). The principles that contribute to building resilience include increasing diversity, creating redundancy, improving connectivity, controlling feedback, handling slow variables, considering socio-ecological interactions, raising engagement, offering opportunities for experimentation and learning, and promoting polycentric governance. These ideas are essential for building systems that are more resistant to problems, such as climate change. For example, the United Nations Agenda 2030, with its 17 Sustainable Development Goals and 169 targets, offers a framework for tackling global issues and fostering sustainable development. The potential, linkages, and synergies between the economic, social, and environmental aspects of development and the SDGs are spelt out in the agenda. Reducing inequality and increasing climate change resilience are two important goals for some of these interconnections and interactions. Africa's Agenda 2063 is a unique opportunity to remove structural inequalities and address poverty; building climate change involves social exclusion and marginalisation. It seeks to enhance resilience for sustainable development by lowering vulnerability to natural disasters.

Climate Change in Matabeleland North Province

Notable climate disasters are the main indicators of climate change in Zimbabwe and have resulted in rising temperatures and decreasing rainfall patterns. The nation is vulnerable to extreme weather events, such as heat waves, flash floods, heavy rains, strong winds, and hailstorms (UNDP, 2017). Climate-related obstacles underscore the need to implement tactics such as *Intwasa/Pfumvudza* to augment mitigation and climate resilience in the agricultural sector, to effectively navigate the consequences of climate change. The Tugwi-Murkosi floods of 2013–2014, the recurring Tsholotsho floods, the ongoing drought, and the most recent Cyclone Idai in 2019 had multiple destructive impacts. They forced population displacement in the affected communities. The climate variability in Matabeleland North Province reflects Zimbabwe because of its location in Agroecological zones IV and V, experiencing extreme weather conditions. The province, which is vulnerable to the consequences of global warming because it is in the country's semi-arid regions, relies mostly on rain-fed agriculture for subsistence of its rural population. Furthermore, a decrease in natural grass during the dry season has been observed in significant areas that produce beef because of the region's recurrent and severe droughts over the previous ten years (Ngara, 2017). Droughts have a detrimental effect on livestock, which are an essential asset in rural and farming communities, as well as low harvests and shortages in food supply (Nyathi, 2024; UNDP, 2017).

Furthermore, the region is characterised by unpredictable and poorly distributed rains, barely exceeding 450 mm per year, coupled with temperatures above 39 °C. The sustainability of crops faces challenges under such conditions, particularly in areas where rainfall is sparsely distributed (Nyathi et al., 2024). However, drought-tolerant plants including cowpeas, millet, and sorghum can withstand severe weather conditions. In regions with little rainfall throughout the farming season, cropping activities and livestock production are significantly affected (Rusinamhodzi, 2015). Owing to the prolonged dry season, smallholder farmers receive poor harvests, and natural grass and browse resources for their livestock are undesirable, as noted above. As a result of climate change, farmers have lost livestock, particularly cattle (Chingarande, 2020).

Study Setting, Methods and Materials

The Umguza district (see Fig. 20.1) is located in the Matabeleland North Province of Zimbabwe. The district is bounded by Matobo and Umzingwane districts in the southeast, Bulilima and Tsholotsho in the west, Kusile (formerly part of the Lupane District) in the north, and Bubi in the north and east. It surrounds the Bulawayo metropolitan area to the south on three sides (Johannes et al., 2017). Like other Districts in the province, Umguza is drought-stricken and receives less than 500 mm of rainfall each season. Farmers in this area always suffer from frequent droughts, water shortages, and perennial household food insecurity (Phiri et al., 2021). According to Zimstat (2022), the district has 28,358 households and a population of 113,265. This study was conducted in four purposively selected wards. The research was conducted in wards 8, 15, 16, and 19. The four wards were selected because of the number of farmers enrolled in the *Intwasa/Pfumvudza* initiative. The wards have documented challenges of poverty and food insecurity (Johannes et al., 2017; Phiri et al., 2021).

This study employed a triangulated approach to data collection, utilising in-depth interviews, key informant interviews, and document analysis. The sampling strategy involved purposive selection of 40 participants for in-depth interviews, six key informants, and relevant documents for analysis. Document analysis was conducted by reviewing Rural District Council food security reports, reports from non-governmental organisations operating in the district, and peer-reviewed journal articles on the subject matter. Key informants included extension service officers, representatives of local authorities, non-governmental field staff, and traditional leaders. The informants were selected based on their extensive knowledge and experience related to the focus of the study (Denzin and Lincoln, 2000).

Informed consent was obtained from all participants prior to data collection. This was achieved by distributing a consent form that clearly outlined the study's purpose

Fig. 20.1 Showing the Umguza District. *Source* Authors

and the participants' rights, including confidentiality and the option to withdraw from the study at any time without consequences. The collected data were analysed using a thematic content approach widely adopted by qualitative researchers to explore participants' experiences and perspectives on a particular subject (Bernard, 1995). This approach allows for the identification and analysis of recurring patterns and themes within the data, providing valuable insights into the phenomenon under investigation (Schensul & LeCompte, 1999; Pope and Mays, 2000). Several measures were implemented to ensure the trustworthiness and credibility of our findings. These included member checking, where participants were allowed to review and validate the researchers' interpretations of their responses, and triangulation of data sources to corroborate the findings from multiple perspectives (Denzin and Lincoln, 2000). The research team maintained a reflexive stance throughout the study, acknowledging and mitigating potential biases and preconceptions that could influence the data collection and analysis processes (Bernard, 1995; Denzin and Lincoln, 2000). This approach ensured that the findings accurately represented participants' perspectives and experiences (Schensul & LeCompte, 1999).

Results and Discussion

Contribution of Intwasa/Pfumvudza Towards Building Smallholder Farmers' Resilience to Climate Change

The *Intwasa/Pfumvudza* concept was designed for drought-proof agriculture and promotes the adoption of research to enhance the productivity of seed varieties. The Government of Zimbabwe (GoZ) approved the program to address issues of low production capacity and low profit margins in smallholder farming communities (GoZ, 2020). One key informant highlighted that *Intwasa/Pfumvudza* is grounded in the principles of conservation agriculture. Applying this concept aims to tackle additional production challenges such as mitigating soil erosion in arable lands and helping farmers boost yields by intensifying production on small plots. Initially, the programme required each farmer to establish three plots: two dedicated to ensuring food self-sufficiency and one for growing soya beans in high-potential areas or sunflower seeds in low-potential areas aimed at generating income. Additionally, grain from the second plot was designated for sale to the Grain Marketing Board (GMB). For the 2021/22 season, the requirement was expanded to five plots per farmer. This programme, rooted in conservation agriculture principles, aims to increase agricultural productivity and resilience, particularly against drought. Farmers are expected to overcome production challenges such as soil erosion and achieve higher yields from small plots by promoting improved seed varieties and sustainable farming practices.

Table 20.1 Yields for *Intwasa/Pfumvudza* beneficiaries and traditional farming households

Farming season	Farming method	Maize per hectare
2019/2020	Traditional farming	1.15t
2020/2021	*Intwasa/Pfumvudza*	5.3 t
	Traditional farming	1.17 t
2021/2022	*Intwasa/Pfumvudza*	1.49 t
	Traditional farming	0.74 t

Source AGRITEX file report (2022)

Despite its goals, weak institutions like the GMB and issues with the partisan supply of farming inputs hinder the program's success. Some smallholder farmers expressed concerns about the reliability of the GMB, as one farmer noted:

> We were told that some of our products would be ferried to GMB. I worry that GMB has been unreliable regarding payments for the past years. They take time to pay for the already delivered maize, and considering our inflationary rates, we end up receiving money less than its original value, unable to plan and prepare for the next season.

However, the benefits of the programme were evident during the 2020/21 farming season in Umguza District, which recorded a bumper harvest due to the adoption of the model and favourable rains. Table 20.1 illustrates the programme yields for the years 2019 and 2022. In the 2019/2020 season, traditional farming yielded an average of 1.15 tonnes of maize per hectare. After introducing *Intwasa/Pfumvudza*, maize production significantly increased to 5.3 tonnes per hectare in the 2020/2021 season. Although the 2021/2022 season faced erratic rainfall and heat waves, *Intwasa/Pfumvudza* farmers still achieved 1.49 tonnes per hectare, compared to traditional farming's 0.74 tonnes per hectare.

Adopting the *Intwasa/Pfumvudza* approach has led to higher maize yields per hectare, even under adverse weather conditions, enhancing food security and increasing farmer incomes. This improvement underscores the importance of the programme in building resilience among smallholder farmers, enabling them to cope better with climate variability and secure their livelihoods.

Table 20.2 compares maize yields between the 2019/20 and 2020/21 agricultural seasons across various districts of Zimbabwe. The data revealed a notable upward trend in maize yield following the implementation of the *Intwasa/Pfumvudza* program. Districts such as Binga, Bubi, Hwange, and Nkayi experienced significant yield increases ranging from 200 to 400%. This substantial improvement can be attributed to the programme's emphasis on conservation agriculture principles and the adoption of improved seed varieties, which have effectively enhanced agricultural productivity in these regions. Notably, the Umguza district stands out, with a remarkable 574% increase in maize yield. This exceptional performance can be directly linked to the successful implementation of the programme. The programme's core tenets, which include optimising small plot sizes and maximising productivity through efficient resource management, have significantly contributed to this substantial yield increase in Umguza.

Table 20.2 Maize yields

District	Maize yields (t/plot)		
	2020/21	2019/20	Percentage
Binga	0.76	0.23	230.96
Bubi	0.99	0.24	313.92
Hwange	0.49	0.1	387.99
Lupane	0.60	0.39	54.02
Nkayi	0.56	0.18	210.63
Tsholotsho	0.65	0.27	139.80
Umguza	1.35	0.2	574.22

Source AGRITEX file report (2022)

The data presented in Table 20.2 provide compelling evidence of the *Intwasa/Pfumvudza* programme's success in boosting maize yields across multiple districts. These substantial yield increases indicate the program's effectiveness in improving agricultural outcomes and underscore its potential to address the longstanding challenges of low production capacity and profitability faced by smallholder farming communities. The programme's positive results suggest that its principles and practices are well suited to the local climatic conditions and resource constraints faced by smallholder farmers in Zimbabwe. By promoting sustainable agricultural practices, such as conservation agriculture, and facilitating the adoption of improved seed varieties, the program has effectively mitigated the adverse effects of climate change and other environmental stressors on crop yields.

Furthermore, the programme's emphasis on efficient resource utilisation and optimising small plot sizes aligns with the needs and capabilities of smallholder farmers, who often operate on limited land resources. By maximising productivity on these small plots, the *Intwasa/Pfumvudza* program has enabled farmers to achieve substantial yield increases while minimising resource inputs. Given the programme's remarkable success in boosting maize yields, it is imperative to consider its broader implementation and expansion to other regions of Zimbabwe. Replicating the programme's principles and practices in areas with similar climatic and socioeconomic conditions could yield comparable positive outcomes, contributing to improved agricultural productivity and enhanced food security at the national level.

Table 20.3 presents the commodity prices offered to smallholder farmers by the Government of Zimbabwe (GoZ) through the Grain Marketing Board (GMB). In line with the *Intwasa/Pfumvudza* concept, a farmer is expected to earn USD378 per tonne of maize, USD1400 per tonne of beans, and USD470 per tonne of sunflower at the end of the farming season. This pricing structure aims to provide smallholder farmers with stable and fair income, thereby incentivising their participation in the program and ensuring its long-term sustainability. According to a consulted informant, the 2020/21 farming season was characterised by above-average rainfall and the absence of prolonged dry spells across the wards. As a result, smallholder farmers participating in the programme achieved good yields, translating into significant income gains.

Table 20.3 Producer prices

Commodity	Producer price (mt)
Maize	USD$378
Sugar beans	USD$1400
Groundnuts-unshelled	USD$705
Sunflower	$USD470

Source GMB (2022) (field research)

This observation aligns with the findings of Tsiko (2021) and Simango (2021), who assert that farmers involved in the Intwasa government initiative have experienced substantial productivity improvements compared to previous years. Simango (2021) further suggests that the programme holds promise for enhancing farmers' incomes owing to its emphasis on sustainable agricultural practices and market linkages.

At the agroecosystem level, the programme promotes sustainable farming practices, such as mulching, composting, and conservation tillage. These practices contribute to soil health and fertility and enhance response diversity, and redundancy within cropping systems. Increased biodiversity through intercropping and agroforestry components helps to buffer against climatic variability by providing alternative response pathways (Lin et al., 2011; Mavesere & Dzawanda, 2023). This diversification of agricultural systems reduces the risk of complete crop failure and ensures a more stable and resilient food-production system. The programme's focus on improving market access and offering competitive commodity prices incentivises smallholder farmers to adopt sustainable agricultural practices and participate in the initiative. By guaranteeing a fair return on investments, the programme addresses the economic barriers that often hinder the adoption of sustainable farming practices among smallholder farmers.

The positive outcomes observed during the 2020/21 farming season, as reported by the consulted informant and supported by Tsiko (2021) and Simango (2021), highlight the programme's potential to enhance agricultural productivity and improve the livelihoods of smallholder farmers. The combination of favourable climatic conditions, sustainable farming practices, and access to stable markets has contributed to the program's success in boosting yields and increasing farmers' income. However, it is crucial to note that the program's long-term sustainability and scalability hinge on continued government support, effective extension services, and smallholder farmers' capacity to adapt to changing climatic conditions. Ongoing monitoring and evaluation of the program's impact and incorporation of lessons learned are essential to ensure its effectiveness in addressing the challenges faced by smallholder farming communities in Zimbabwe.

Challenges Presented by the Intwasa/Pfumvudza Programme

Labour Constraints

The *Intwasa/Pfumvudza* initiative involves digging, planting holes, and maintaining permanent soil cover using an organic mulch. This method is labour-intensive and primarily relies on hand-held hoes, posing significant challenges for smallholder farmers, particularly women, female-headed households, and community segments led by children or the elderly. One of the main obstacles preventing smallholder farmers from adapting to climate change is the unequal access to profitable and productive resources. This inequality hinders their capacity to adjust successfully to the effects of climate change (Antwi-Agyei, 2021). For instance, the labour-intensive nature of *Intwasa*, which requires digging over a thousand planting stations in each plot, presents a significant challenge. Farmers needing help digging the necessary planting stations often miss the presidential input scheme. Larger households are more likely to adapt to climate change because to the labour demands of many agricultural adaptation options.

Similarly, younger farmers are more likely to adapt to climate change because they can easily embrace new methods and access contemporary information and technology. Research conducted in 2017 by Jiri, Mafongoya, and Chivenge supports this observation. An AGRITEX officer confirmed this trend,

> The rectangular Intwasa plot, 16 metres wide by 39 metres long, requires the farmer to dig 1,456 planting stations. Each station is 15 cm deep, 15 cm wide, and 15 cm long. Three maize seeds are evenly distributed in each station, ensuring two maize plants remain after thinning at germination. The planting stations are spaced 75 cm between rows and 60 cm within rows across the slope.

A study by Tanyanyiwa et al. (2022) in Muzarabani revealed that despite the popularity of the *Intwasa/Pfumvudza* initiative among smallholders, participants found it comprehensible, yet physically taxing. Kassim and Nkomoki (2023) support this finding, arguing that although the *Intwasa/Pfumvudza* initiative promotes sustainable farming, female-headed households experienced labour shortages, limitations in access to inputs and technical know-how, and societal and cultural norms that compromised their involvement in the programme. The physical demands of the *Intwasa/Pfumvudza* method, coupled with resource inequalities, underscore the need for more inclusive and accessible agricultural strategies. Addressing these challenges is crucial for ensuring that all smallholder farmers, regardless of their household composition or demographic characteristics, can benefit from climate-smart agricultural practices and improve their resilience to climate change.

Financing

Financing agriculture in Zimbabwe remains a significant challenge for smallholder farmers, directly impacting their ability to adopt and benefit from programs, such as *Intwasa/Pfumvudza*. While the programme shows promise in climate-proofing agriculture and enhancing food security, it also faces substantial financial constraints. Access to finance and the ability to own livestock have been shown to significantly improve farmers' adaptability to climate change (Jiri et al., 2017). To address climate change challenges effectively, it is crucial to provide smallholder farmers with access to innovative financial adaptation mechanisms.

The goal of adopting the *Intwasa/Pfumvudza* farming model in Zimbabwe is to commercialise agriculture and bolster food security among smallholder farmers. Transitioning from these farmers to Climate-Smart Agriculture (CSA) involves embracing new farming techniques, relying on new knowledge and advisory support systems, and engaging with new markets. The government provides inputs to smallholder farmers who participate in the programme through the Ministries of Agriculture and Rural Development and Economic Development and Finance. This assistance is vital for the successful implementation of the model and promotion of sustainable farming practices (Roop & St. Martin, 2021). However, during the 2020/21 farming season, some districts experienced issues regarding the distribution of farming inputs. Specifically, the absence of nitrogenous fertilisers led to problems such as waterlogging and leaching, which manifested as the yellowing of crops and stunted growth during wet spells in December and January. This highlights a significant shortcoming of GoZ's *Intwasa/Pfumvudza* policy, particularly concerning financing and resource allocation.

In the context of climate change, financial and technological limitations often restrict sustainable management of soil and water resources. These constraints can impede the implementation of effective strategies for climate change resilience and mitigation in agricultural planning. Overcoming these challenges requires innovative solutions and increased investments in the agricultural sector. Although wealthier agricultural households might not be as motivated to change due to already optimised practices (Jiri et al., 2017), future interventions must involve private sector participation. The private sector has the resources and expertise to finance smallholder agriculture and enhance resilience and productivity. Mavesere and Dzawanda (2023) recognise the potential of the *Intwasa* farming initiative, but note several challenges, including issues with input access, limited post-harvest management skills, and delays in payments from the Grain Marketing Board (GMB). According to these scholars, substantial efforts are needed across social and ecological dimensions to enhance the initiative's impact on household farming productivity and resilience to climate variability and change.'

Opportunities Presented by the Intwasa/Pfumvudza Programme

Household Food Security

Intwasa/Pfumvudza is primarily focused on ensuring household food security, which can be considered both a strength and weakness. Although its emphasis on household food security is commendable, the model may not directly contribute to national export efforts. However, by ensuring that smallholder farmers achieve food self-sufficiency, the programme indirectly supports commercial farmers by directing their efforts towards export-oriented agriculture. The interviewed smallholder farmers attested that their household food security had significantly improved because of the programme. The findings indicate that successful institutional and policy interventions should focus on alleviating the resource limitations that hinder farmers' ability to embrace climate-smart agricultural practices. These initiatives should be contextually specific and mindful of gender dynamics (Makate et al., 2018). One key informant argued that, under the Intwasa plot model, which involves 52 planting rows with 28 planting stations, every maize plant produces a yield (amounting to 56 maize plants per row). This resulted in approximately 20 kg of shelled maize grain being harvested from each row. Once milled, this quantity was sufficient to provide meals for a family of six people for one week. Consequently, the total yield from all 52 rows, equating to 52 buckets of grain, would be sufficient to feed a six-member family for the entire year.

Enhancing Smallholder Farmers' Income

Improving smallholder farmers' income ensures access to inputs, farm services, and mechanisation. The *Intwasa/Pfumvudza* programme aims to commercialise smallholder farming, as highlighted by a consulted key informant. The objective of this initiative is to enhance the productivity levels of smallholder farmers, which will consequently lead to higher incomes and improved food security. Achieving these goals will make significant strides in alleviating poverty and bolstering the livelihoods of small-scale agricultural communities. The engagement revealed that the government program has significantly stimulated the commercialisation of smallholder agriculture, which is critical in increasing its resilience to shocks and stressors. The study results substantiate Timmer's (1997) assertion that the commercialisation of smallholder agriculture plays a crucial role in promoting economic growth, especially in underdeveloped nations, where the agricultural industry is their main source of income.

Our findings are corroborated by Mavesere and Dzawanda (2023), who argue that the Ggovernment's *Pfumvudza* programme has played an important role in improving household well-being among smallholders in the Mutare district. These two scholars indicate that farmers enrolled in the programme for the 2020/2021 rain season reported improvements in yields and household food security compared to previous years. This resilience allows communities and ecosystems to adapt, recover, and maintain their integrity in the face of changing environmental conditions, helping them thrive despite the adversities brought about by climate change (Engle, 2011; IPCC, 2007). Families can mitigate the risks associated with climate shocks by cultivating various crops and engaging in multiple income-generating activities. This strategy spreads risk across different sectors, ensuring a more stable and resilient livelihood, even when faced with unpredictable environmental disruptions (Nyathi et al., 2024).

The Role of Government in Promoting Intwasa/Pfumvudza

Within this framework, government intervention should focus on mitigating obstacles and capitalising on the programme's potential benefits. Providing smallholder farmers access to essential resources, including knowledge, agricultural supplies, and genetic resources, increases their resistance to climate change. This could include locally adapted seeds, planting materials, and livestock, among other essential resources (FAO, 2017). Box 20.1 below highlights some of the roles played by the government in the programme.

Box 20.1: Government's Role in *Intwasa*
It is difficult for smallholder farmers to plan for the upcoming season because of their difficulties, which include growing input costs for veterinary chemicals, fertilisers, and seeds, unsustainable market transportation costs and shifting market pricing. This necessitates strategic interventions such as subsidies, improved infrastructure, and market information systems to support their sustainability and growth in agriculture. In these conditions, as a government agency, we support smallholder farmers through the government's farming input programmes *Intwasa/Pfumvudza* programme, which supports 1.8 million farmers through training and provision of inputs promotion of locally available seed varieties. In collaboration with local NGO agencies, we have created seed bank projects, village loans and savings projects such as Beekeeping, Rabbit keeping and fisheries for diversity

Source: Government Key Informant.

The contribution of AGRITEX, Zimbabwe's Agricultural Extension Services, was similarly highlighted by another participant in the study area. Underscore was on the crucial role of AGRITEX in providing agricultural extension services, disseminating knowledge, and offering technical assistance to address the challenges that smallholder farmers face. This highlights the importance of government-run agricultural extension programmes to bolster the capacity and resilience of smallholder farmers against various obstacles. The participant hinted that:

> AGRITEX personnel play a critical role in the distribution of inputs under the Intwasa/Pfumvudza programme. The department also trains farmers on how to prepare our fields as

part of the initiative. Despite the importance of AGRITEX, our district needs more resources and capacity, which hinders its ability to serve farmers effectively.

This statement highlights the need for increased investments in training, infrastructure, and resources to empower AGRITEX officers. Enhanced support would enable them to better assist smallholder farmers in overcoming challenges and improving their agricultural productivity. Most interviewed smallholder farmers indicated that AGRITEX is making significant progress in empowering them by providing expertise in implementing Climate-Smart Agriculture (CSA) practices. However, resource constraints pose significant challenges. Additionally, some farmers expressed concerns about the lack of markets to sell their produce. Sikwila and Mashunje (2013) argue that several government initiatives aim to help smallholders escape poverty by enhancing their agricultural productivity. Unfortunately, overwhelming evidence indicates that smallholder farmers are constrained by limited institutional support (Dick-Sagoe et al., 2023). This lack of support makes it challenging for smallholders to access market information and acquire technical skills (Guidoboni et al., 2023; Nyathi et al., 2024).

Conclusion and Policy Implications

This study aimed to evaluate the viability of the *Intwasa/Pfumvudza* initiative implemented by the Government of Zimbabwe to enhance the resilience of smallholder farmers to climate change in selected wards in the Umguza District. The findings indicate that by providing agricultural inputs and extension services, *Intwasa/Pfumvudza* strengthens absorptive, adaptive, and transformative capacities at both farm plot and livelihood scales. Optimum plant density, intercropping, and soil water conservation techniques facilitate production intensification within confined areas. This response diversity buffers against sensitivity to dynamic rainfall and temperature patterns, whereas crop diversification distributes risk, thereby supporting food and economic security objectives. Government initiatives have positively impacted farmers' resilience to climate change, as evidenced by increased yields and improved household food security. However, the programme's success depends on timely access to farming inputs and availability of labour at the household level.

To enhance the *Intwasa/Pfumvudza* initiative, it is essential to recalibrate agricultural policy and public finance management to prioritise this programme. Mainstreaming climate change considerations into strategic planning and adaptive governance frameworks is necessary. Strengthening coordination between research, extension services, education, and farmers will promote knowledge co-creation and social learning, which are critical to adaptation. Encouraging private sector involvement and establishing multi-stakeholder partnerships will ensure comprehensive support. Civil society participation should be promoted to enhance transparency and accountability regarding resource allocation. Developing innovative financing solutions through public–private partnerships can address barriers such as high costs, short tenures, and post-harvest losses.

Regularising land tenure removes constraints on accessing agricultural loans and insurance, thereby incentivising adaptation investments. It is crucial to facilitate continued capacity building and implement early warning systems to support responsive decision making in volatile conditions. Adopting a holistic systems approach that integrates actors across various scales with revised policies, institutions, and coordinated support services will foster resilient climate-smart smallholder agriculture. An effective monitoring and evaluation system is necessary to measure the short-term and long-term impacts of such initiatives on smallholder productivity, food security, and income sustainability. Implementing these recommendations will enhance the effectiveness of the *Intwasa/Pfumvudza* initiative, ensuring that it meets its goals and contributes to building a resilient and climate-adaptive agricultural sector in Zimbabwe.

References

Action on Climate. (2018). *Mitigation, adaptation, and resilience: Climate terminology explained.* Available online at: http://www.actiononclimate.today/act-oninformation/mitigation-adaptation-and-resilience-climate-terminologyexplained/

Alemaw, B. F. (2020). Framework of best practice for climate change adaptation in Africa: The water–development Nexus. In J. I. Matondo et al. (Eds.), *Climate variability and change in Africa.* Springer Nature.

Alemaw, F. B., & Matondo, I. J. (2020). Overview of climate variability and change in Africa: Perspectives and experiences. In J. I. Matondo et al. (Eds.), *Climate variability and change in Africa.* Springer Nature.

Antwi-Agyei, P. (2021). Building smallholder farmers' resilience to climate change for food security in a future climate in Ghana. Available online from Building Smallholder Farmers' Resilience to Climate Change for Food Security in a Future Climate in Ghana Climate & Development Knowledge Network.

Berbés-Blázquez, M., Mitchell, C. L., Burch, S. L., & Wandel, J. (2017). Understanding climate change and resilience: Assessing strengths and opportunities for adaptation in the Global South. *Climatic Change, 141*(2), 227–241.

Bernard, H. R. (1995). *Research methods in Anthropology* (2nd ed.). Sage Publications.

Chingarande. (2020). *Zimbabwe market study: Matabeleland North Province report.* Research Technical Assistance Centre.

CIAT and World Bank. (2017). *Climate-smart agriculture in Zimbabwe. CSA country profiles for Africa series.* CIAT.

C2ES. (2019). *What is climate resilience, and why does it matter?* C2ES.

Denzin, N. K., & Lincoln, Y. S. (2000). *Handbook of qualitative research.* Sage Publications.

Dick-Sagoe, C., Hope, K. N., & Asare-Nuamah, P. (2023). Perceived impact of climate variability and change on livelihoods of smallholder farmers in Lesotho. *African Journal of Science, Technology, Innovation and Development, 15*(2), 175–184. https://doi.org/10.1080/20421338.2022.2058339

Dube, T., Mlilo, C., Moyo, P., Ncube, C., & Phiri, K. (2018). Will adaptation carry the future? Questioning the long-term capacity of smallholder farmers' adaptation strategies against climate change in Gwanda District, Zimbabwe. *Journal of Human Ecology, 61*(1–3), 20–30. https://doi.org/10.1080/09709274.2018.1452866

England, M. I., Dougill, A. J., & Stringer, L. C. (2018). Climate change adaptation and cross-sectoral policy coherence in southern Africa. *Regional Environmental Change, 18*, 2059–2071. https://doi.org/10.1007/s10113-018-1283-0

Engle, N. L. (2011). Adaptive capacity and its assessment. *Global Environmental Change, 21*(2), 647–656.

FANRPAN. (2017). *Climate-smart agriculture in Zimbabwe*. FANRPAN.

FAO. (2017). *Climate smart agriculture sourcebook*. http://www.fao.org/climate-smart-agriculture-sourcebook/concept/module-a1-introducing-csa/chapter-a1-4/en/

FAO. (2020). *Climate proofing: Zimbabwe's key to rural resilience and food security*. Food and Agricultural Organisation of the United Nations.

Folke, C., Carpenter, S. R., & Walker, B. (2010). Resilience thinking: Integrating resilience, adaptability and transformability. *Ecology and Society, 15*(4), 1–7.

Gbegbelegbe, S., Serem, J., Stirling, C., Kyazze, F., Radeny, M., Misiko, M., Tongruksawattana, S., Nafula, L., Gakii, M., & Sonder, K. (2018). Smallholder farmers in eastern Africa and climate change: A review of risks and adaptation options with implications for future adaptation programmes. *Climate and Development, 10*(4), 289–306. https://doi.org/10.1080/17565529.2017.1374236

GMB. (2022). *Producer prices for 2020/2021 season*. Government Printers. Harare.

GoZ. (2020). *National development strategy 1*. Government Printers. Harare.

Guidoboni, M. V., Duparque, A., Boissy, J., Mouny, J.-C., Auberger, J., & van der Werf, H. M. (2023). Conservation agriculture reduces the climate change impact of popcorn and wheat crop rotation. *PLoS ONE, 18*(5), e0285586.

IPCC. (2007). *Climate change 2007—Impacts, adaptation and vulnerability contribution of working group II to the fourth assessment report of the IPCC*. Intergovernmental Panel on Climate Change.

IPCC. (2014). *Climate change 2014: Impacts, adaptation and vulnerability. IPCCWGIIAR5 technical summary*. Intergovernmental Panel on Climate Change.

IPCC. (2019). *Climate change and land: An IPCC special report on climate change, desertification, land degradation, sustainable land management, food security, and greenhouse gas fluxes in terrestrial ecosystems*. Intergovernmental Panel on Climate Change.

Jiri, O., Mafongoya, P. L., & Chivenge, P. (2017). Building climate change resilience through adaptation in smallholder farming systems in semi-arid Zimbabwe. *International Journal of Climate Change Strategies and Management, 9*(2), 151–165.

Johannes, B., Moyo, S., & Abiodun, O. A. (2017). Assessing communal farmers' preparedness to drought in the Umguza District, Zimbabwe. *International Journal of Disaster Risk Reduction, 22*, 194–203.

Kassie, B. T., Asseng, S., Rotter, R. P., Hengsdijk, H., Ruane, A. C., & Van Ittersum, M. K. (2015). Exploring climate change impacts and adaptation options for maize production in the Central Rift Valley of Ethiopia using different climate change scenarios and crop models. *Climatic Change, 129*(1–2), 145–158. https://doi.org/10.1007/s10584-014-1322-x

Kassim, S. C., & Nkomoki, W. (2023). *Promoting food security in Zimbabwe by addressing adoption challenges for female farmers: Pfumvudza agricultural*. Tropentag Conference 2023. https://www.tropentag.de/2023/abstracts/posters/891.pdf

Lin, B. B., Chappell, M. J., Vandermeer, J., Smith, G., Quintero, E., Bezner-Kerr, R., & Perfecto, I. (2011). Effects of industrial agriculture on climate change and the mitigation potential of small-scale agro-ecological farms. *Cabi Reviews*, (2011), 1–18.

Makate, M., Makate, C., Mango, N., & Siziba, S. (2018). Increasing resilience of smallholderfarmers to climate change through multiple adoptions of proven climate-smart agriculture innovations. Lessons from Southern Africa. *Journal of Environmental Management, 231*(2019), 858–868.

Matondo, I. J., Alemaw, F. B., & Sandwidi, P. J. W. (2020). *Climate variability and change in Africa*. Springer Nature.

Mavesere, F., & Dzawanda, B. (2023). Effectiveness of *Pfumvudza* as a resilient strategy against drought impacts in rural communities of Zimbabwe. *GeoJournal, 88*(3), 3455–3470. https://doi.org/10.1007/s10708-022-10812-3

Mujere, N. (2021). *Assessing the potential contribution of Pfumvudza towards climate smart agriculture in Zimbabwe: A review*. Preprint. https://doi.org/10.20944/preprints202101.0619.v1. https://www.preprints.org/frontend/manuscript/7c927741 1edd4296732ba780b45ad3a4/download_pub

Ndlovu, S., Mathe, B., Phiri, K., & Nyathi, D. (2020). Factoring water harvesting into climate change adaptation: Endogenous responses by smallholder farmers in Gwanda district, Zimbabwe. *Cogent Social Sciences, 6*, 1. https://doi.org/10.1080/23311886.2020.1784652

Ngara, T. (Ed.). (2017). *Climate-smart agriculture manual for agriculture education in Zimbabwe*. UNEP DTU Partnership. Available at: https://www.uncclearn.org/wp-content/uploads/library/climate-smart_agriculture_manual_0.pdf

Nyasimi, M., Amwata, D., Hove, L., Kinyangi, J., & Wamukoya, G. (2014). *Evidence of impact: Climate-smart agriculture in Africa*. CCAFS Working Paper No. 86. CGIAR Research Program on Climate Change, Agriculture and Food Security (CCAFS).

Nyathi, D. (2024). Motives and constraints of rural livelihoods diversification in dry-land Agrarian settings of Matabeleland, Zimbabwe. https://doi.org/10.1177/00219096241257696

Nyathi, D., Ndlovu, J., Mare, A., Dzvimbo, M. A., & Ndlovu, M. (2024). Women's vulnerability and adaptive capacity to climate change in Agrarian settings of Zimbabwe. *Climate Change Management*, 541–559. https://doi.org/10.1007/978-3-031-58261-5_23

Oldreive, B. (2006). *Foundations for farming trainers' manual*. Foundations for Farming.

Phiri, K., Nhliziyo, M., Madzivire, S. I., Sithole, M., Nyathi, D., & Read, R. (2021). Understanding climate smart agriculture and the resilience of smallholder farmers in Umguza district, Zimbabwe. *Cogent Social Sciences, 7*(1). https://doi.org/10.1080/23311886.2021.1970425

Pope, C., & Mays, N. (2000). *Qualitative research in health care*. BMJ Books.

Roop, R., & St. Martin, C. C. (2021). Building climate resilience of smallholder family farms by implementing integrated soil and water management strategies in Trinidad and Tobago. In *Handbook of climate change management: Research, leadership, transformation* (pp. 3711–3739). Springer.

Rusinamhodzi, L. (2015). Tinkering on the periphery: Labour burden not crop productivity increased under no-till planting basins on smallholder farms in Murehwa district, Zimbabwe. *Field Crops Research, 170*, 66–75.

Schensul, J., & LeCompte, M. (1999). *Ethnographer's toolkit*. Altamira Press.

Sikwela, M. M., & Mushunje, A. (2013). The impact of farmer support programmes on household income and sustainability in smallholder production: A case study of the Eastern Cape and KwaZulu Natal farmers, South Africa. *African Journal of Agricultural Research, 8*(21), 2502–2511.

Simango, A. (2021). *Zimbabwe young farmers lead the charge in agriculture*. Accessed December 25, 2023 from https://www.un.org/africarenewal/magazine/september-2021/zimbabwean-young-farmers-lead-charge-agriculture

Tanyanyiwa, V. I., Kanyepi, T., & Katanha, A. (2022). Zimbabwe's Pfumvudza agriculture programme—reality or rhetoric? In W. Leal Filho, M. Kovaleva, & E. Popkova (Eds.), *Sustainable agriculture and food security*. Springer.

Timmer, C. P. (1997). Farmers and markets: The political economy of new paradigms. *American Journal of Agricultural Economics, 79*(2), 621–662.

Tsiko, S. (2021). *Farmers reap benefits of Pfumvudza/Intwasa concept*. Accessed March 25, 2024 https://www.herald.co.zw./farmers-reap-benefitsof-pfumvudza/intwasa-concept

UN. (2016). *Climate change resilience: An opportunity for reducing inequalities*. United Nations.

UNDP. (2017). *Zimbabwe human development report climate change and human development: Towards building a climate resilient nation*. UNDP.

UNDP. (2018). *Annual report 2018*. United Nations Development Programme. New York.

Unganai, L. S., & Murwira, A. (2010). *Challenges and opportunities for climate change adaptation among smallholder farmers in southeast Zimbabwe in 2nd international conference: Climate, sustainability and development in semi-arid regions* (pp. 16–20). Ceará Convention Center.

Vincent, K., Cull, T., Chanika, D., Hamazakaza, P., Joubert, A., Macome, E., & Mutonhodza-Davies, C. (2013). Farmers' responses to climate variability and change in southern Africa—Is it coping or adaptation? *Climate and Development, 5*(3), 194–220.

Wilbanks, J. T. F., Abeysinghe, A. C., Burton, I., Gao, Q., Lemos, C. M., Masui, T., O'Brien, L. K., & Warner, K. (2014). *Climate-resilient pathways: Adaptation, mitigation, and sustainable development*. Cambridge University Press.

Zimstat. (2022). *Population census report*. Government Printers. Harare.

Zolnikov, T. (2019). *Global adaptation and resilience to climate change*. Palgrave McMillan.

Douglas Nyathi is a Post-Doctoral Fellow at the University of Johannesburg, South Africa. He holds a Ph.D. in Public Policy from the University of KwaZulu Natal. His thesis focused on livelihood diversification in agrarian settings. Dr Nyathi's research focuses on the intersection of livelihoods, climate change and poverty amongst smallholder farmers in the Global South. His research and teaching interests explore how climate change affects vulnerable populations, particularly in relation to their livelihoods and economic well-being. Dr Nyathi has considerable acquaintance in monitoring, evaluation and documentation of success stories, having been consulted by local and international organisations.

Mbongeni Nhliziyo holds a Master of Science Degree in Public Policy and Development Management from Lupane State University and is currently a lecturer in the Department of Development Studies at Lupane State University. He has previously lectured at Zimbabwe Open University in the Department of Development Studies. He is a research fellow at the Public Policy and Research Institute of Zimbabwe, a leading research organisation on democracy, governance, and public policy advocacy in Zimbabwe. His research and teaching interests are in Africa's development. Nhliziyo has published peer-reviewed articles in accredited academic journals.

Joram Ndlovu is an Associate Professor and Academic Leader for Research in the School of Social Sciences at the University of KwaZulu-Natal. Professor Ndlovu has over two decades of teaching and research at different Universities in Southern Africa. He has published extensively and supervised Masters and Ph.D. students. His research interest is exploring the intersection of livelihoods, climate change, and indigenous knowledge. He aims to investigate how indigenous communities adapt to environmental challenges, leveraging traditional knowledge to sustain their livelihoods. By understanding these dynamics, he seeks to develop more effective strategies for climate resilience and community empowerment. His research focuses on environmental governance, sustainable development, eco-tourism and climate change.

Manase K. Chiweshe is a renowned sociologist in the Department of Social and Community Development at the University of Zimbabwe and a Research Associate at Rhodes University in South Africa. His research focuses on everyday life in African spaces, emphasising gender, football, and sports studies. He won the 2015 Gerti Hesseling Prize for best paper published in African Studies. He has authored books on football and the agrarian question in Zimbabwe. Chiweshe also has some interest in rural transformation and climate change.

MIX
Papier aus verantwortungsvollen Quellen
Paper from responsible sources
FSC® C105338

If you have any concerns about our products,
you can contact us on
ProductSafety@springernature.com

In case Publisher is established outside the EU,
the EU authorized representative is:
Springer Nature Customer Service Center GmbH
Europaplatz 3, 69115 Heidelberg, Germany

Printed by Libri Plureos GmbH
in Hamburg, Germany